Advances in
Aircraft Flight Control

Advances in Aircraft Flight Control

Edited by

MARK B. TISCHLER

Taylor & Francis
Publishers since 1798

UK Taylor & Francis, Ltd., 11 New Fetter Lane, London EC4P 4EE
USA Taylor & Francis, 325 Chestnut Street, Philadelphia, PA 19106

British Library Cataloguing in Publication Data

A catalogue record for this book is available from the British Library.

ISBN 07484 0479 1

Library of Congress Cataloguing in Publication Data are available

Cover design by Amanda Barragry

Typeset in Times 10/12pt by Santype International Limited, Salisbury, Wilts

Printed by Great Britain by T. J. Press (Padstow) Ltd, Cornwall

Contents

Contents

vi

Series Introduction

Control systems has a long and distinguished tradition stretching back to nineteenth-century dynamics and stability theory. Its establishment as a major engineering discipline in the 1950s arose, essentially, from Second World War driven work on frequency response methods by, amongst others, Nyquist, Bode and Wiener. The intervening 40 years has seen quite unparalleled developments in the underlying theory with applications ranging from the ubiquitous PID controller widely encountered in the process industries through to high-performance/fidelity controllers typical of aerospace applications. This development has been increasingly underpinned by the rapid developments in the, essentially enabling, technology of computing software and hardware.

This view of mathematically model-based systems and control as a mature discipline masks relatively new and rapid developments in the general area of robust control. Here intense research effort is being directed to the development of high-performance controllers which (at least) are robust to specified classes of plant uncertainty. One measure of this effort is the fact that, after a relatively short period of work, 'near world' tests of classes of robust controllers have been undertaken in the aerospace industry. Again, this work is supported by computing hardware and software developments, such as the toolboxes available within numerous commercially marketed controller design/simulation packages.

Recently, there has been increasing interest in the use of so-called intelligent control techniques such as fuzzy logic and neural networks. Basically, these rely on learning (in a prescribed manner) the input–output behaviour of the plant to be controlled. Already, it is clear that there is little to be gained by applying these techniques to cases where mature mathematical model-based approaches yield high-performance control. Instead, their role is (in general terms) almost certainly going to lie in areas where the processes encountered are ill-defined, complex, nonlinear, time-varying and stochastic. A detailed evaluation of their (relative) potential awaits the appearance of a rigorous supporting base (underlying theory and implementation architectures for example) the essential elements of which are beginning to appear in learned journals and conferences.

Elements of control and systems theory/engineering are increasingly finding use outside traditional numerical processing environments. One such general area in which there is increasing interest is intelligent command and control systems which are central, for example, to innovative manufacturing and the management of advanced transportation systems. Another is discrete event systems which mix numeric and logic decision making.

It is in response to these exciting new developments that the present book series of Systems and Control was conceived. It will publish high-quality research texts and reference works in the diverse areas which systems and control now includes. In addition to basic theory, experimental and/or application studies are welcome, as are expository texts where theory, verification and applications come together to provide a unifying coverage of a particular topic or topics.

The book series itself arose out of the seminal text: the 1992 centenary first English translation of Lyapunov's memoir *On the General Problem of the Stability of Motion* by A. T. Fuller, and was followed by the 1994 publication of *Advances in Intelligent Control* by C. J. Harris. The current volume in the series is *Intelligent Control in Biomedicine* by D. A. Linkens. Forthcoming titles in the series include *True Digital Control* by P. C. Young, A. Chotai and W. Tych, *Sliding-Mode Control* by V. Utkin and *A Knowledge-Based Approach to Systems Control* by G. K. H. Pang.

E. ROGERS
J. O'REILLY

Foreword

The highly interdisciplinary nature of the modern aircraft flight control problem presents a substantial challenge to control engineering. This challenge starts at the conceptual control system design phase, where the desired system response characteristics required for good handling qualities are specified as a function of aircraft configuration, mission, flight condition, failure states, and conditions of visibility including the influence of vision aids. These handling-qualities characteristics together with the servo-loop requirements for disturbance rejection and stability/performance robustness across large variations in dynamic characteristics with flight condition define a formidable set of interdisciplinary design specifications. The control law architecture and design methodology is selected to meet these complex requirements. But the practical considerations such as the control engineer's (or company's) design philosophy, and the ease of flight test development and optimization can also have an important influence.

Modern tools for flight control system development are also wide ranging, requiring an appreciation of several engineering disciplines. Hardware-in-the-loop simulation, software development/validation facilities (DF), and advanced ground and in-flight piloted simulation are used to evaluate the complete system performance for the many flight conditions and control system modes. The nonreal-time and real-time models used in these tools include complex descriptions of the aircraft aerodynamics, flight dynamics, propulsion systems, and structural dynamics characteristics to ensure that the simulation results will be representative of expected flight behavior. A detailed representation of the digital control system implementation is also necessary to capture aeroservoelastic and pilot–airframe coupling phenomena. System identification tools are used to document dynamic characteristics during development, and later to support control law and handling-qualities optimization during flight test.

The complexity of modern digital flight control systems and the trend toward increased technical specialization makes it difficult for one person to acquire the needed interdisciplinary perspective. More important is that the long time interval between modern aircraft projects means that new projects may not fully benefit from the experience gained in earlier flight control programs. Key lessons learned, such as the critical importance of minimizing time delay and actuator saturation, have often been forgotten or ignored only to be re-encountered with costly consequences on later programs. Further, there is a wealth of knowledge in one specialized aircraft community that can be effectively leveraged by another. For example, the fixed-wing community has over a decade of experience with production fly-by-wire control system development that is a technology resource for the more recent fly-by-wire application to rotary-wing aircraft. Conversely, in the rotorcraft community, frequency-domain methods for system identification are effective for analyzing complex and unstable system dynamics. These methods can be more extensively leveraged by the fixed-wing community to reduce the cost of control system flight test development and optimization. Finally, the fixed-wing and rotary-wing testing organizations have each developed specialized techniques for flight control and handling-qualities evaluation that also can be effectively leveraged.

Excellent textbooks are available which cover the full range of flight control design and analysis methods. However, owing to the proprietary nature of aircraft development programs, the available references generally present few flight results creating a dependency on analytical and simple simulation studies to illustrate and validate the proposed methods. Such studies may not expose robustness and practical implementation problems of some of the proposed control system methods. Furthermore, there is little insight into the actual control system development and flight test optimization process. Open-literature publications describing the valuable flight experience that has been obtained in recent years for a wide variety of fly-by-wire projects are generally diffused throughout the many aerospace technical and flight-test-oriented conferences in Europe and the United States.

This collection of papers seeks to address the need to provide the flight control engineer with a single comprehensive resource that reviews many of the current aircraft flight control programs from the perspective and hindsight of experienced practitioners. The programs described herein have widely exploited modern interdisciplinary tools and techniques, and the discussions include extensive flight test results. There are many important 'lessons learned' from the experience gained when design methods and requirements were tested and optimized in actual flight demonstration. This book is organized into four parts. Following the discussion of control system design specifications and development tools in Part One, the remaining three parts are organized according to aircraft type: rotorcraft and V/STOL (Vertical/Short Takeoff and Landing) aircraft; transport aircraft; and high-performance aircraft.

Part One reviews the key flight control design specifications and validation methods for modern aircraft. The first chapter is an overview of US handling-qualities specifications for fixed-wing and rotary-wing military aircraft. These specifications are important design drivers for setting flight control system architecture and component performance. The second chapter explores the key role of system identification in modern flight control development. This technology provides an integrated flow of dynamic response data around the entire life-cycle of aircraft development from initial specifications, through simulation and bench testing, and into flight test optimization. A critical 'paper trail' is thus established to track system performance and avoid costly changes late in the development process.

Part Two reviews four advanced flight programs for rotorcraft and V/STOL configurations. The control problem for these aircraft is made especially demanding by unstable bare-airframe dynamic characteristics, high levels of measurement noise and vibration, strong interaxis coupling, and lightly damped structural and rotor blade dynamic modes within the control bandwidth. The first chapter is by the DLR (Germany) and describes the development and flight research applications of a rotorcraft in-flight simulator capability 'ATTHeS' based on a Bo-105 helicopter equipped with a fly-by-wire control system. The model-following control laws were based directly on bare-airframe state-space models extracted from flight data using frequency-domain system identification. The second chapter reviews the evolution of advanced flight control technology for rotorcraft at the Boeing Helicopter Company from the TAGS experimental test bed to the newest US scout–attack helicopter, the RAH-66 Comanche. Comprehensive aeroservoelastic analysis and advanced piloted hardware-in-the-loop simulation are used to validate the end-to-end rotorcraft control system design as extensively as possible before first flight. In the third chapter of this section, researchers from NASA Ames describe a nonlinear inverse controller implementation for the QSRA powered-lift aircraft and an advanced conceptual STOVL (Short Takeoff and Vertical Landing) fighter aircraft. This controller approach has the advantage of eliminating the need for complex gain scheduling over the large configuration changes of the QSRA and STOVL flight envelopes. The last chapter presents the development and testing of the VAAC experimental aircraft, a Harrier I trainer equipped with an advanced fly-by-wire flight control system. This work by DRA Bedford demonstrates the potential of digital flight control for greatly improving the handling qualities of jet-lift V/STOL aircraft configurations.

Part Three discusses advanced flight control technology for military and commercial transport aircraft. Active gust suppression, auto trimming, multiaxis sidestick controllers, and highly augmented responses are featured on modern transports to reduce pilot workload and to improve ride qualities. A full-time multiply-redundant stability-augmentation system allows the relaxation of bare-airframe static-stability margins which in turn reduces trim drag and thereby improves fuel performance. The first chapter presents design and flight test results for the US advanced tactical transport aircraft, C-17, which features a full-authority, quad-redundant, digital flight control system. Special emphasis is given to the importance of control system time delays on aircraft handling qualities. The second chapter presents an overview of the Airbus fly-by-wire control laws, systems, certification and development methods. The chapter emphasizes the extensive use of simulation tools in the development, validation, and certification of civil transport control systems, including a turbulence suppression system for the A340. In the third chapter, authors from Boeing Defense and Space Group used the multivariable linear quadratic regulator (LQR) design method to improve significantly the lateral autopilot performance for the Boeing 767 as illustrated using flight test data. The design method is also applied to an advanced fighter in a conceptual study that includes detailed handling-qualities analyses using Gibson's template criteria.

Part Four presents flight control development methods and flight test results for six advanced high-performance military aircraft programs. Multiply-redundant nonconventional control surfaces and relaxed static-stability concepts are utilized to achieve increased maneuverability and agility even at the edges of the flight envelope, while at the same time minimize radar cross-section. The extensive use of piloted simulation, system identification, and detailed handling-qualities analysis is emphasized as critical to the successful achievements of the difficult flight control goals for this class of aircraft. In the first chapter of this part, authors from Israel Aircraft Industries describe the control design, development, and flight testing for the Lavi aircraft. Extensive ground and in-flight piloted-simulation evaluations of the proposed handling qualities and control laws were effectively used to refine Lavi characteristics prior to first flight, and reduce flight test development costs. The control law design approach for combat aircraft development at Alenia is presented in the second chapter. The author discusses the development history of the AMX autopilot, including flight test results, and emphasizes the control law architecture for supporting the AMX multimode capability. The author also reviews the control law structure and design approach for the new Eurofighter EF 2000 aircraft. In the third chapter, authors from BAe Defence Ltd present the control law design philosophy and extensive flight test results for the Experimental Aircraft Programme (EAP). An important aspect of this program was the demonstration of carefree maneuvering through automatic angle-of-attack and normal-acceleration limiting. Another important contribution of this chapter is the review of the analysis methodology and flight tests for evaluating aeroservoelastic coupling and its alleviation. The fourth chapter presents the control law design and flight test lessons learned on the X-29A forward-swept wing demonstrator program at the NASA Dryden Flight Research Center. Low- and high-angle-of-attack control laws were developed and evaluated. The authors have made extensive use of on-line system identification to validate control system stability margins during envelope expansion. In the fifth chapter, authors from Wright Laboratory and McDonnell Douglas Aerospace–East describe in detail the system requirements and control law design methods for the F-15 STOL and Manoeuvre Technology Demonstrator (S/MTD). A unique aspect of this program was the combined use of both classical and modern multivariable (LQG/LTR) control law design methods. The classical design method used a novel inverse approach directly to tune control law parameters to achieve desired frequency-domain handling-qualities characteristics. In the last chapter of this part, authors from Daimler-Benz Aerospace AG present the advanced control law design and implementation for the X-31A post-stall experimental aircraft. The multivariable LQG approach for the control law design makes effective use of the many sensors and unconventional aerodynamic and thrust vectoring controls on the aircraft. Other special features of the program include compensation for gravity effects, and inertial and gyroscopic coupling.

This book was developed as a follow-on project to a collection of papers that was published in a Special Issue of the *International Journal of Control* (IJC) on the same topic (Tischler 1994). Many of the papers in the IJC collection were updated by the authors to include the latest flight test results for inclusion in this volume. Some of the additional contributions to the current volume were updated from papers at an excellent recent symposium 'Active Control Technology: Applications and Lessons Learned' (Anon. 1994) held by the Flight Mechanics Panel (FMP) of the Advisory Group of Aerospace Research and Development (AGARD). The support and help of my fellow colleagues on the AGARD Flight Mechanics Panels (FMP) (now evolved into the current Flight Vehicle Integration Panel (FVP)) in organizing the earlier IJC Special Issue and this current volume was invaluable and is gratefully acknowledged. The AGARD panels represent an important scientific and technical resource throughout the NATO nations. Preparation of the contributed papers in this book represented a considerable effort and commitment on the part of each author and their parent organization. Their efforts in bringing together this unique collection of aircraft flight control experience are greatly appreciated.

MARK B. TISCHLER
Rotorcraft Group Leader
Flight Control Branch
Aeroflightdynamics Directorate
US Army Aviation and Troop Command
Ames Research Center
Moffett Field, California 94035-1000
USA

REFERENCES

ANON., 1994, *Proceedings of the AGARD Flight Mechanics Panel Symposium on 'Active Control Technology: Applications and Lessons Learned'*, AGARD-CP-560, Turin, Italy, 9–13 May.

TISCHLER, M. B., Guest Editor, 1994, *International Journal of Control, Special Issue: Aircraft Flight Control*, **59**(1).

List of Contributors

MOSHE ATTAR
Israel Aircraft Industries, Ben Gurion International Airport, Israel

JEFFREY E. BAUER
NASA Dryden Flight Research Center, Edwards, CA 93523, USA

H. BEH
Daimler-Benz Aerospace AG, Military Aircraft LME12, 81663 München, Germany

PIER LUIGI BELLUATI
Alenia System Technology Group, Corso Marche, 41 Turin, Italy

JAMES D. BLIGHT
Boeing Defense & Space Group, PO Box 3707, Seattle, WA 98124, USA

JOHN T. BOSWORTH
NASA Dryden Flight Research Center, Edwards, CA 93523, USA

GERD BOUWER
Deutsche Forschungsanstalt für Luft- und Raumfahrt e.V. (DLR), Institut für Flug-mechanik, D-38108 Braunschweig, Germany

JOHN J. BURKEN
NASA Dryden Flight Research Center, Edwards, CA 93523, USA

B. CALDWELL
Aerodynamics Dept, BAe Defence Ltd, Warton, Lancs PR4 1AX, UK

KEVIN D. CITURS
McDonnell Douglas Aerospace–East, St Louis, Missouri, USA

List of Contributors

ROBERT CLARKE
NASA Dryden Flight Research Center, Edwards, CA 93523, USA

CHARLES DABUNDO
Boeing Defense and Space Group, Helicopters Division, Philadelphia, PA 19142, USA

R. LANE DAILEY
Boeing Defense and Space Group, PO Box 3707, Seattle, WA 98124, USA

JAMES M. DAVIS
Boeing Defense and Space Group, Helicopters Division, Philadelphia, PA 19142, USA

ELI ERENTHAL
Israel Aircraft Industries, Ben Gurion International Airport, Israel

C. FAVRE
Aerospatiale, 316 Route de Bayonne, 31060 Toulouse Cedex 03, France

C. FIELDING
British Aerospace Defence Ltd, Military Aircraft Division, Warton, Lancs, PR4 1AX, UK

JAMES A. FRANKLIN
NASA Ames Research Center, Moffett Field, CA 94035-1000, USA

S. L. GALE
Defence Research Agency, Bedford, UK

DAGFINN GANGSAAS
Boeing Defense and Space Group, PO Box 3707, Seattle, WA 98124, USA

D. V. GRIFFITH
Experimental Flying Squadron, Aircraft & Armaments Evaluation Establishment, Boscombe Down, UK

WOLFGANG VON GRÜNHAGEN
Deutsche Forschungsanstalt für Luft- und Raumfahrt e.V. (DLR), Institut für Flugmechanik, D-38108 Braunschweig, Germany

FROHMUT HENSCHEL
Deutsche Forschungsanstalt für Luft- und Raumfahrt e.V. (DLR), Institut für Flugmechanik, D-38108 Braunschweig, Germany

G. HOFINGER
Daimler-Benz Aerospace AG, Military Aircraft LME12, 81663 München, Germany

ROGER H. HOH
Hoh Aeronautics, Inc., Lomita, California, USA

P. HUBER
Daimler-Benz Aerospace AG, Military Aircraft LME12, 81663 München, Germany

JÜRGEN KALETKA
Deutsche Forschungsansalt für Luft- und Raumfahrt e.V. (DLR), Institut für Flug-mechanik, D-38108 Braunschweig, Germany

JAMES F. KELLER
Boeing Defense and Space Group, Helicopters Division, Philadelphia, PA 19142, USA

ERIC KENDALL
C-17 Avionics/Flight Controls, McDonnell Douglas Corporation, Long Beach, California, USA

KENNETH H. LANDIS
Boeing Defense and Space Group, Helicopters Division, Philadelphia, PA 19142. USA

A. McCUISH
Aerodynamics Dept, BAe Defence Ltd, Warton, Lancs PR4 1AX, UK

DAVID G. MITCHELL
Hoh Aeronautics, Inc., Lomita, California, USA

DAVID J. MOORHOUSE
Wright Laboratory, Wright–Patterson AFB, Ohio, USA

HEINZ-JÜRGEN PAUSDER
Deutsche Forschungsanstalt für Luft- und Raumfahrt e.V. (DLR), Institut für Flug-mechanik, D-38108 Braunschweig, Germany

G. T. SHANKS
Defence Research Agency, Bedford, UK

MENAHEM SHMUL
Israel Aircraft Industries, Ben Gurion International Airport, Israel

MARK B. TISCHLER
Aeroflightdynamics Directorate, US Army Aviation and Troop Command, Ames Research Center, Moffett Field, CA 94035-1000, USA

ALDO TONON
Alenia Flight Mechanics Group, Corso Marche, 41 Turin, Italy

PART ONE

Specification and Validation Methods

1

Handling-qualities specification – a functional requirement for the flight control system

ROGER H. HOH and DAVID G. MITCHELL

1. Introduction

The objective of any flight control system is to provide the pilot with a means to control the aircraft safely and effectively throughout its flight envelope. That is, to provide good handling qualities.† Therefore, the handling-qualities specification represents a logical source of functional requirements to guide the development of the flight control system. Experience has shown that, with a few exceptions, this viewpoint is confined to the handling-qualities community. The use of the handling-qualities specification as a source of functional requirements for the flight control system is rarely, if ever, carried out in practice. In the 'real world', flight control design is accomplished as a portion of the avionics system, and handling qualities are studied as a subset of aerodynamics. The lack of high-quality communication between these disciplines has been formally recognized as a problem since 1982 (see Reference 2). The flight control engineer tends to focus on issues related to sensor and actuation hardware, and related fault detection and isolation logic. The control laws are usually relegated to the category of a relatively minor issue to be resolved on the simulator. It is common for the control law selection to be based on the past experience of the engineers and pilots in charge, as opposed to the mission requirements of the subject aircraft.

Handling qualities are usually first considered explicitly during early flight testing of the prototype aircraft. Historically, this approach has led to aircraft that are reasonably safe as long as the pilots are well trained and current. However, with the advent of active control technology (ACT) flight control systems, it has led to serious problems that include divergent pilot-induced oscillations (PIOs) that, in some cases, have led to loss of the aircraft. In most cases these PIOs can be traced directly to handling-qualities deficiencies that are predicted by criteria that exist in current handling-qualities specifications.

The primary goal of this chapter is to familiarize flight control system designers with the latest concepts and criteria that have been implemented into, or are

---

† Handling qualities are defined in Reference 1 as 'those qualities or characteristics of an aircraft that govern the ease and precision with which a pilot is able to perform the tasks required in support of an aircraft role'.

planned for, the handling-qualities specifications. We intend to present these concepts and criteria in the context of functional requirements for the flight control system.

2. Status of handling-qualities specifications

The current fixed-wing handling-qualities specification in the United States is MIL-STD-1797A. This specification supercedes the MIL-F-8785 series (A through C). The differences between MIL-F-8785C and MIL-STD-1797A are summarized below.

- MIL-STD-1797A is in the approved military standard format. The basic specification is a list of criteria with blanks in place of numbers for the quantitative boundaries. Appendix A of MIL-STD-1797A consists of a collection of criteria. The user is expected to select the appropriate criterion values from Appendix A to build a custom standard for a given application.

- MIL-STD-1797A contains criteria that are specifically oriented towards ACT-type control systems.† The database that supports the criteria in MIL-STD-1797A contains substantially more data for highly augmented aircraft than that supporting previous specifications.

The development of MIL-STD-1797 was accomplished during the period between 1980 and 1982, resulting in a proposed specification that is documented in Reference 3. That report was converted to the actual MIL-Standard during the period 1982 through 1987. Since that time, work has been accomplished to develop a more mission-oriented specification. Most of this work was done to develop a flying-qualities specification for US Army helicopters, resulting in an Aeronautical Design Standard, ADS-33C. ADS-33C (Reference 4) was used to specify the handling-qualities requirements for the Boeing/Sikorsky RAH-66 ACT rotorcraft. It has since been refined and subjected to the tri-service review process, see Reference 5. Reference 5 is expected to supersede MIL-H-8501A as the military rotorcraft specification in the United States within a year. In this chapter, we shall refer to Reference 5 as the rotorcraft MIL-Standard under the assumption that final acceptance is imminent.

While there has been considerable work to upgrade MIL-STD-1797A to include the now-accepted mission-oriented concepts, that work has not been incorporated into the specification as of this writing. Plans are to accomplish a major upgrade from MIL-STD-1797A to MIL-STD-1797B in 1997. It is recommended that control system designers consult References 6 and 7 to augment MIL-STD-1797A as a source of functional requirements.

3. Handling-qualities specifications – are they necessary?

There is substantial resistance to specifications in the name of cost cutting and paperwork reduction. What is overlooked is the fact that most specification require-

---

† Active control technology (ACT) is the currently agreed-on terminology for aircraft that have flight control systems that do not use a mechanical link between the control surfaces and the cockpit controls. Such aircraft usually incorporate flight control computers that allow modification of the response to control inputs. Another generic term for this type of flight control system is 'fly-by-wire'. ACT is currently the favored term as it does not imply that the signals are electrical and transmitted through wires. Some ACT aircraft use fiber optics, for example.

ments are a result of lessons that have been learned at great cost. Nearly all serious handling deficiencies that have occurred can be traced to noncompliance with existing specification criteria. The YF-22 PIO, which led to a crash on the runway at Edwards AFB, is a good example. This aircraft fails the Equivalent System Time Delay and Bandwidth Phase Delay criteria in MIL-STD-1797A by a wide margin. Such all-too-common cases of noncompliance with specification criteria are an indication that these criteria are not commonly used as functional requirements for the flight control system.

One probable reason for this situation is the complexity of the handling-qualities criteria. Handling qualities are difficult to specify because they inherently involve the quantification of pilot workload. The tools available to measure pilot workload require considerable experience and background knowledge to apply. For example, the Cooper–Harper subjective rating scale (Reference 1) is the basis for the criteria in all current handling-qualities specifications. Cooper–Harper handling-qualities ratings (HQRs) have proven to be reasonably reliable under certain test conditions, but subject to serious shortcomings in others (Reference 8). Specifically, the scale has been shown to produce consistent and reliable results when used properly in a research experiment where the pilot subjects have no vested interest in the outcome. However, when there is a vested interest, experience has shown that subjective opinions and ratings can be unreliable, even with the most experienced and conscientious test pilots (e.g. see Reference 8). Those not familiar with this phenomenon (or who do not believe that it exists) tend to place excessive reliance on piloted simulation to design empirically the flight control system. This approach often has appeal to management because it makes use of a very expensive company asset. Unfortunately, the limited visual and motion cues that are available in even the most advanced simulators, in combination with the potential for basing control laws on the preconceived notions of a few pilots, have a long history of failure.

Quantitative handling-qualities criteria are based on the results of research experiments that include variable-stability flight tests using pilots with no vested interest. This is the fundamental reason why the quantitative specification criteria should be met, even if this does not seem intuitively necessary.

Consider the following scenario. During the preliminary design review of a new aircraft, the engineers note that it does not meet certain handling-qualities criteria in the handling-qualities MIL-Standard. It is noted that the work required to modify the flight control system would involve considerable additional cost, and would have an adverse impact on the schedule. Management is not sure that the problem is real, and does not want to spend money and time unless it is absolutely necessary. The company test pilots are asked their opinion based on simulator and/or flight evaluations. They indicate that the flying qualities are satisfactory. In this environment, it is not practical (or a good career decision) for the engineers to question the validity of evaluations made by the test pilots. If there is no contractual requirement to meet the flying-qualities specification, the decision to ignore the 'problem' is inevitable. And this in spite of the fact that such a decision is based on the opinion of pilots with a strong vested interest, and often with experience that is limited to unstructured testing using *ad hoc* tasks.

In this all-too-common scenario, pilot opinion is judged to have precedence over specification criteria. This implicitly ignores the fact that the quantitative criteria are based on the collective opinions of test pilots, with no vested interest, performing highly structured tasks, usually in a variable-stability flight test environment. The only way to avoid this pitfall is to *make the most current handling-qualities specification a contractual requirement*. Feedback from engineers on a recent ACT rotorcraft development indicated that having the ADS-33C handling-qualities specification as a contractual requirement was extremely valuable in obtaining management approval to deal with identified problems.

The need for good handling qualities is subtle because most handling-qualities deficiencies can be compensated for by a competent pilot (albeit with training and practice). Once a pilot has managed to develop the skill required to overcome a handling-qualities deficiency, he or she may even consider it an enhancement. The personality factors that result in this phenomenon are complex. They are almost certainly related to the competitive nature of an individual drawn to a profession that involves continuous testing of his or her mental and physical skills. Implicit in this 'good handling qualities are for sissies' attitude is a synergy with management where the primary goal is to limit costs and to stay on 'success-oriented' schedules. However, we are learning that even the most skilled pilots make fundamental errors when confronted with highly stressful real-life scenarios, e.g. tired crew, short runway, turbulence, windshear, night, etc. While such an unfortunate combination is rare, it is precisely when safety dictates good handling qualities.

Manufacturers of large aircraft have often taken a position that the boundaries for short-period frequency in MIL-STD-1797A are too stringent for large aircraft in the power-approach flight condition (Category C). They argue that many existing aircraft that do not meet these boundaries are routinely landed safely by average pilots. Few operational pilots will admit that under some conditions, significant recent training and skill is required to accomplish the required tasks successfully. The vested-interest implications are obvious in the context of this discussion, and very subtle in the real world of flight control design.

In summary, the need for a flying-qualities specification is subtle, but of primary importance to maintain the desired level of safety in critical flight conditions. In military operations, the need for the pilot to divide his or her attention between flying and the inherent high stress and high workload associated with tactical operations implies an even greater need for good flying qualities.

4. Flying-qualities specifications for active control technology (ACT) aircraft

ACT has become the basis for the flight control system design on many new commercial and military aircraft. Ideally ACT technology eliminates the compromise between good handling (large tail, forward c.g.) and good performance (small tail, aft c.g.). Such ideal handling qualities may require the use of very high gains to make the aircraft respond 'naturally' regardless of the configuration. Experience has shown that there are three factors that prevent this ideal situation: (1) what constitutes a 'natural' response is not well understood, (2) filtering necessitated by flexible modes, noise, controller characteristics, and digital processing limits the use of very high gains, and (3) high gains can only be achieved with adequate control power (large tail). Recent work on the handling-qualities specification for rotary-wing aircraft has been heavily oriented towards developing requirements that take these factors into account. Such upgrades are planned for the fixed-wing specification (MIL-STD-1797A) at the next major revision, in 1997.

For conventional (non-ACT) aircraft, the handling qualities are established primarily by the configuration. This results in a compromise with performance (primarily tail sizing). Final refinements are made through the use of aero-mechanical flight control devices such as bobweights, downsprings, q-bellows, servo-tabs and spring tabs. Properly designed, none of these devices significantly affects dynamic stability characteristics such as the short-period frequency and damping. The shape of the response of these aircraft to control inputs does not vary significantly for viable configurations, and hence simply specifying the values of a few parameters is sufficient (e.g. short-period frequency and damping)†. Generally speaking, this holds true even for aircraft with a limited-authority stability-augmentation system.

With ACT, the shape of the response to a control input can be drastically modified from a conventional aircraft. To cope with this, the specification criteria can no longer be based on a few parameters that define the frequency response of a classical airplane (phugoid mode, short-period mode, and n/α). The current MIL-STD-1797A represents a first cut at making the transition from the specification of handling qualities of classical airplanes to those designed around the ACT concept. For example, the Lower-Order Equivalent System (LOES) criterion (Reference 3 or 9) is based on the concept of using equivalent values of the classical parameters, with the addition of an equivalent time delay factor to account for the filtering noted above. Because the lower-order model used in the specification is the short-period approximation, this criterion is restricted to ACT designs that result in response shapes that emulate 'a conventional airplane'. This gives rise to a need to define, in precise terms, what is a conventional response. Further, it raises the question: is a conventional response the best response?

It is important to recognize that historically the response of airplanes was not something that could be significantly altered. A great deal of effort was expended to extend the center-of-gravity (c.g.) range, or to decrease the required tail size just a few percent. In the end, the flying qualities resulted from a compromise with performance, and by today's standards were not good. At forward c.g. the stick forces tended to be very heavy and the gust response excessive. At aft c.g., the stability was often near neutral. Hence, the answer to the question of what is the best response is not simply to make the aircraft fly 'naturally'. In fact the best response shape has been found to be dependent on the task. The concepts of Response-Type and Mission–Task–Element (MTE) have been introduced into the handling-qualities specifications to allow a distinction to be made regarding the shape of the aircraft response to control inputs as a function of task. Furthermore, the best Response-Type has also been found to be dependent on visual cuing in some cases. To account for this, the usable cue environment (UCE) concept has been developed. These new handling-qualities specification concepts are discussed in this chapter.

5. The mission-oriented handling-qualities specification

Developers of handling-qualities specifications have always been concerned with tying the specification criteria to the aircraft missions. This resulted in the Category

† The 'shape' of a response is best defined on frequency-response or 'Bode' plots of aircraft attitude and flight path. This defines the magnitude and phase of the aircraft attitude and flight path response to cockpit control inputs at all frequencies. Experience has shown that a consideration of the frequency range between 0·10 and 20·0 rad s^{-1} is generally adequate for the study of manual control.

and Class definitions found in the early flying-qualities specifications such as MIL-F-8785 (A through C). The reader is referred to References 11, 12 and 13 for background material related to the B and C versions. The concepts developed in these references have continued to evolve, as documented in References 3, 6, 7, 10, and in Appendix A of the MIL-Standard for handling qualities (Reference 9). The currently recognized elements of a mission-oriented flying-qualities specification are briefly summarized below, and are discussed in detail in this chapter.

- *Handling-qualities Levels.* The pass/fail criteria are defined in terms of Levels. Compliance with the Levels must be accomplished by meeting a comprehensive set of quantitative criteria. Because those criteria are not perfect, they are spot checked through highly constrained piloted evaluations where Levels are assigned by the evaluation pilots (Demonstration Maneuvers).

- *Response-Types.* A classification of the shape of the response of an aircraft to control inputs.

- *Mission–Task–Elements (MTEs).* Elements of an aircraft mission that are defined as flying-qualities tasks.

- *Usable cue environment (UCE).* A rating scale that allows the effect of degraded visual cuing on handling qualities to be quantified.

- *Quantitative criteria.* Parameters whose values predict the handling-qualities Levels. The parameters are obtained from analysis during the control system development, and from open-loop control inputs from flight testing for specification compliance.

- *Demonstration Maneuvers.* MTEs that are performed by evaluation pilots to assign handling qualities Levels.

All of these have been implemented in the rotorcraft MIL-Standard (Reference 5), and work is in progress to upgrade MIL-STD-1797A to implement these mission-oriented flying-qualities specification elements (e.g. see Reference 6).

The way in which these elements are utilized in the mission-oriented flying-qualities specification is illustrated by the flowchart in Fig. 1. The user activities are identified above the dotted line. These consist of a definition of the missions and environment in which the aircraft is being designed to operate. The user is required to sort through the list of MTEs in the specification, and select those that apply. If new MTEs are required, they should be defined in detail. Typical environmental specifications include atmospheric effects such as wind velocity and density altitude, and visual cuing effects such as lighting, temperature, fog, visible moisture, etc. The now common use of night vision aids (usually forward-looking infrared (FLIR) and night vision goggles (NVGs)) results in the need to include light levels (drives NVG performance) and temperature gradients such as might be affected by periods of cold soaking (degrades FLIR operation). With this information and a knowledge of current technology, it is possible to estimate the performance of the vision aid.

A UCE value can be estimated (and verified if the display device is available) to allow the user to determine the Response-Type required for each of the MTEs. The methodology to accomplish this is discussed in subsequent sections of this chapter. The definition of the operational missions also allows the user to define the required flight envelopes (limit of performance parameters such as altitude, airspeed, etc.). The aircraft configuration and sizing results directly from the mission requirements, and in the basic airframe dynamics that must be controlled.

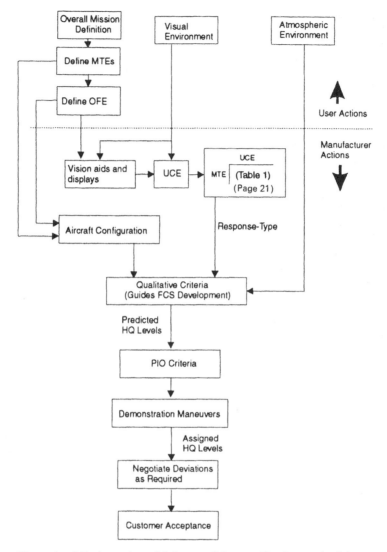

Figure 1.   Mission-oriented flying-qualities specification methodology.

The quantitative specification criteria should be used as functional requirements for the design of the flight control system. This is a crucial step, and one that tends to be overlooked. Some of the current MIL-STD-1797A criteria are out of date and should not be used. Most notably, the Bandwidth, Phase Delay and Dropback criteria are updated in Reference 6. These versions of the criteria will be included in MIL-STD-1797B, when it becomes available in 1997.

The next step in the process is to perform demonstration maneuvers for those MTEs selected by the user. The HQRs assigned by the pilots from this exercise should be consistent with those predicted from the quantitative criteria. If the assigned HQRs are Level 2 when Level 1 was predicted, the root cause should be identified and corrected. If this process is being accomplished on a ground-based simulator, the lack of good visual cuing must be accounted for as discussed in the section (12) titled 'Demonstration Maneuvers'. Once a prototype aircraft is available, both the quantitative criteria and Demonstration Maneuvers must be verified. Data for the quantitative criteria is obtained from open-loop flight testing (e.g. steps, pulses, frequency sweeps), whereas the data for the Demonstration Maneuvers is obtained from closed-loop testing using procedures that are consistent with a well-performed handling-qualities experiment.

As noted above, and in Fig. 1, the user must provide certain input data for the mission-oriented specification. While such data should be available, experience has shown that the user often does not have the information in a usable format for handling qualities. One advantage of a mission-oriented specification is to encourage the user to develop a precisely defined set of requirements in a format that can be used in the handling-qualities specification, and therefore as functional requirements for the flight control system. Armed with this information, and by using the specification methodology, it is possible to define a required Response-Type, and to define preliminary flight control system design in the pre-proposal stage. If these results show excessive complexity and cost, the user may choose to relax the severity of the required environment, consider less aggressive MTEs, or relax the required Level of flying qualities.

The mission-oriented specification, when properly used, puts all of these factors into a manageable framework, and therefore it can be used for preliminary design and associated cost–benefit studies. The downside of this is that the answers are often not consistent with the budget and schedule planned for the program. For example, modern military rotorcraft must be able to operate at night and in bad weather. The handling-qualities specification shows that this capability along with a low level of pilot workload requires considerable additional augmentation. There is a tendency for less-informed team members to blame the specification, i.e. the 'shoot the messenger syndrome'.

In summary, without a mission-oriented handling-qualities specification, accounting for the many factors that result in the aircraft handling qualities throughout the flight envelope, and for all environmental conditions, was impossible. As a result, deficiencies tended to show up as 'unpleasant surprises' during flight testing or, even worse, as operational problems. An overview of the details of the elements of modern mission-oriented handling-qualities specifications is presented in the remainder of this chapter.

6. Handling-qualities Levels

The basic premise of all recent handling-qualities specifications is that a 'pass' implies that the aircraft is satisfactory without improvement for the designated MTEs. This is quantified in terms of Cooper–Harper handling-qualities ratings (HQRs) as shown in Fig. 2. In practice, the quantitative criteria boundaries are a result of fairing lines through HQR data (from research experiments) according to the Level limits defined in Fig. 2. The underlying premise is that the flying-qualities tasks used to obtain the basic HQR data in the research environment are a valid representation of the MTEs that occur in the operational environment. In that context, the criterion boundaries are said to be a prediction of the ratings that would be given by pilots in accomplishing the MTEs.

Levels of flying qualities obtained from the quantitative criteria are based on the *prediction* of HQRs, whereas flying-qualities Levels obtained from the Demonstra-

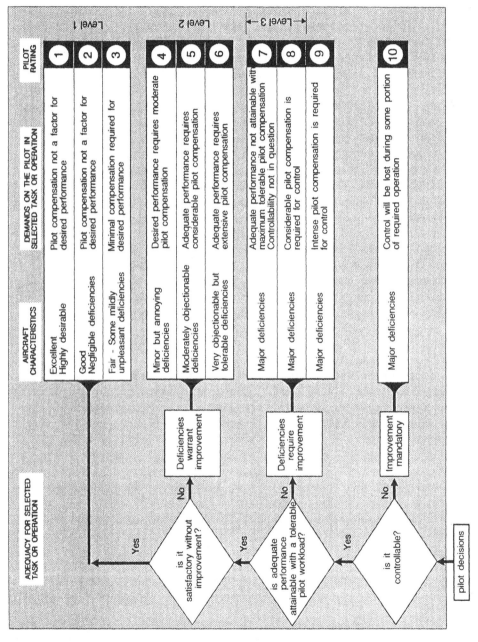

Figure 2. Definition of Levels based on Cooper–Harper handling-qualities rating (HQR) scale.

11

tion Maneuvers are based on HQRs that are *assigned* to the actual aircraft. When these two methods of specification compliance are in disagreement, the fundamental reasons must be identified to determine which is correct. Such disagreements are a warning flag, and indicate that either the quantitative criteria have been misapplied (or are incorrect), or the Demonstration Maneuvers are not being correctly performed, or are not appropriately defined. This type of scrutiny often exposes fundamental errors in the control law development and evaluation process, and should not be overlooked for the sake of meeting a schedule.

At present the rotorcraft flying-qualities MIL-Standard uses the HQR scale in Fig. 2 to define Levels, whereas the fixed-wing standard technically does not. MIL-STD-1797A 'officially' defines Levels through a separate set of definitions. A synopsis of the MIL-STD-1797A Level definitions is as follows: Level 1 implies that desired performance is achievable with no more than minimal pilot workload, Level 2 allows the mission to be completed with an increased workload and/or some degradation in performance, and Level 3 implies that the aircraft can be controlled, but that the mission cannot be successfully completed. Since all of the quantitative boundaries in MIL-STD-1797A are obtained by fairing lines of constant HQR, the 'official' definitions are never invoked. Defining the flying-qualities Levels directly in terms of HQRs results in a more straightforward connection with the data used to develop the criteria. In addition, using the HQR scale as the Level definition allows the specification to take advantage of many lessons learned with respect to Cooper–Harper pilot ratings, when testing compliance using the Demonstration Maneuvers. There are currently no Demonstration Maneuvers associated with MIL-STD-1797A, so that assigned ratings are not given. In practice, the methodology for obtaining predicted Levels is identical for the fixed- and rotary-wing specifications, i.e. the HQR data is averaged at the 3·5, 6·5 and 8·5 values to define the Level 1, 2, and 3 boundaries.†

Using the quantitative criteria, a predicted level of handling qualities can be obtained. Ideally, that will be Level 1. However, it may turn out that the costs associated with augmenting the airframe to Level 1 are outside the available budget. A decision to accept Level 2 with some negotiated limits on the quantitative criteria would be a potential way to use the specification criteria and supporting data. A modest degradation in handling qualities might be quite acceptable if the cost savings were significant. Such modifications to the specification would be handled on a case-by-case basis, with mutual agreement between the manufacturer and customer/user. If degradations from Level 1 exist, the PIO criteria will take on special importance, although they should be checked in any event (further discussed in § 13).

7. Response-Types

The mission-oriented flying-qualities specification takes into account the capability of ACT aircraft to exhibit characteristic response shapes that are entirely

† Whether or not it is correct to average HQRs has been the subject of much debate. Intuitively, it seems incorrect to average ratings that are based on the semantic meanings of phrases. However, analyses have shown that the Cooper–Harper phrases are indeed linear, at least for ratings between 1 and 7, and that averaging produces negligible errors (see Reference 14). This is fortunate, since essentially all handling-qualities criteria are based on fairing lines between the rating data that falls in the Levels as defined in Fig. 2. Such 'fairing' of the data is, of course, averaging.

different from classical aircraft. Examples are rate-command and attitude-command systems. Examples of Response-Types for fixed-wing aircraft are shown in Fig. 3. The Response-Types are identified primarily by the characteristic shape of their frequency response to control input for attitude and flight path. Such an output/input definition does not dictate a specific control system architecture, but does provide very specific guidelines for the control law designer.

The roles of the Response-Type classification in the mission-oriented flying-qualities specification are summarized below.

(a) Certain MTEs, sometimes in combination with a degraded visual environment (DVE), may require a specific Response-Type to achieve Level 1.

(b) Some Response-Types have unique requirements that the flight control system designer must be aware of.

(c) Some criteria are valid for certain Response-Types and not for others.

An example of the first of these Response-Type roles is the need for an attitude-command–attitude-hold (ACAH) Response-Type for low-speed and hover tasks in a DVE. Operational experience has shown that precision low speed and hover involves considerable additional pilot workload when the visual cuing is degraded because of a lack of fine-grained texture. This is usually found with night vision goggles (NVGs) or forward-looking infrared (FLIR) devices in critical conditions such as low-light levels (for NVGs) or a cold soak (for FLIR). There have been numerous accidents as a result of the inability of pilots to stabilize the helicopter, and still have sufficient excess workload capacity available to maintain an adequate level of situational awareness (Reference 15). Simulation, and flight testing, have shown that the use of an ACAH Response-Type in combination with altitude hold

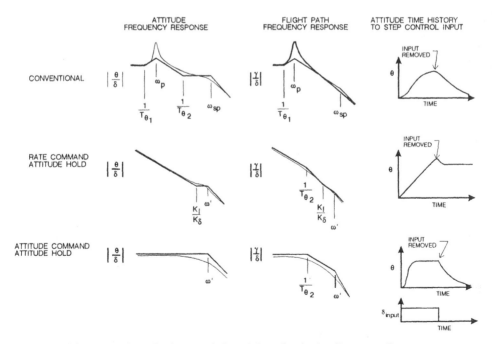

Figure 3. Generic characteristics of three fixed-wing Response-Types.

greatly reduces pilot workload in the DVE (References 16 and 17). These tests have shown that ACAH plus height hold is necessary to achieve Level 1 in conditions of moderately degraded visual cuing. For severely degraded visual cuing, a translational-rate-command (TRC) Response-Type has been shown to be necessary for Level 1.

There has not been very much research in this area for fixed-wing aircraft. However, there is strong evidence that ACAH, when properly mechanized, provides excellent handling qualities for landing (e.g. see Reference 18). Rate Response-Types have a tendency for larger longitudinal touchdown dispersions (a tendency to float), and the conventional Response-Type tends to be excessively responsive to turbulence. When properly mechanized, all of these Response-Types can be made to produce Level 1 flying qualities, but ACAH tends to produce the most consistent touchdown precision in critical conditions of wind and turbulence. That is probably because ACAH exhibits more phase margin in the flight path response in the frequency range of piloted crossover (approximately 0.3 to 1.0 rad s^{-1}) than conventional or rate Response-Types.

An understanding of the strengths, weaknesses and idiosyncrasies of each Response-Type is essential for a successful design of the flight control system (item (*b*) in the above list). For example, the ACAH Response-Type is highly dependent on trim, and therefore the trim system must be carefully optimized. In addition, the very same stability that produces excellent touchdown performance for the ACAH Response-Type causes problems for an aggressive maneuver such as a go-around in fixed-wing aircraft, and a quickstop in a helicopter. Some provision must be made to allow the pilot to make large attitude changes (such as a forward-loop parallel integrator that switches in at larger control deflections). The nature of ACAH is such that excessive phase lag will result in pitch bobbling. For this reason, it is more important to keep the phase delay parameter (see § 11) at, or preferably below, the Level 1 limit for this Response-Type, than for rate or conventional Response-Types. Finally, the ACAH Response-Type may require more control power than rate or conventional Response-Types. If adequate control power is not available, rate limiting may occur that would cause pitch bobbling at best, and possibly a PIO. The control system designer must be aware of these factors, both during the preliminary design when the Response-Type is being selected, and during the detailed development of the FCS. For example, if control power is limited, the ACAH Response-Type might not be a viable candidate for a fixed-wing aircraft. (Since ACAH is not required for Level 1 pilot ratings, the designer has the option to use other Response-Types. However, for a helicopter, this luxury would not be available if the mission includes operation in a DVE. In that case, more control power would be required resulting in a potential tradeoff between handling qualities and performance.)

For the rate Response-Type for fixed-wing aircraft, it is important that the pure rate response in pitch is not achieved at the expense of the flight path response as illustrated in Fig. 4. This example shows the flight path to stick response for a typical rate Response-Type. Two values of the integral-to-proportional gain ratio (K_I/K_δ) are illustrated. The value of 4.6 is representative of the F-16, which is known to be difficult to land consistently with precision. The plot with $K_I/K_\delta = 1.0$ is representative of the X-29, which did not exhibit such problems. It can be shown that for rate Response-Types, increasing values of K_I/K_δ tend to augment the pitch attitude response, but degrade the flight path response. In Fig. 4, the F-16 is seen to

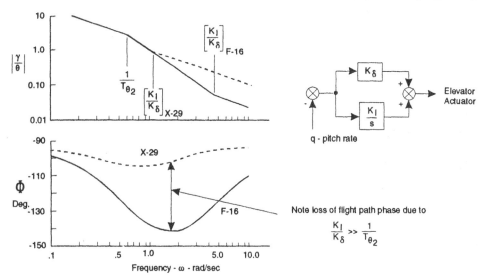

Figure 4.   Illustration of important characteristics associated with the rate Response-Type for fixed-wing aircraft.

exhibit considerably more phase lag than the X-29 in the flight path to stick response, which would be expected to degrade the handling qualities for landing. Generically, this result occurs because the lead at $1/T_{\theta_2}$ in the attitude response becomes a lag in the flight path response, and the region between $1/T_{\theta_2}$ and K_I/K_δ is K/s^2 (i.e. the lag is directly proportional to the ratio of K_I/K_δ to $1/T_{\theta_2}$). For the rate Response-Type, it is important to set K_I/K_δ close to $1/T_{\theta_2}$.

Conventional Response-Types with adequate values of equivalent short-period frequency suffer from a lightly damped phugoid mode and excessive response to turbulence and windshear. Complementary filtering of the angle-of-attack feedback is essential so that atmospheric turbulence does not excite the system. A possible alternative is to feed back inertial angle of attack ($\alpha_I = \theta - \gamma$), with a separate stall-warning (or envelope-limiting) system, similar to what would be used for ACAH. It is notable that the combination of a flight control system that holds angle of attack constant with pitch, and airspeed constant with an autothrottle, results in a flight path angle rate-command system. Experience has shown that flight path angle rate Response-Types experience similar touchdown precision problems as more conventional pitch-rate-command systems. It is therefore important either to turn off the autothrottle prior to flare, or to allow the pilot to override the autothrottles in the flare.

The above examples serve to illustrate that a knowledge of the fundamental characteristics of each Response-Type is essential for any tradeoff study that might be conducted to determine the proper Response-Type for the specified mission. For example, a trade study to determine the proper Response-Type for landing should include significant low-frequency turbulence and windshear on short final approach to expose potential weaknesses of the conventional Response-Type, an aggressive go-around (for the ACAH Response-Type), and precision touchdown requirements (for the rate Response-Type).

As noted by (*c*) in the above list, the Response-Type category also can be beneficial as a metric to define when certain criteria are applicable. The Lower-Order

Equivalent System (LOES) criterion in MIL-STD-1797A is a good example. This criterion is based on the special case of an equivalent system that uses the classical airplane Response-Type as a lower-order model (see Fig. 3). Hence it implicitly assumes that the shape of the response is that of a classical airplane. The fitting routine adjusts the equivalent short-period and phugoid frequencies and damping ratios to fit this model. Any phase lag that is left over is accounted for by the equivalent time delay τ_e. If the ACT control laws do not result in a classical Response-Type (i.e. frequency-response plots do not look like the classical airplane Response-Type in Fig. 3), it is simply not correct to fit the higher-order system to this model. The use of angle-of-attack feedback to augment the short period typically results in a classical Response-Type. Inertial feedbacks, such as pitch rate and pitch attitude, result in nonclassical response shapes such as the rate-command–attitude-hold (RCAH) and attitude-command–attitude-hold Response-Types in Fig. 3. Comparison of the Response-Type shapes in Fig. 3 clearly shows that it is not appropriate to attempt to fit the RCAH and ACAH shapes to the classical airplane shape. Unfortunately, this mistake continues to made. It is a good example of why a user's guide for compliance is needed. As an aside, note that the *general* concept of lower-order equivalent systems is valid for all Response-Types. Incorporation of this type of criterion would require the development of a database for each lower-order classification (RCAH, ACAH, etc.). The RCAH and ACAH Bode asymptotes shown in Fig. 3 would be the obvious choices for the LOES models for these Response-Types. However, in the context that the Bandwidth criterion applies to the short-term response of all Response-Types, it does not seem necessary to accomplish the considerable work and experimentation required to develop lower-order models for every conceivable ACT Response-Type.

8. Mission–Task–Elements (MTEs)

The concept of the MTE is to start with the total mission as defined by the user, and to break it down into elements that consist of handling-qualities tasks. As discussed earlier in this chapter, good handling qualities are only required for critical combinations of task and environment, and fall in the category of 'nice to have, but not essential' the rest of the time. Therefore, the MTEs must be defined to represent the critical tasks, where good handling is a necessity. The tasks that result are by definition not representative of normal operational activity. On the other hand, it is important to insure that MTEs do not require agility or precision that will never be required. Examples of MTEs for fixed- and rotary-wing aircraft are given below.

- *Offset landings.* Offset landings are commonly used (e.g. Reference 19) to force the pilot to be 'in the loop'. This is intended to expose problems that would show up in turbulence and windshear where the pilot is unable to enter the flare in a stabilized condition. The required touchdown precision should be based on the shortest runways that would be used by the aircraft when at maximum landing weight.

- *Accel/decel for helicopters.* This MTE is taken from the helicopter MIL-Standard and is described in detail in Fig. 5. It is an aggressive maneuver and probably would not be required for civilian rotorcraft.

The MTE in Fig. 5 illustrates several points that should be considered when making evaluations. First, it is defined in substantial detail to insure that all pilots will

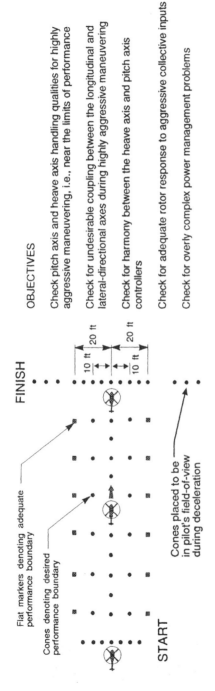

OBJECTIVES

Check pitch axis and heave axis handling qualities for highly aggressive maneuvering, i.e., near the limits of performance

Check for undesirable coupling between the longitudinal and lateral-directional axes during highly aggressive maneuvering

Check for harmony between the heave axis and pitch axis controllers

Check for adequate rotor response to aggressive collective inputs

Check for overly complex power management problems

Suggested Course for Acceleration-Deceleration MTE.

DESCRIPTION OF MANEUVER

Starting from a stabilized hover, rapidly increase power to approximately maximum, and maintain altitude constant with pitch attitude. Hold collective constant during the acceleration to an airspeed of 50 knots. Upon reaching the target airspeed, initiate a deceleration by aggressively reducing the power and holding altitude constant with pitch attitude. The peak attitude should occur just before reaching the final stabilized hover.

DESIRED PERFORMANCE

Complete the maneuver over the reference point at the end of the course. The longitudinal tolerance on the final hover position is plus zero and minus a distance equal to one half of the helicopter (positive forward).

Maintain altitude below 50 ft.

Maintain lateral track within ±10 feet of centerline.

Maintain heading within ±10 degrees.

Within 1.5 seconds from initiation of the maneuver, achieve at least the greater of 95% maximum continuous power or 95% maximum transient limit that can be sustained for the required acceleration, whichever is greater. If 95% power results in objectionable pitch attitudes, use the maximum nose down pitch attitude that is felt to be acceptable. This pitch attitude shall be considered as a limit of the Operational Flight Envelope.

Decrease power to full down collective within 3 seconds to initiate the deceleration. Significant increases in power are not allowable until just before the final stabilized hover.

Figure 5. Example of Demonstration Maneuver from the rotorcraft MIL-Standard.

17

perform it in the same way. Second, there are very specific desired and adequate performance boundaries. This is done because one use of the MTE is as a Demonstration Maneuver where the level of flying qualities is achieved via assigned Cooper–Harper HQRs. Lessons learned over the years have shown that the HQR scale only produces meaningful and repeatable results if the task and performance limits are well defined. This is further discussed in § 12.

The MTEs are used in three ways in the mission-oriented flying-qualities specification:

(a) MTEs are used to make a distinction between criterion boundaries for aircraft with different missions. This is illustrated in § 11 (e.g. see Fig. 7 in that section).

(b) The MTEs are used as Demonstration Maneuvers as shown in Fig. 5.

(c) The MTEs are used in combination with the UCE to determine the proper Response-Type (discussed further in the next section of this chapter).

9. Usable cue environment

It has become common practice to employ vision aids such as forward-looking infrared (FLIR), night vision goggles (NVGs), or even millimeter-wave radar to allow continued operation during night and bad weather. These vision aids have been highly successful as a means to see and avoid objects. However, under some adverse conditions they are not able to produce the high values of spatial frequency (fine-grained texture) necessary for the human operator to perceive translational rates that are required for tasks such as landing or low-speed and hover tasks in rotorcraft. The UCE scale was developed for the rotorcraft MIL-Standard to provide a minimum Response-Type requirement for conditions of degraded visual cuing (see Reference 17). The specification includes a method to quantify the visual environment in terms of the ability of the pilot to stabilize the rotorcraft (see the following section). It has not been determined whether such a requirement is necessary for fixed-wing aircraft.

Ideally, it should be possible to specify the display characteristics necessary for the human pilot to accomplish precision hover with an acceptable workload. This might be done in terms of the modulation transfer function (MTF – defined as spatial frequency vs. depth of modulation) of the display device. Unfortunately, the technology is not sufficiently mature to allow the specification of the MTF limits required for the pilot to accomplish low-speed and hover tasks with an acceptable level of workload. The methodology provided in the helicopter MIL-Standard provides an experimental determination of the quality of the display.

To understand the basis for this procedure, it is important to make a distinction between visual cues required to see objects vs. those required to stabilize the aircraft. It is assumed that the display devices provide imagery of sufficient quality to allow the pilot to see objects well enough to use them as a tracking reference, or to avoid them. The ability to use the available visual cuing to stabilize the aircraft is a less understood concept. Testing has shown that some level of fine-grained texture is required to provide the cues that a pilot uses to accomplish a precision hover with respect to some object (see Reference 16). For example, the current generation of NVGs produce sufficient texture with an illumination that occurs when the moon is greater than one-half. As the available light decreases, so does the ability of the NVGs to reproduce fine-grained texture (blades of grass, etc.).

Tests have shown that the effect of insufficient fine-grained texture in the pilot's field-of-view is identical to degrading the aircraft handling qualities. The specification methodology uses this finding as follows. The test aircraft consists of a Rate Response-Type with known Level 1 handling qualities. Certain MTEs are performed, and the pilots are required to rate their ability to be aggressive and precise according to the visual cue rating (VCR) scale shown in Fig. 6. At least three pilots are required to fly the tasks and the ratings are averaged across those pilots, but not across the tasks. The averaged VCRs are plotted on the UCE scale (Fig. 6) to determine a UCE for each task. Test data has shown (Reference 16) that for UCE = 1, no additional stabilization is required for Level 1 (e.g. NVGs with moon more than one-half). For UCE $\geqslant$ 2, additional stabilization is required. This is discussed in more detail in the following section of this chapter.

As of this writing, there has been very little work done to determine the applicability of the UCE methodology for fixed-wing aircraft. Some potential scenarios that might be amenable to providing workload relief with additional stabilization in fixed-wing aircraft are listed below.

- Air-refueling requires considerable precision, and has characteristics similar to hover in that the pilot must perceive and react to very small translational rates before they become large rates. Anecdotal comments have indicated that pilots must keep the tanker in the field-of-view to provide the necessary cuing. There is evidence that the addition of attitude stabilization might

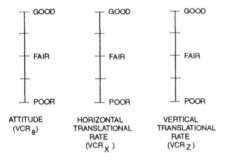

VISUAL CUE RATING SCALE

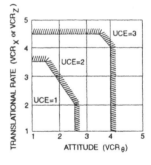

USABLE CUE ENVIRONMENT DEFINITION

Pitch, roll, and yaw attitude and lateral-longitudinal and vertical translational rate shall be evaluated for effectiveness for stabilization according to the following definitions.

Good: Can make aggressive and precise corrections with confidence and precision is good

Fair: Can make limited corrections with confidence and precision is only fair.

Poor: Only small and gentle corrections are possible, and consistent precision is not attainable

Figure 6. Definition of usable cue environment (UCE) scale from References 4 and 5.

reduce pilot workload for the basic air-refueling task (Reference 20), and could be especially valuable in conditions of poor visual cuing (e.g. in clouds).

- Very low-visibility landings (700 RVR and less) using head-up displays and synthetic vision have been demonstrated in the research environment (Reference 21). The head-up display (HUD) used in these tests included the usual symbology with the addition of superimposed imagery of the runway environment from sensors such as millimeter-wave radar. If that imagery does not provide the necessary texture, flare guidance using a flight director symbol on the HUD is required. It may be possible to relax the need for flight director guidance through additional stabilization (e.g. with an ACAH Response-Type).

10. Specification of Response-Type as a function of Mission–Task–Element and Usable Cue Environment

The Response-Types required to achieve Level 1 flying qualities are specified as a function of the UCE and MTEs in the helicopter MIL-Standard in Reference 5 as illustrated in Table 1. It is important to emphasize that the definitions of the Response-Types are based on fundamental properties that have been found to be required to achieve Level 1. To enhance physical interpretation, the Response-Types have been labeled with the augmentation that would typically be used to achieve the necessary response properties, i.e. attitude-command–attitude-hold (ACAH). A cursory review of Table 1 might lead one to conclude that the flying-qualities specification defines the control system architecture, thereby eliminating the possibilities for innovation. In fact, the Response-Type definitions have been developed in terms of first-principle requirements developed from flight and simula-tion test data as interpreted by the use of pilot–vehicle analysis techniques (Reference 22). For example, empirical test results have shown that attitude stabili-zation and altitude hold improve the ability to hover in the DVE. Pilot–vehicle analysis procedures were used to determine that this is a result of the fact that the pilot lead-generation requirements to accomplish a precision hover are substantially less with ACAH than for rate (workload is known to be a strong function of the required lead). Lead generation in hover requires that the pilot be able to perceive very small translational and angular rates, and it was concluded that these rates are much more difficult to perceive when the fine-grained texture is removed from the visual field.

These analysis results were used to generalize the definition of the Response-Type to include all responses that do not require excessive pilot lead generation to accomplish hover. As an example, the requirement for attitude-command Response-Types from the helicopter MIL-Standard is given below.

4.2.5 Character of Attitude Command Response-Types. If Attitude Command is speci-fied as a required Response-Type in Table 1, a step cockpit pitch (roll) controller force input shall produce a proportional pitch (roll) attitude change within 6 seconds. The attitude shall remain essentially constant between 6 and 12 seconds following the step input. However, the pitch (roll) attitude may vary between 6 and 12 seconds following the input, if the resulting ground referenced translational longitudinal (lateral) acceler-ation is constant, or its absolute value is asymptotically decreasing towards a con-

	UCE = 1		UCE = 2		UCE = 3	
	Level 1	Level 2	Level 1	Level 2	Level 1	Level 2
Required Response-Type for All MTEs. Additional Requirements for Specific MTEs are Given Below.	**RATE**	**RATE**	**ACAH**	**RATE + RCDH**	**TRC+RCDH+ RCHH+PH**	**ACAH**
Precision MTEs						
Hover			RCHH + RCDH			RCHH + RCDH
Hovering turn			RCHH			RCHH
Precision Landing			RCDH			RCDH
Pirouette			RCHH			RCHH
Slope landing			RCDH			RCDH
Tasks requiring divided attention (see 3.6.2)	PH+RCDH+RCHH		RCHH + RCDH			RCHH + RCDH
Aggressive MTEs						
Acceleration and deceleration			RCDH+RCHH			RCDH + RCHH
Rapid sidestep			RCDH+RCHH			RCDH+ RCHH
Vertical remask			RCDH			RCDH
Slalom			RCHH			RCHH
Aggressive turn to target			RCDH+RCHH			RCDH + RCHH
Aggressive bobup/down			RCDH			RCDH

Notes:

1. TRC is not recommended for pitch pointing tasks

2. The task descriptions for UCE=1 are found in Sections 5.8.1 and 5.8.2

3. The task descriptions for UCE-2 and 3 are found in Sections 5.8.4 and 5.8.5

4. The rank-ordering of combinations of Response-Types from least to most stabilization is defined as:
   1. Rate
   2. ACAH+RCDH
   3. Rate+RCDH+RCHH+PH
   4. ACAH+RCDH+RCHH
   5. ACAH+RCDH+RCHH+PH
   6. TRC+RCDH+RCHH+PH

5. A specified Response-Type may be replaced with a higher rank of stabilization. For UCE=1, it is important to insure that the higher rank of stabilization does not preclude meeting the Moderate and Large Amplitude requirements.

Definitions:

Rate => Rate or Rate Command Attitude Hold (RCAH) Response-Type (4.2.4 and 4.2.5)

ACAH => Attitude Command Attitude Hold Response-Type (4.2.5 and 4.2.6)

RCHH => Vertical-Rate Command with Altitude (Height) Hold Response-Type (4.2.8.1)

RCDH => Rate-Command with Heading (Direction) Hold Response-Type (4.2.9.2)

PH => Position Hold Response-Type (5.1.11)

TRC => Translational-Rate-Command Response-Type (5.1.12)

Table 1. Illustration of use of MTE and UCE to determine proper Response-Type (from Reference 5).

stant. A separate trim control must be supplied to allow the pilot to null the cockpit controller forces at any achievable attitude.

The obvious way to achieve this Response-Type is through the feedback of pitch attitude, but it is defined in such a way that any control system that provides the required translational acceleration profile to a step input would be acceptable. As an example, during the RAH-66 Comanche competition, one manufacturer successfully complied with this Response-Type definition by means of a linear acceleration command system without attitude feedback. That flight control system was highly innovative, and is an example of the flexibility of the Response-Types in Table 1.

11. Quantitative criteria

Prior to the implementation of Response-Types and UCE, the flying-qualities specification consisted entirely of a collection of quantitative criteria. These criteria continue to play a major role in the specification of flying qualities.

11.1. *Short-term small-amplitude attitude changes*

The short-term, small-amplitude pitch, roll and yaw response criteria are of primary importance because they are a measure of the ability of the pilot to control the aircraft with precision. Not surprisingly, many existing criteria are focused on the short-term pitch response of aircraft. MIL-Standard 1797A contains no less than six short-term pitch response criteria (LOES/CAP, Bandwidth, Time Response, Dropback, Nichols Chart and Neal–Smith). In addition a currently planned revision will include the Smith–Geddes criterion, a short-term small-amplitude requirement for PIO prevention. Our recommendation at this time is to select one criterion for MIL-STD-1797B, and to use the others for supporting analyses, if additional insight is desired. A detailed discussion of the rationale for this recommendation is given in References 6 and 7. The existence of multiple criteria for short-term small-amplitude maneuvering in MIL-STD-1797A requires that all users be handling-qualities experts who understand the strengths and weaknesses of each. This is simply not practical, and may explain why control systems engineers have avoided the use of the handling-qualities specification criteria. Since MIL-STD-1797B is not scheduled for completion until 1997, we recommend using the latest version of the Bandwidth/Phase Delay/Dropback criteria set for short-term small-amplitude maneuvering because they (1) consistently correlate the entire database better than other criteria, (2) are easy to apply both for analysis of flight test data and for control system design, (3) apply to all Response-Types and (4) explicitly account for attitude and flight path control. These criteria are contained in References 6 and 7, and are summarized in Fig. 7†. This set of criteria has undergone continual refinement by a number of researchers throughout the United States and Europe since the late 1970s (see Reference 6).

The rotorcraft MIL-Standard (Reference 5) uses the Bandwidth/Phase Delay criteria exclusively for short-term small-amplitude maneuvering. As a result there has been a substantial amount of experience gained with testing for compliance for a number of rotorcraft, most notably the Boeing/Sikorsky RAH-66 Comanche. The US Army Aeroflightdynamics Directorate at NASA Ames Research Center in coordination with the Airworthiness Qualification Test Directorate at Edwards Air Force Base have developed flight test techniques and a computer program for generating high-quality frequency-response data (e.g. see References 23 and 24).

The original version of the Bandwidth criterion for precise up-and-away tasks (Category A) was excessively stringent. It is still in MIL-STD-1797A, so it is important to use the criteria in Fig. 7 as a functional requirement for a flight control

---

† The boundaries in Fig. 7 are not complete, and the user should consult References 6 and 7 before applying them.

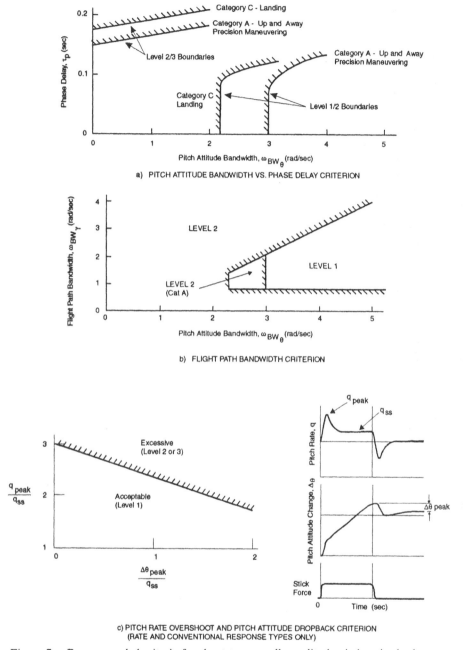

Figure 7. Recommended criteria for short-term small-amplitude pitch attitude changes.

system design, not the MIL-Standard. The relaxed boundaries shown in Fig. 7 resulted from the incorporation of a version of the Dropback criterion into the short-term small-amplitude criteria set. It was found that the fundamental deficiency of configurations that supported the high-bandwidth limit was dropback, not bandwidth. The boundaries shown in Fig. 7 successfully correlate most of the existing handling-qualities database (see Reference 6).

As noted above, there are numerous criteria for small-amplitude short-term control. The recommendation to use Bandwidth/Phase Delay/Dropback for the handling-qualities specification requires some justification, which is summarized below.

- *LOES/CAP†*. This is an excellent criterion for classical Response-Types, and should be used as a supporting criterion for the design of flight control systems with an architecture that results in a conventional Response-Type. It is not selected as a specification criterion because the version of the equivalent system criterion in MIL-STD-1797A is based on the short-period approximation for conventional Response-Types, and therefore does not apply to other Response-Types. When using this criterion, it is very important to keep $1/T_{\theta_2}$ fixed during the matching process, or to match the attitude and normal acceleration responses simultaneously (as required by the specification). This is because CAP is a measure of maneuver margin, which is a flight path parameter (stick-force/g is directly proportional to maneuver margin).

$$\text{CAP} = \frac{\bar{c}G_e W}{I_{yy}} \left(-\frac{dC_m}{dC_L} - \frac{\rho S \bar{c}}{4m} C_{m_q} \right)$$

$$\text{Maneuver margin} = N_m = -\frac{dC_m}{dC_L} - \frac{\rho S \bar{c}}{4m} C_{m_q}$$

The first term in the equation for CAP is the mean aerodynamic chord × elevator gearing × aircraft weight over the pitch moment of inertia.

Fixing $1/T_{\theta_2}$ and matching the pitch and normal acceleration responses simultaneously have the identical effect of insuring that the integrity of the flight path response is not lost in the matching process. This can be seen by noting that the relationship between flight path and pitch attitude is well approximated as $\gamma/\theta \approx 1/(T_{\theta_2}s + 1)$.

- *Time-Response*. The Time-Response criterion in MIL-Standard 1797A requires that the critical phase rolloff parameter be measured in the time domain. The problem is one of sensitivity since 0.05 seconds of delay is worth approximately one HQR (see Reference 9). Experience has shown that it is impossible to measure the delay from a time history with the necessary accuracy. Problems include obtaining an ideal step, turbulence and repeatability. The fundamental connection between time delay and HQRs is the phase rolloff which is clearly better measured in the frequency domain (e.g. see References 7 and 10).

Finally, this criterion was obtained by mapping the CAP criterion into a pitch attitude response in the time domain. As noted above, CAP is a measure of maneuver margin, which is a flight path (stick-force/g) parameter. Since CAP is directly related to the flight path response, it is probably not correct to map this parameter into a pitch attitude response, and to use that pitch response as the criterion.

- *Neal–Smith criterion*. This is a highly insightful criterion that is based on closed-loop pilot modeling. It has been successfully employed by a number

---

† CAP is the control anticipation parameter (see Reference 9).

of researchers to identify important trends in flying qualities within a given experiment. It is not recommended as a specification criterion for the following reasons. The assumed pilot model includes a 3 dB droop that can cause one of two things to happen: (1) the solution does not converge, or (2) the predicted pilot model contains excessive phase lag. These results have been observed for conventional aircraft, known to have Level 1 flying qualities.† The assumed pilot model also includes a time delay to account for the neuro-muscular lag in the pilot. Experience has shown that the results are sensitive to the value of delay, and several values have been assumed by various researchers. The criterion requires a computer program to iterate on a solution making it awkward to use (i.e. given a frequency-response plot, it is not possible to determine directly if the configuration meets the criterion). For the analysis of flight test data, it is necessary to develop a transfer function model of the aircraft before the criterion can be applied. This requires the assumption of a task bandwidth, and the results are very sensitive to the value selected. Finally, the criterion only characterizes the attitude response so that deficiencies related to the flight path response will not be identified. In spite of these shortcomings as a flying-qualities specification requirement, the Neal–Smith criterion provides good insight for comparisons between competing configurations. In that context it represents a structured way to conduct pilot–vehicle analyses, with the criterion boundaries providing a basis for comparisons between configurations.

- *Gibson Nichols Charts.* This criterion characterizes the pitch attitude frequency response on a plot of gain vs. phase. Envelopes are defined on this grid to separate satisfactory and unsatisfactory responses, but there has not been any effort to correlate pilot ratings to define Levels of handling qualities (e.g. see Reference 9). The primary drawback of envelope-based criteria for a specification is that it is difficult to specify a pass/fail criterion without being overly stringent (i.e. any excursion outside the envelope is a failure). In addition, this criterion does not account for flight path control. It is applicable only to fighters, and predicts poor handling qualities for ACAH and similar Response-Types. The phase-rate part of the criterion is identical to the Bandwidth Phase Delay.

The above criteria are described in detail in Appendix A of MIL-STD-1797A (Reference 9).

Numerous other criteria have been developed, but not generally adopted by the user community for various reasons. The most familiar of these are the C* criterion (Reference 26) and the Smith–Geddes criterion (Reference 27). The C* (C-star) criterion was not included in MIL-Standard 1797A, but has been used rather extensively by flight control system designers. It is based on envelopes of the C* parameter which is a combination of pitch rate and normal acceleration vs. time. The C* criterion has been shown to be ineffective as a metric to correlate HQRs (Reference 28), and has therefore received very little attention in the handling-qualities literature for the past 15 years. However, it is mentioned here because it tends to be

† Several researchers have developed alternative pilot models, e.g. Reference 25. These should be employed if the criterion is selected for use as guidance during the development of the flight control system.

adopted by control system designers, especially if state-space optimal control techniques are employed. The time-response envelopes of this criterion tend to be ideally suited as a cost function metric for these techniques. Unfortunately, time-response metrics have been shown to be unsuitable as a measure of short-term small-amplitude handling qualities. C* has been adopted by some large airframe manufacturers as a name for control laws that use pitch rate and normal acceleration as feedbacks. It is important not to confuse the criterion with the control law, since all they have in common is pitch rate and normal acceleration as parameters.

The Smith–Geddes handling qualities and PIO criteria are extremely simple to apply (Reference 27), but have several shortcomings. The most significant is that they exhibit a strong tendency to predict poor flying qualities for aircraft with known good flying qualities, especially for the power-approach flight condition. For example, Reference 7 shows that the Smith–Geddes criteria correctly predicted PIO in only 17 of 51 cases for precision-landing tasks. In all cases, the failure of the criteria was to predict a PIO when no such tendencies were noted by the pilots. The prediction of PIO is always due to a low neutral-stability frequency. It is not possible to increase this frequency substantially without increased feedback gains. Therefore, compliance with the criteria tends to result in excessively high gains that can lead to actuator rate limiting (and actually induce a PIO). The criteria do not include a parameter that is sensitive to the very important phase curve shape, and use a straight-line fit to the amplitude plot (of the frequency response) to calculate the criterion frequency (Reference 7). Both of these are important deficiencies that must be resolved before either the handling qualities or PIO versions of the criterion should be used as a guide for flight control system design. There is essentially no data to support claims of a high success rate, and considerable data to show that it consistently predicts Level 2 and 3 flying qualities for aircraft that are rated Level 1, and PIO tendencies where none have been found. For example, the full-up F-15 is predicted to be PIO prone, whereas pilots are essentially unanimous in their opinion (a rare event) that the aircraft has excellent handling qualities and no PIO tendencies.

11.2. *Moderate-amplitude attitude changes*

For many years, handling-qualities research focused almost exclusively on the short-term small-amplitude response characteristics of aircraft. Large-amplitude (full-control input) maneuvering has been confined almost entirely to control sizing. Moderate-amplitude maneuvering has become more important for aircraft that employ ACT flight control systems, because actuator rate limiting tends to occur during such maneuvers. Rate limiting can result in a sudden change in the aircraft response, especially if the unaugmented aircraft has poor stability characteristics. There have been several instances of PIOs that can be traced to actuator rate limiting.

An Attitude Quickness criterion was developed for the helicopter MIL-Standard, specifically to identify flight control system problems associated with moderate-amplitude maneuvering (Reference 5). A proposed Attitude Quickness requirement for fixed-wing aircraft is shown in Fig. 8, taken from Reference 6. It is recommended that this boundary be used as part of the functional requirements for fixed-wing flight control systems until MIL-STD-1797B becomes available.

The $p_{pk}/\Delta\phi$ parameter (Fig. 8) is equal to the bandwidth for small amplitudes. Attitude Quickness therefore represents a logical extension to the short-term small-

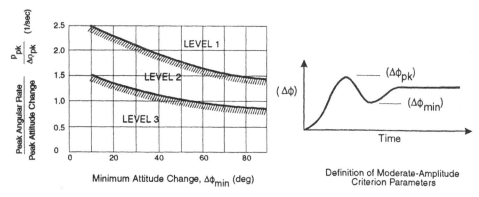

Figure 8.   Attitude Quickness criterion proposed for fixed-wing aircraft roll control (Class IV, Category A).

amplitude criteria shown in Fig. 7(*a*), to account for more aggressive maneuvering. It is also suggested that it represents an ideal parameter to define the 'agility' of the aircraft.

Flight test data for this criterion is generated with open-loop, pulse-like control inputs of varying magnitudes. It is very important that the test pilots do not obtain data for this criterion using closed-loop control to capture an attitude. The proper flight test technique involves a rapid stick displacement away from trim and back to trim with as little overshoot as possible. If the pilot compensates to achieve a target attitude, he or she is acting like a stability augmenter, which would cause the data to be overly optimistic.

12. Demonstration Maneuvers

The Demonstration Maneuvers were developed for the rotorcraft MIL-Standard in Reference 5 in recognition of the fact that the quantitative criteria are not perfect. For example, there is little useful guidance for sidestick controllers. Satisfactory characteristics for such controllers depend on a large number of variables such as the required maneuvers and the Response-Type, so it is unlikely that there will be a valid general set of criteria in the near future. New aircraft developments that utilize unusual controllers (i.e. not a centerstick or wheel) must rely on internal testing to obtain desirable characteristics.

The Demonstration Maneuvers are intended to provide for an overall check on the quantitative criteria, where those criteria do not have adequate supporting data or are nonexistent. As a word of caution, it is important not to ignore the quantitative criteria and to test only the Demonstration Maneuvers. There is a strong temptation for flight test organizations to treat the Demonstration Maneuvers as if they were the entire specification since they are easily tested and do not require supporting analysis. Furthermore, the testing is more interesting since the pilots can make a direct evaluation of the aircraft as opposed to simply generating data to be used for criteria.

If the visual environment is a factor, the Demonstration Maneuvers should be conducted in the degraded visual environment (DVE). This requires that this environment be defined, and that the display device be tested in a known Level 1 aircraft to establish the UCE (see Fig. 6). In some cases it is possible to make a valid estimate of the UCE. For example, it is reasonably well known that NVGs with a quarter-moon or less illumination result in UCE = 2.

Initial evaluations of a flight control system are always made using a piloted simulator. The Demonstration Maneuvers provide excellent handling-qualities tasks for these evaluations. However, experience has shown that the ability of digital image generators to produce fine-grained texture is consistent with UCE = 2 (i.e. the visual environment is degraded for the purpose of a handling-qualities evaluation). This provides a dilemma, in that there is no way for a pilot to make a distinction between poor handling qualities and the DVE. This problem is restricted to tasks where the visual cuing is an essential element of the closed-loop piloting task. Examples are low-speed and hover tasks for rotorcraft and precision-landing tasks for fixed-wing aircraft. Experience has shown that the touchdown sink rate of an aircraft will be approximately 3 ft s^{-1} higher in ground-based simulation than in flight.

It is easily seen that the use of the simulator to make changes in the flight control system can lead to very misleading results. The US Army Aero-flightdynamics Directorate has developed a procedure to quantify the simulator visual system by obtaining UCE ratings in the simulated day environment. These are referred to as SIMDUCE (simulated day UCE) ratings (see Reference 29). Demonstration Maneuvers conducted on the simulator with a SIMDUCE of 2 or 3 are only useful to spot deficiencies that are not related to precision maneuvering with respect to outside references. The HQRs obtained in this environment would be expected to be Level 2, for a Level 1 aircraft.

The available alternatives to ground-based simulation for control system design are: (1) make changes after the prototype is flying (expensive) and (2) use a variable-stability aircraft to augment ground-based simulation during the control system development. The latter approach has proven to be a wise decision many times over, and aircraft that are developed without such testing nearly always exhibit problems. These problems are much more difficult and expensive to resolve late in the program, when the prototype is flying. If variable-stability testing is not possible, strict adherence to the quantitative criteria is essential.

At this time Demonstration Maneuvers are included in the rotorcraft MIL-Standard in Reference 9 for the following categories:

- Precision tasks in the good visual environment (GVE)
- Aggressive tasks in a good visual environment
- Precision tasks in the degraded visual environment (DVE)
- Moderately aggressive tasks in the degraded visual environment

The MTEs are identical for the GVE and the DVE, but the performance standards are relaxed in the DVE to account for the inability of the pilots to see well enough to be aggressive. Furthermore, it is not practical to require the aggressive maneuver standards for the more stable flight control system architecture that is necessary to reduce pilot workload in the DVE (e.g. ACAH).

There are currently no Demonstration Maneuvers in the fixed-wing MIL-Standard. Work has been ongoing to develop such maneuvers under an Air Force Contract (Standard Evaluation Maneuvers or STEMS, see Reference 30). Unfortunately, these maneuvers have not been developed in the context of a rigorous handling-qualities experiment (to produce valid HQRs), and therefore would require refinement to be included as fixed-wing Demonstration Maneuvers.

12.1. *Instructions for the conduct of the Demonstration Maneuvers*

Years of experience with handling-qualities flight testing have provided many lessons. These have been incorporated into the conduct of the Demonstration Maneuvers to insure valid results. Excerpts from the helicopter MIL-Standard are provided below to illustrate the methodology.

> Only those Mission-Task-Elements designated by the procuring activity must be accomplished. The Demonstration Maneuvers must be accomplished by at least three pilots. These pilots shall assign subjective ratings using the Cooper–Harper Handling Qualities rating scale. The arithmetic average of the Cooper–Harper ratings must be $\leqslant 3.5$ when operating in the Operational Flight Envelopes, and $\leqslant 6.5$ in the Service Flight Envelopes. All individual ratings and associated commentary must be documented and supplied to the procuring activity. It is desired that the maneuvers be performed at the Normal States within the Operational Flight Envelope that are most critical from the standpoint of flying qualities. It is emphasized that performance is not an issue in these tests, and that the flight conditions should be selected accordingly. It is not necessary to test in a high density altitude environment unless the aircraft has been shown to not meet the quantitative requirements for such a condition, and satisfactory results in this section are being used to support a deviation.
>
> The use of the Cooper–Harper Handling Qualities Rating Scale [see Fig. 2] requires the definition of numerical values for desired and adequate performance. These performance limits are set primarily to drive the level of aggressiveness and precision to which the maneuver is to be performed. Compliance with the performance standards may be measured subjectively from the cockpit or by the use of ground observers. It is not necessary, or even desirable, to utilize complex instrumentation for these measurements. Experience has shown that lines painted on the aircraft and markers on the ground are adequate to perceive whether the rotorcraft is within desired or adequate performance parameters. In any event, the contractor shall develop a scheme for demonstrating compliance which uses at least outside observers and in-cockpit observations. This plan will be subject to approval by the government.
>
> The evaluation pilot is to be advised any time his performance fails to meet the desired or adequate limits, immediately following the completion of the maneuver, and before the pilot rating is assigned. In cases where the performance does not meet the specified limits, it is acceptable for the evaluation pilot to make as many repeat runs as necessary to insure that this is a consistent result. Repeat runs to improve performance may expose handling qualities deficiencies that arise when the pilot increases his gain. Such deficiencies should be an important factor in the assigned pilot rating.
>
> If the inability to meet a performance standard in good visual conditions is due to a lack of visual cuing, the test course should be modified to provide the required pilot cues. This is allowed in the context that the purpose of these maneuvers is to check aircraft handling, not problems associated with a lack of objects on the test course. However, once the test course is established for the GVE, it should not be modified to conduct the maneuvers in the degraded visual environment (DVE), unless such modifications are specifically noted in those sections.

Experience with the Demonstration Maneuvers has resulted in lessons that should be taken into account for future testing.

- The MTEs are specified in considerable detail to insure that each evaluation pilot flies the maneuver in the same way. This level of detail causes the maneuvers to require some practice to learn. New users may object to this, and will tend to develop new maneuvers that they understand (but are usually no less complex). It is important to avoid this tendency since the specification maneuvers usually are designed with specific objectives that may not be immediately obvious to the testing activity.

- It is tempting to save resources by cutting corners on the setup of the test course. This produces invalid results, since the test course provides the cues that drive the pilot's level of aggressiveness.

- It is easy to become overly preoccupied with the measurement of performance against the desired and adequate standards, probably because it is easier to judge pass/fail on the basis of performance than pilot workload. However, the pilot workload is intended to be the pass/fail criterion, and the performance standards are only intended to drive the level of pilot aggressiveness. If discussions relative to excessively expensive instrumentation occur, this is probably the culprit.

- As noted above, the intent of these maneuvers is to identify excessive pilot workload. Therefore, the pilots must use the Cooper–Harper scale in a way that emphasizes workload, not performance. For example, if desired performance was obtained, but the pilot felt that there were moderately objectionable deficiencies, the proper HQR would be 5, not 4 (see the scale in Fig. 2). Comments to the effect that 4 is the proper rating because desired performance was achieved indicates that workload is not receiving the proper emphasis.

13. PIO – a special case?

Experience has shown that schedule and cost constraints nearly always result in compromises during the development of the flight control systems of modern aircraft. These compromises typically result in degradations in aircraft handling qualities, and therefore failure to meet the stringent Level 1 limits. Most (if not all) current highly augmented aircraft do not meet all of the Level 1 limits of MIL-STD-1797A, or the helicopter MIL-Standard in Reference 5 and are not considered to be Level 1 for all mission tasks. Typical problems are slow actuators, stick filters necessitated by the use of force transducers, bending-mode filters, anti-aliasing filters, computational time delay, excessive parallel integrator gains, and improper flight control system design.

The problem is amplified by the fact that the use of active control technology often results in a significant modification of bare-airframe response characteristics. For example, the unaugmented X-29A doubles amplitude in well under 1 second, and is completely uncontrollable by the pilot. The augmented aircraft is well damped with a crisp response characteristic. This requires considerable pitch control power that is normally achieved with increased control surface sizing. Note that this is in direct conflict with the primary performance objectives that led to active control design in the first place. The result is smaller surfaces that must move very rapidly, leading to actuator rate limiting as an inherent problem in the design. In that context, it is not surprising that we have seen a strong tendency for PIO in highly augmented aircraft. On the other hand, strict adherence to the flying-

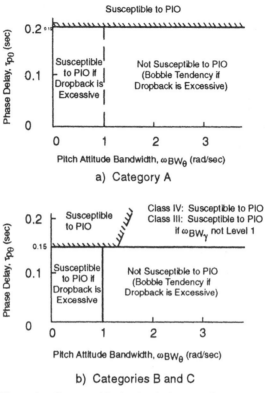

Figure 9. Proposed PIO criteria from Reference 7.

qualities specification could result in excessively large control surfaces. A criterion is needed that allows the tradeoff between performance and handling qualities to be made while accounting for the inviolable PIO limits.

If we accept the basic premise that Level 2 handling qualities are good enough to accomplish the mission, albeit with increased pilot workload, then it is inevitable that in the context of the above tradeoffs between performance and handling, plus the usual budget and schedule crises, Level 2 can be acceptable. This has been a fact of life for all modern aircraft development programs. Given the adaptability of the human pilot, and the proper 'can-do' attitude of professional and military pilots, the Level 2 aircraft is an acceptable compromise and gets the job done. When the deficiency that causes the aircraft to fall in a Level 2 region has a potential for a catastrophic PIO, however, it takes on a far more important role than simply a requirement for increased pilot workload. Therefore, it is important to make a distinction between deficiencies that cause a flying-qualities parameter to predict Level 2, and one that can result in a dangerous PIO.

Work is currently in progress to develop a standalone PIO criterion. At this time the most promising criterion is a version of the Bandwidth/Phase Delay boundaries shown in Fig. 7. This PIO criterion successfully correlated a large percentage of known PIOs (see Reference 7), and is shown in Fig. 9. The boundaries in the figure should be considered tentative since work is in progress to continue investigating other criteria for the prediction of PIO.

31

14. Summary

The methodology for mission-oriented handling-qualities specifications is reasonably mature. The rotorcraft MIL-Standard is up to date and should be used directly to develop functional requirements for flight control system design. The fixed-wing MIL-STD-1797A has not had a major revision since its inception in 1982, and is out of date in some areas. Therefore, functional requirements for fixed-wing flight control systems should be based on MIL-STD-1797A, with the recommended revisions from Reference 6 (Figs 7 and 8). In addition, the PIO criteria from Reference 7 should be incorporated (Fig. 9). These caveats will no longer be required after the planned major upgrade to MIL-STD-1797B in 1997.

REFERENCES

1 COOPER, GEORGE E. and HARPER, ROBERT P. JR., 1969, The use of pilot rating in the evaluation of aircraft handling qualities. NASA TN D-5153, Apr.

2 BERRY, DONALD T., 1982, Flying qualities: a costly lapse in flight control design? *Aeronautics and Astronautics*, Apr.

3 HOH, ROGER H. *et al.*, 1982, *Proposed MIL Standard and Handbook – Flying Qualities of Air Vehicles, Volume II: Proposed MIL Handbook*. AFWAL-TR-82-3081, Nov.

4 Handling qualities requirements for military rotorcraft. US Army Aviation Systems Command, ADS-33C, Aug. 1989.

5 HOH, ROGER H., MITCHELL, DAVID G., KEY, DAVID L. and BLANKEN, CHRIS L., 1993, Military standard – rotorcraft flight and ground handling qualities, general requirements for. US Army Aviation and Troop Command, MIL-STD-XXXX, 24 Nov. (draft).

6 MITCHELL, DAVID G., HOH, ROGER H., APONSO, BIMAL L. and KLYDE, DAVID H., 1994, Proposed Incorporation of mission-oriented requirements into MIL-STD-1797A. WL-TR-94-3162, Oct.

7 MITCHELL, DAVID, G. and HOH, ROGER H., 1995, Development of a unified method to predict tendencies for pilot-induced oscillations. WL-TR-95-3049, June.

8 HOH, ROGER H., 1990, Lessons learned concerning the interpretation of subjective handling qualities pilot rating data. AIAA 90-2824, presented at the AIAA Atmospheric Flight Mechanics Conference, Portland, OR, Aug.

9 Military standard, flying qualities of piloted aircraft. MIL-STD-1797A, 30 Jan. 1990.

10 HOH, ROGER H., 1988, Unifying concepts for handling qualities criteria. AIAA Paper No. 88-4328, AIAA Atmospheric Flight Mechanics Meeting, Minneapolis, Minnesota, Aug.

11 CHALK, C. R. *et al.*, 1969, Background information and user guide for MIL-F-8785B(ASG), Military specification – flying qualities of piloted airplanes. AFFDL-TR-69-72, Aug.

12 CHALK, CHARLES R., DiFRANCO, DANTE A., LEBACQZ, J. VICTOR and NEAL, T. PETER, 1973, Revisions of MIL-F-8785B (ASG) proposed by Cornell Aeronautical Laboratory under Contract F33615-71-C-1254. AFFDL-TR-72-41, Apr.

13 MOORHOUSE, DAVID J. and WOODCOCK, ROBERT J., 1982, Background information and user guide for MIL-F-8785C, Military specification – flying qualities of piloted airplanes. AFWAL-TR-81-3109, July.

14 MITCHELL, DAVID G. and APONSO, BIMAL L.,1990, Reassessment and extensions of pilot ratings with new data. AIAA 90-2823, Aug.

15 HOH, ROGER H., 1990, The effects of degraded visual cuing and divided attention on obstruction avoidance in rotorcraft. Hoh Aeronautics, Inc., TR 1003-1, Dec.

16 HOH, ROGER H., 1986, Handling qualities criterion for very low visibility rotorcraft NOE operations. Presented at the AGARD Flight Mechanics Panel Meeting, Rotorcraft Design for Operations, Amsterdam, Oct.

17 HOH, ROGER H., BAILLIE, STEWART W. and MORGAN, J. MURRAY, 1987, Flight investigation of the tradeoff between augmentation and displays for NOE flight in low visibility. Presented at the Mideast Region National Specialists' Meeting for Rotorcraft Flight Controls and Avionics, American Helicopter Society, Cherry Hill, NJ, 13–15 Oct.

18 MITCHELL, DAVID G., APONSO, BIMAL L. and McRUER, DUANE T., 1990, Task-Tailored Flight Controls: Volume 1 – Ultra Precision Approach and Landing System. NASA CR-182066, Nov.

19 BERTHE, C. J., CHALK, C. R. and SARRAFIAN, S., 1984, Pitch rate flight control systems in the flared landing task and design criteria development. NASA CR-172491, Oct.

20 TASCHNER, MICHAEL J., 1993, A limited handling qualities investigation of rate command/attitude hold and attitude command/attitude hold response-types in the probe and drogue air refueling task. AFFTC-TLR-93-38, Dec.

21 BURGESS, MALCOLM A. *et al.*, 1993, *Synthetic Vision Technology Demonstration – Volume 1 of 4 – Executive Summary*. DOT/FAA/RD-93/40, I, Dec.

22 McRUER, D. T., and KRENDEL, E. S., 1974, Mathematical models of human pilot behavior. AGARD AG-188, Jan.

23 HAM, JOHNNIE A., GARDNER, CHARLES K. and TISCHLER, MARK B., 1993, Flight testing and frequency domain analysis for rotorcraft handling qualities characteristics. American Helicopter Specialists Meeting, 'Piloting Vertical Flight Aircraft, A Conference on Flying Qualities and Human Factors', San Francisco, CA, Jan.

24 TISCHLER, MARK B., and CAUFFMAN, MAVIS G., 1994, CIFER Version 2.1, Comprehensive Identification From FrEquency Responses – An interactive facility for system identification and verification. Aeroflightdynamics Directorate, US Army Aviation and Troop Command, Ames Research Center, Moffett Field, CA, NASA CP10149, USAATCOM TR94-A-017.

25 CHALK, C. R., 1980, Calspan recommendations for SCR flying qualities design criteria. NASA CR-159236, Apr.

26 TOBIE, H. N., ELLIOTT, E. M. and MALCOLM, L. G., 1966, A new longitudinal handling qualities criterion. National Aerospace Electronics Conference, Dayton, Ohio, 16–18 May.

27 SMITH, RALPH H., 1977, A theory for longitudinal short-period pilot induced oscillations. AFFDL-TR-77-57, June.

28 SMITH, ROGERS E., 1978, Effects of control system dynamics on fighter approach and landing longitudinal flying qualities. AFFDL-TR-78-122, Mar.

29 KEY, DAVID L. and HOH, ROGER H., 1987, New handling-qualities requirements and how they can be met. 43rd Annual Forum Proceedings, American Helicopter Society, St. Louis, MO, May, pp. 975–990.

30 CORD, T. J., 1989, A standard evaluation maneuver set for agility and the extended flight envelope – an extension of HQDT. AIAA Atmospheric Flight Mechanics Conference, Aug., pp. 92–95.

2

System identification methods for aircraft flight control development and validation

MARK B. TISCHLER

1. Introduction

System identification is a procedure for accurately characterizing the dynamic response behavior of a complete aircraft, subsystem or individual component from measured data. This key technology for modern fly-by-wire flight control system development and integration provides a unified flow of information regarding system performance around the entire life-cycle from specification and design through development and flight test (Fig. 1). A similar 'roadmap' for application of system identification methods to rotorcraft development was previously proposed by Schrage in a comprehensive report dedicated to this topic (Hamel 1991). An excellent historical summary and overview of system identification is given by Hamel and Jategaonkar (1995). System identification has been widely utilized in recent aircraft programs including many of those described in the present volume. Common applications for flight control system development include: definition of

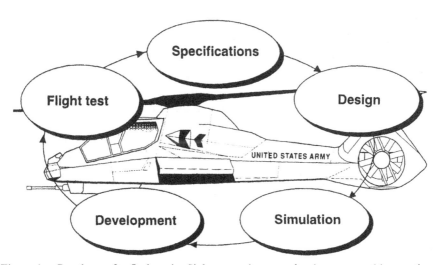

Figure 1.   Road map for fly-by-wire flight control system development and integration.

system requirements, specification and analysis of handling qualities, evaluation of proposed control law concepts, validation and improvement of complex simulation models, validation of subsystem components and development facilities, and flight test optimization of control laws.

Frequency-domain identification approaches are especially well suited to the development and validation of flight control systems. Feedback stability and noise amplification properties are determined from the broken-loop frequency response, and characterized by metrics such as crossover frequency, and associated gain and phase margins. Command tracking performance is determined from the closed-loop frequency response, and characterized metrics such as bandwidth and time delay, and equivalent system eigenvalues. Frequency-domain identification approaches allow the direct and rapid (including real-time) identification of these frequency responses and metrics, without the need to first identify a parametric (state-space) model structure such as is required in applying time-domain methods. Careful tracking of the broken-loop and end-to-end closed-loop frequency-response behavior from the preliminary design studies through detailed design and simulation and into flight test provides an important 'paper trail' for documenting system performance and solving problems that may appear in the later phases of development.

The availability of comprehensive and reliable computational tools has substantially enhanced the acceptability of frequency-domain techniques in the flight control and flight test community. Benefits from applying these techniques include the reduction of flight test time required for control system optimization and handling-qualities evaluation, especially for complex control law architectures, and improvements in the final system performance. Frequency-domain methods offer a transparent understanding of component and end-to-end response characteristics that can be critical in solving system integration problems.

This chapter reviews frequency-domain system identification methods for development and integration of aircraft flight control systems. These methods were developed under a long-term research activity at the Ames Research Center by the Army Aeroflightdynamics Directorate (AFDD), NASA and Sterling Software. Many of the flight applications have been to rotorcraft, which pose an especially difficult challenge to system identification (Tischler 1990). The dynamics of these aircraft are highly coupled, and unstable. Additionally, the rotorcraft dynamics include lightly damped fuselage and rotor modes. Vibration and low excitation signal content, especially near hover, results in typically low signal-to-noise ratios. Experience in developing and applying system identification methods to the rotorcraft problem has produced a set of tools that has proven highly reliable for the broad scope of applications reviewed in this chapter. The first section presents a summary of the frequency-domain approach and the *C*omprehensive *I*dentification from *FrE*quency *R*esponses (CIFER®) comprehensive analysis facility. The remainder of the chapter is organized into five sections following the flight control development flowchart of Fig. 1 from specifications and design through flight test optimization. Each section illustrates important techniques with examples based on fixed- and rotary-wing projects at the Ames Research Center.

2. Overview of AFDD/NASA system identification techniques

The AFDD/NASA frequency-domain system identification procedure is shown in Fig. 2, and is reviewed in this section. Details of the procedure are found in

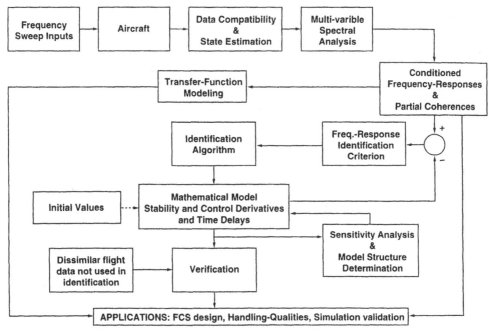

Figure 2.   Frequency-domain system identification procedure.

Tischler and Cauffman (1992). System identification methods are requirements for specific application to flight control system design of high-bandwidth rotorcraft are given by Tischler (1990).

Aircraft or subsystem component dynamics are excited by a pilot-generated or computer-generated frequency-sweep input. The dynamic responses are generally measured by dedicated sensors, and the data is either recorded on board or tele-metered to the ground for processing. Kalman filtering techniques (or simple numerical integration) are used to check data compatibility and eliminate spurious instrumentation system biases, scale factors and dropouts. Here, unmeasured signals may be estimated from the available measured states.

The foundation of the AFDD/NASA approach is the high-quality extraction of a complete multi-input/multi-output (MIMO) frequency-response database. These responses fully characterize the linearized coupled characteristics of the system without *a priori* assumptions. Advanced multivariable spectral analysis using the Chirp-Z transform and composite optimal window techniques have been developed and exercised over many flight applications. These methods provide significant improvement in identification quality relative to standard fast Fourier transform (FFT) methods. The frequency-response database directly supports important flight control system applications including: handling-qualities analysis and specification compliance testing, simulation validation and servo-loop stability analysis.

Transfer function fitting is a rapid procedure for extracting simple single-input/ single-output parametric models of specific frequency-response pairs. These transfer function models define the lower-order equivalent system (LOES) of the fixed-wing handling-qualities specifications (MIL-STD-1797) and directly support root-locus techniques for flight control system design.

Accurate MIMO state-space models are often needed to support multivariable control law design, simulation model validation and improvement, and validation of aerodynamic theory or wind tunnel results. Here, sophisticated nonlinear search algorithms are used to extract a general state-space model that matches the complete MIMO input/output frequency-response database. A significant advantage of identifying parametric models from frequency responses is the capability to define individually the appropriate frequency range for each response pair based on the associated coherence function – a valuable accuracy and linearity metric. The coherence function is also useful for automatically selecting error weighting in the cost function independent of the model structure. A methodical and integrated model structure determination procedure simplifies the model to a minimum set of reliable parameters that accurately characterizes the MIMO frequency-response database. Finally, the identified state-space model is validated by comparing predicted time responses with the actual flight responses for test inputs not used in the identification procedure.

The frequency-domain system indentification procedure is incorporated in a sophisticated interactive computational facility known as CIFER®. Integrated databasing and extensive user-oriented utilities are distinctive features of CIFER® for organizing and analyzing the large amounts of data which are generated for flight test identification projects. A screen-driven interface is tied to the database for rapid user interaction. Previous program setups and analysis results are retrieved by simply referencing case names. Then, changes can be easily made by moving the cursor around on the user screens and modifying the default or previously saved program parameter values. The changes are then updated in the database with a single key stroke. Utilities are available for quick inspection, searching, plotting or tabulated output of the contents of the database. Extensive analysis modules within CIFER® support: (1) rapid identification of transfer function models; (2) signal spectral analysis; (3) handling qualities and classical servo-loop analysis; and (4) time- and frequency-domain comparisons of identification and simulation model predictions with flight data. Aircraft applications of CIFER® have included the full life-cycle of flight control system development depicted in Fig. 1. The Deutsche Forschungsanstalt für Luft und Raumfahrt (DLR) Institute for Flight Mechanics (Braunschweig, Germany) has also developed and widely applied excellent methods for frequency-domain system identification. Applications to flight mechanics and flight control studies at the DLR include rotorcraft, transport aircraft and high-performance aircraft (Kaletka and von Grunhagen 1989, Kaletka and Fu 1993).

3. Design specifications and specification acceptance testing

Formulating design specifications is the starting point for flight control system development, while validating the achievement of these design goals is the concluding step in the process (Fig. 1). Dynamic models of expected system behavior are determined in the design process using system identification and are tracked and updated throughout the aircraft development and flight testing. This documentation provides an important 'paper trail' that minimizes flight control development time and reduces the need for costly flight test tuning. This section presents system identification methods for defining and verifying design specifications. Flight test examples illustrate the analysis of handling qualities and servo-loop stability characteristics.

Early handling-qualities specifications for fixed-wing aircraft (MIL-F-8785A, Anon. 1954), and for rotary-wing aircraft (MIL-H-8501A, Anon. 1961) were based on simple dynamic modeling concepts and time-domain metrics. These specifications were suitable to aircraft in which stability-augmentation systems (SAS) did not significantly alter the character of the (classical) bare-airframe flight mechanics responses. Compliance testing techniques depended on standard step and doublet inputs long used in the flight test community, with little requirement for sophisticated post-flight data processing.

Modern fly-by-wire aircraft employ high-bandwidth digital flight control systems to achieve greatly increased agility and disturbance rejection across a significantly widened operational flight envelope as compared with the older generation of aircraft. The flight control includes complex feedback and feedforward shaping and advanced control moment devices that profoundly alter the bare-airframe characteristics and invalidate the classical stability and control modeling concepts and testing methods. For example, modern combat aircraft achieve independent pitch pointing and flight path control with direct lift devices and vectored thrust, rather than the coupled attitude–path response to elevator for conventional aircraft. This capability greatly enhances weapon pointing and air-to-air combat maneuvering. Another common feature of advanced aircraft is sidestick controllers which reduce weight, space and cockpit complexity compared to standard center sticks. Classical static stick-stability testing is an invalid method for determining speed stability since the side sticks possess automatic trimming at neutral stick position and feedback loops provide the required stability independent of the trim gradient.

A new concern that arises for modern fly-by-wire aircraft is the potential for the accumulation of effective time delays due to digital flight control computations, flight control systems filters and fly-by-wire actuators. Actuator rate limiting can also contribute large equivalent time delays in modern aircraft (Buchholz *et al.* 1995). Excessive delays have been repeatedly cited as a key cause for handling-quality problems and stability-loop margin degradation in modern aircraft, yet equivalent time delay cannot be reliably measured using the standard testing techniques. Clearly the dynamics modeling concepts, specifications and testing techniques must be appropriate to the unique characteristics of modern highly augmented aircraft.

System identification provides an accurate, rapid and reliable approach for defining design specifications and for validating aircraft flight performance for highly augmented flight control systems. The modern US fixed-wing specification (MIL-STD-1797, Anon. 1987) and rotorcraft specification (ADS-33C, Anon. 1989) are based on extensive frequency-domain system identification analyses of flight test and simulation responses. Numerous examples from these and comparable European handling-qualities specifications are presented in this volume and in the references of this chapter. Two common handling-quality specifications are the Bandwidth/Phase Delay criteria and the LOES criteria. The former is checked directly from frequency-response identification, and the latter is checked from a transfer function fit of the frequency-response result. An illustration of flight test and handling-qualities analyses based on these specifications is now presented.

The Advanced Digital Optical Control System (ADOCS) demonstrator (Fig. 3), developed by Boeing Helicopters under contract to the US Army, was a UH-60A helicopter highly modified with redundant processors, instrumentation, and side-

Figure 3. Advanced Digital Optical Control System (ADOCS) demonstrator.

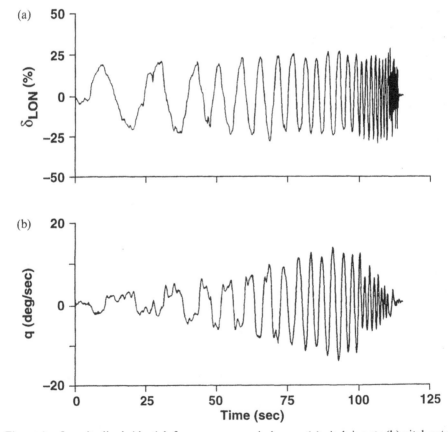

Figure 4. Longitudinal sidestick frequency-sweep in hover: (a) pitch input; (b) pitch rate.

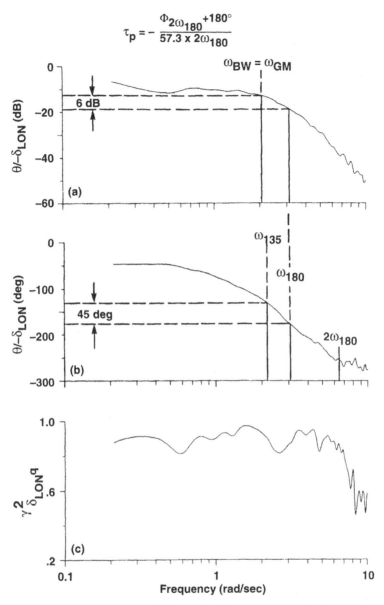

Figure 5. Identification of ADOCS pitch-rate response in hover.

stick controllers (Glusman *et al.* 1987). The overall program objective of the ADOCS was to provide the technology base for the engineering development of an advanced battlefield-compatible flight control system that: (1) enhanced aircraft mission capability; (2) improved handling qualities; and (3) decreased pilot workload. System identification flight tests and analyses using CIFER® were conducted to document the response characteristics and to compare handling-qualities characteristics with the (proposed at that time) ADS-33 design specifications (Hoh *et al.*

41

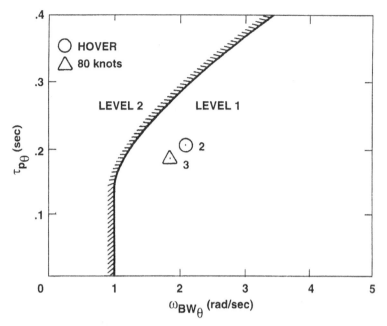

Figure 6. Handling-qualities correlation of ADOCS small-amplitude pitch response in hover and 80 knots; average pilot ratings are shown next to the data.

1988). Aircraft excitation was achieved via piloted frequency sweeps using the side-stick controller as shown in Fig. 4. Real-time telemetry of pilot inputs and aircraft responses ensured that pre-established aircraft flight limits were not exceeded.

The ADOCS frequency response for pitch response due to longitudinal input is shown in Fig. 5, along with the determination of bandwidth and phase delay as required by the ADS-33 specification. The value of the coherence function (Fig. 5) is consistently above 0.8 for the frequency range of 0.2–8 rad s^{-1} indicating excellent identification. At higher frequencies, the coherence drops, which reflects the intentionally reduced piloted inputs. The pitch bandwidth and phase delay values obtained from the hover identification results are shown on the ADS-33 specification boundary in Fig. 6. Level 1 (satisfactory) handling qualities for moderate pilot-gain tasks such as the helicopter 'bob-up' are predicted, which is consistent with the Cooper–Harper pilot rating displayed next to the data symbol. A good correlation of pilot rating and the predicted handling-qualities result was also indicated in pitch for the 80 knot flight condition.

Transfer function (LOES) models of the key on-axes responses were generated from single-input/single-output fits of the identified ADOCS frequency responses. The LOES pitch response in hover is

$$\frac{\theta}{\delta_{LON}} = \frac{-0.876(s + 0.229)e^{-0.238s}}{s[0.539, 1.82]} \tag{1}$$

The response is dominated by a well-damped second-order mode with a frequency of 1.8 rad s^{-1}. The LOES handling-qualities specification boundaries of Fig. 7 have

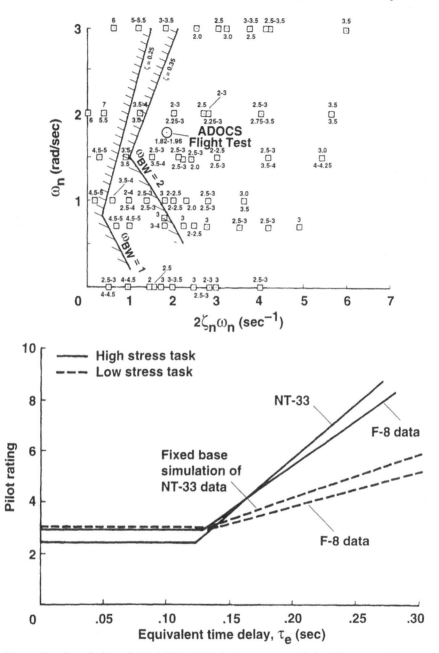

Figure 7. Correlation of ADOCS LOES pitch response with handling-qualities data.

been established based on system identification analyses of an extensive flight test and simulation database (Hoh and Ashkenas 1979). The ADOCS characteristics are seen to lie well within the Level 1 region. A second important characteristic of the LOES pitch response is the relatively large equivalent time delay, $\tau = 238$ ms. Handling-qualities experience indicates that the equivalent time delay should not

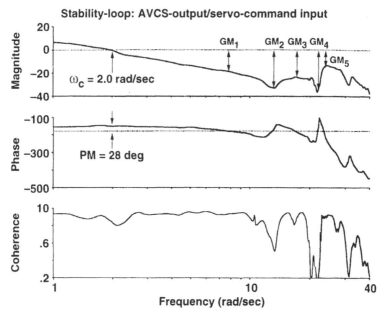

Figure 8.   Rotorcraft servoelastic stability margin testing.

greatly exceed $\tau = 120$ ms, thereby suggesting ADOCS handling-qualities degradation for 'high-gain' tasks. Comparable levels of equivalent time delay in the roll axis
were considered to be a key contributor to pilot-induced oscillations (PIOs) for
'high-gain' piloting tasks such as slope landing (Tischler *et al.* 1991). Additional
examples of handling-qualities analyses using lower-order equivalent-system modeling are presented in this volume.

The stability characteristics of aircraft rigid-body and structural dynamics may
be greatly affected by the feedback loops of the flight control system. Feedback
may degrade the flutter margin stability at the same time as it improves the rigid-
body stability and handling qualities. Military specification 9490D (Anon. 1975)
defines minimum levels of control system gain and phase margin as determined
from a broken servo-loop frequency-response analysis. The specifications are given
as a function of frequency range, with larger gain margins required for the structural elastic modes (Table 1). Figure 8 shows the broken servo-loop frequency
response of a large single-rotor helicopter as obtained by computer-generated
frequency-sweep flight test procedures. The rigid-body response crossover frequency
is $2 \cdot 0$ rad s^{-1} with an associated phase margin of 28°. This margin is slightly below
the recommended specification value. The gain margin is checked at each crossing
of the 180° ($\pm n360$) phase line as shown in the figure. The critical margin is the
minimum value (GM$_5$), which is 15 dB at a frequency of $23 \cdot 5$ rad s^{-1}. This frequency corresponds to the first vertical bending mode for the tail boom of this
aircraft. Reference to Table 1 indicates this gain margin to be well within accepted
design specifications. The coherence function of Fig. 8 shows excellent identification
accuracy for this flight test across the broad frequency range of interest (1–30 rad
s^{-1}). Sharp drops ('holes') or peaks in the coherence function reflect structural
anti-nodes and nodes respectively. Examples of fixed-wing programs using

Airspeed mode freq. Hz	Below $V_{0\,min}$	$V_{0\,min}$ to $V_{0\,max}$	At limit speed V_L	At $1.15 \times V_L$
$f_M < 0.06$	GM = 6 dB No phase reqt. below $V_{0\,min}$	GM = ± 4.5 PM = ± 30	GM = ± 3.0 PM = ± 20	GM = 0.0 PM = 0.0 Stable at nominal phase and gain
$0.06 \leqslant f_M < $ 1st ASE		GM = ± 6.0 PM = ± 45	GM = ± 4.5 PM = ± 30	
1st ASE $< f_M$		GM = ± 8.0 PM = ± 60	GM = ± 6.0 PM = ± 45	

Table 1. MIL-F-9490D gain and phase margin requirements (dB, deg) from Caldwell (1994).

frequency-domain system identification for elastic-mode stability-margin evaluation include the X-29 (Clarke *et al.* 1994) and EAP (Caldwell 1994) aircraft.

4. Design

The design process establishes control system loop architecture and associated control law parameters that achieve desired handling-qualities and servo-loop stability specifications. During the conceptual design phase, system identification procedures are applied to simple linear-design models to establish a baseline description of the proposed control system approaches and an initial check of specification compliance. At the detailed design stage, system identification methods can extract highly accurate linear-control system design models from complex simulation models or wind tunnel data. These applications of system identification in the design process are illustrated in this section.

Conceptual control system design studies are commonly based on simple stability and control derivative descriptions and the transfer function of the airframe dynamics as obtained from first-principles aerodynamic theory. Control law architectures are conceived and the initial system is modeled in a computer-aided design (CAD) environment such as MATLAB® (1992). System identification provides LOES models which are very useful in characterizing the end-to-end system dynamics and delays, and for an initial check against the design specifications. Flight control system design parameters are then adjusted until the identified LOES characteristics satisfy the design requirements. In the F-15 S/MTD demonstrator project, a numerical optimization design tool was developed to adjust control law parameters automatically to meet LOES specifications (Moorhouse and Citurs 1994).

Detailed flight control design efforts are based on very complex high-order and nonlinear simulation models. Force and moment descriptions are developed for each of the aircraft elements such as the wings, propulsion system and flight control systems based on wind tunnel look-up tables, component bench-test data and analytical theory. The simulation of multiple rigid-body systems, or flexible bodies, involves sets of dynamic equations of motion linked by constraint conditions. In many simulations these sets of equations are numerically integrated in serial form to reduce the complexity of deriving a fully coupled multibody simulation. The distributed or serial nature of these complex simulations may thus preclude the extraction of an integrated high-order linear model of the fully coupled system as is needed for accurate control design studies.

Even when the simulation architecture allows for the direct extraction of higher-order linear models using classical numerical perturbation methods, the assumption of independent perturbations results in incorrect phasing of the state variables within the multidimensional look-up tables. For example, the look-up table for aerodynamic pitching moment may depend both on angle of attack and pitch rate, so $C_{m_q} = f(\alpha)$. Thus, the correct determination of phugoid dynamics depends on maintaining representative phasing of q and α within the linearization process. Selection of perturbation size can also strongly influence the linearization results. These effects can significantly degrade the predictive accuracy of the extracted linear model. Much more accurate linear models are obtained by simulating piloted frequency-sweep inputs and extracting state-space models using system identification just as if from flight test data.

Engelland extracted accurate stability and control derivative models of a conceptual A/STOVL aircraft from a complex nonlinear off-line simulation (Engelland *et al.* 1990) to support control system design studies. The excitation input consisted of computer-generated frequency sweeps and white noise. In a procedure described by Ballin and Dalang-Secretan (1991), artificial feedback control loops were included to keep the aircraft flight condition near the reference trim point during the inputs. Starting from the perturbation derivative results CIFER® was used to identify a more accurate 6 DOF bare-airframe model. The perturbation and CIFER® derivatives are compared in Table 2. Longitudinal frequency responses of the two linear models are compared with the complete simulation responses in Fig. 9 for a flight condition of 120 knots. The linear model obtained using system identification is seen to be much more accurate than the numerical perturbation model for the high-frequency (3.0–20 rad s^{-1}) pitch-rate response q/δ_θ, and for the low-frequency (0.1–1.0 rad s^{-1}) longitudinal acceleration response a_x/δ_θ. The models are essentially identical in the mid-frequency range. A time-domain comparison of the two linear models with the nonlinear simulation response is shown in Fig. 10 for a small ($1°$) pitch-doublet input. The system identification model is seen to track the nonlinear behavior much more closely than the numerical perturbation model. The improvements are most noticeable for the long-term response (low-frequency) behavior, which is consistent with the frequency-response comparison of Fig. 9.

These results show that system identification provides an A/STOVL linear model that will be much more accurate than models extracted using numerical perturbation methods. The improvement obtained by 'flying' the frequency-sweep input is especially apparent at low frequencies where the dynamic responses are

Derivative	Perturb. value	CIFER value	C.R. (%)	Insens. (%)
X_u	-0.03471	-0.03602	-5.662	2.289
X_w	0.03958	0.02852	6.910	2.840
$X_{\dot{w}}$	$6.764E-04$	$6.764E-04†$	—	—
X_q	0.2451	$0.2451†$	—	—
X_{PCD}	$-7.690E-03$	$-8.303E-03$	-7.504	3.584
X_{PLA}	0.02270	0.02229	3.731	1.835
X_{Θ_N}	-0.5150	-0.5586	-2.353	1.005
Z_u	-0.04596	-0.03312	-13.62	4.579
Z_w	-0.3704	-0.2817	-4.386	1.377
$Z_{\dot{w}}$	-0.01023	$-0.01023†$	—	—
Z_q	-3.754	$-3.754†$	—	—
Z_{PCD}	0.1389	0.1551	5.571	2.698
Z_{PLA}	-0.3800	-0.3305	-2.254	1.016
Z_{Θ_N}	-0.01724	-0.03055	-4.646	2.242
M_u	$1.661E-04$	$-1.059E-03$	-6.016	1.745
M_w	$1.222E-03$	$3.715E-03$	5.263	1.283
$M_{\dot{w}}$	$-1.286E-03$	$-1.286E-03†$	—	—
M_q	-0.4971	-0.6852	-5.561	1.873
M_{PCD}	0.02494	0.02818	2.517	0.9822
M_{PLA}	$4.993E-04$	$4.993E-04†$	—	—
M_{Θ_N}	$2.502E-04$	$4.953E-04$	10.16	4.257

† Perturbation value used.

Table 2. Comparison of A/STOVL perturbation derivatives and CIFER® results.

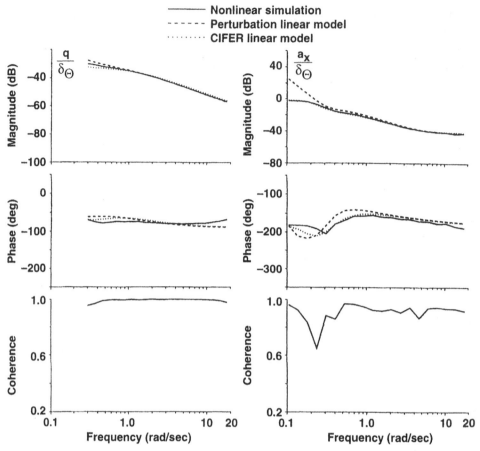

Figure 9. Frequency-response comparison of perturbation and identification models with nonlinear A/STOVL simulation.

larger, and correct phasing of the representative motion variables for entry into the multidimensional look-up tables is important. A cursory time-domain comparison of the numerical perturbation results with the nonlinear simulation response would suggest the presence of strong nonlinearities in the A/STOVL aircraft dynamics. However, the very close agreement of the system identification model with the nonlinear simulation shows that the method of linear model extraction is much more important in this case than the nonlinear characteristics of the simulation.

Success in achieving maximum control system performance and robustness in flight depends heavily on the predictive accuracy of the linear-design models. The system identification approach provides highly accurate design models for design at specific flight conditions, but it is clearly more time intensive than the simple numerical perturbation method. This is not a practical approach for checking control system behavior at the tens or hundreds of off-nominal conditions.

5. Simulation

The detailed implementation of the control system design is evaluated in comprehensive real-time piloted-simulation trials. System identification techniques are

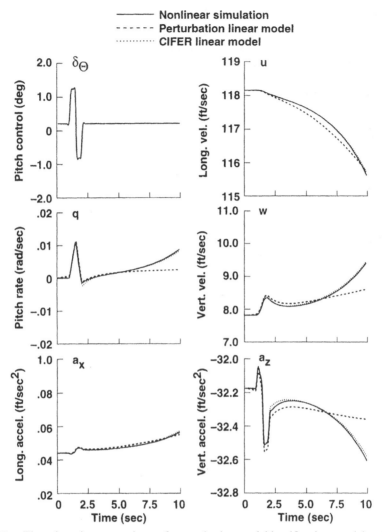

Figure 10. Time-domain comparison of perturbation and identification models with non-linear simulation of A/STOVL aircraft.

exercised to validate the real-time math model implementation of nonlinear digital control laws. Also, these techniques are used to document simulator motion and visual systems. Once flight test data is available, system identification provides an important tool for validating and updating the simulation math models. This section illustrates system identification techniques for validating simulation math models and simulator validation using XV-15 tilt-rotor (airplane mode) and UH-60A helicopter results. A companion frequency-domain format is proposed for specifying simulation model fidelity for the on-axis responses. Finally, an analysis based on an A/STOVL piloted-simulation study illustrates the use of system identification for determining actuator authority requirements.

Direct frequency-response comparison of the end-to-end performance of the complex simulation model with the conceptual design models and specifications constitutes an important 'dynamic check' which often exposes unexpected processing delays such as in the numerical integration procedures, or errors in the digital (Z-plane) implementation of control laws. This technique is also useful in exposing degradation in control system performance due to high-order structural or other hardware dynamics modeled in the advanced design simulation model that may not have been taken into account in the conceptual studies.

Simulator visual and motion systems should track the math model response as accurately as possible to ensure that the pilot's cuing environment is correct and that the handling-qualities evaluation obtained in the simulator reflects what may be expected in flight. Nonlinear compensation algorithms have been developed by McFarland (1988) that offset visual system delays, thereby minimizing the mismatch between the simulator visual system response and the math model. In work reported by Atencio (1993) and illustrated in Fig. 11, there is nearly perfect agreement of the DIG-1 visual system image (with McFarland compensation) and the UH-60A helicopter simulation math model. Math model commands to the simulator motion drive are attenuated using wash-out logic. The wash-out parameters are

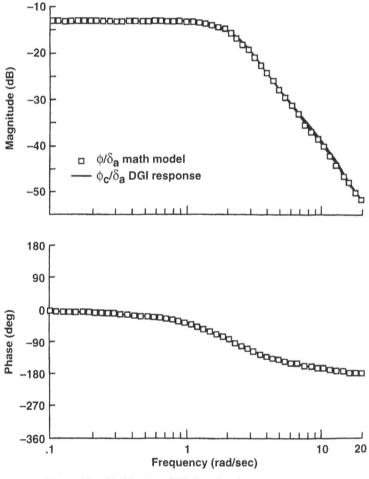

Figure 11. Validation of DIG-1 visual system response.

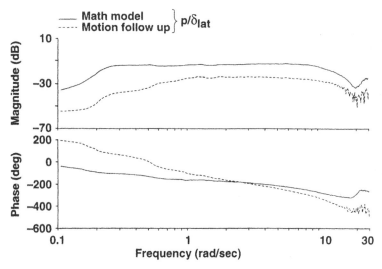

Figure 12.   Documentation of VMS motion follow-up for UH-60A roll response in hover.

selected to preserve the dynamic behavior in the frequency range of most concern to the pilot (e.g. 1–10 rad s^{-1} for pitch and roll tasks), while accommodating the restricted motion environment of the simulator. Figure 12 from Atencio (1993) compares the (washed-out) cab roll motion to stick input with the math model response for the UH-60A simulation in the Ames Vertical Motion Simulator (VMS). In the 1–10 rad s^{-1} frequency range, the simulator motion drive response follows the math model, although the motion is less than one to one as seen by the vertical shift in the magnitude curves. The motion drive wash-out logic is designed to minimize phase distortions in this frequency range as can be seen in the figure. At low frequencies the large motion is washed out, and considerable errors are encountered in the magnitude and phase response as expected. At high frequency, the motion drive is unable to follow rapid commands to the aircraft model, resulting in larger phase lags of the motion follow-up as seen in the figure.

Once flight test data is available, system identification tools are exercised to validate and update the simulation math models. The direct comparison of frequency-response behavior provides a clear picture of model fidelity as a function of frequency. This is critical for validating piloted simulations since the requirements on pilot cuing accuracy are also frequency dependent. The separate display of the magnitude and phase responses allows the sources of simulation discrepancies to be more easily determined. For example, an excessive time delay (τ) in the simulation math model or hardware causes a linear phase shift with frequency ($\phi = -\tau\omega$). Scaling errors in the simulation model appear as a clear vertical shift (in dB) in the magnitude curve. These effects are all combined in the time domain and therefore are not easily discernible in the traditional time-response comparison methods for validation. Further, the procedure of overlaying time histories is often not very accurate since the flight responses rarely begin in a trim-quiescent condition.

Tischler (1987) conducted an extensive flight test program and simulation math model validation study on the XV-15 tilt-rotor aircraft shown in Fig. 13. This tilt-rotor math model is based on comprehensive look-up tables of full-scale wind

(a)

(b)

Figure 13.   XV-15 tilt-rotor aircraft: (a) hover configuration; (b) cruise configuration.

tunnel test data, and detailed theoretical models of the rotor-system behavior and rotor-on-airframe aerodynamic interference effects. Figure 14 compares the flight and simulation roll responses for a flight condition of 170 knots. Excellent dynamic response fidelity is seen in the close match of the simulation prediction and the measured flight response. Figure 15 replots these results in terms of magnitude and phase errors as a function of frequency. Here 0 dB magnitude and 0° phase indicate perfect tracking of the flight and simulation results. Also shown in the figure are math model mismatch boundaries proposed herein for the highest-fidelity training simulations (FAA Level D). These boundaries correspond to the LOES mismatch criteria from the fixed-wing handling-qualities criteria (Hoh *et al.* 1982). The XV-15

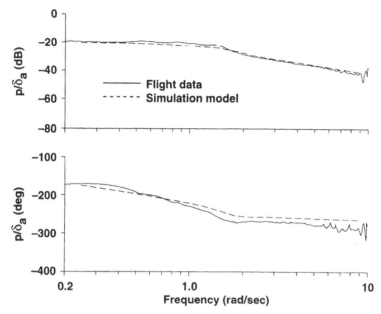

Figure 14.   XV-15 tilt-rotor simulation model validation for 170 knots.

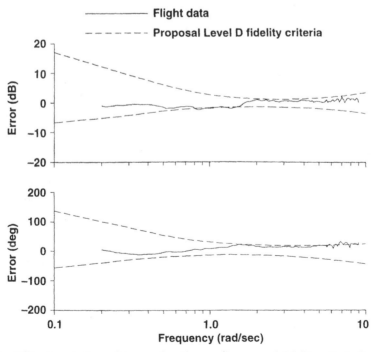

Figure 15.   Tilt-rotor math model error functions and proposed fidelity criteria for Level D simulators.

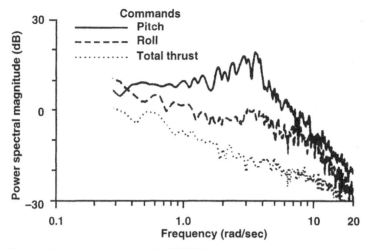

Figure 16.   Power spectra of A/STOVL propulsion system commands.

simulation math model complies with the proposed Level D (high-fidelity) criteria. This result is consistent with the very favorable pilot comparison of simulator and flight behavior (Churchill and Dugan 1982). The same approach of mismatch boundaries in the frequency domain has also been independently proposed and applied by DLR researchers to detect the effects of unnoticeable dynamics in the case of helicopters (Hamel and Jategaonkar 1995) and for evaluating the fidelity of in-flight simulation (Buchholz *et al.* 1995).

Direct comparisons of stability and control derivatives identified from flight tests with values identified from simulation math models can be used to derive correction factors for significantly improving the model fidelity. For example,

$$L_{\text{corrected}} = f(\text{nonlin. sim. eqns}) + [(L_p)_{\text{flight}} - (L_p)_{\text{sim}}]p$$

$$+ [(L_{\delta_{\text{lat}}})_{\text{flight}} - (L_{\delta_{\text{lat}}})_{\text{sim}}]\delta_{\text{lat}} + \cdots \tag{2}$$

Identification tools provide a systematic and accurate approach to determine these correction factors which are routinely used by the simulator industry to improve dynamic fidelity.

Comprehensive simulation studies are often used to define flight control system hardware requirements such as actuator and sensor filter bandwidths. Franklin *et al.* (1991) used CIFER® to determine actuator bandwidth requirements for a conceptual A/STOVL aircraft. The spectral characteristics of the stabilization and command augmentation system (SCAS) commands to the aircraft control surfaces were obtained for ensemble analyses of simulated flight tasks, and are shown in Fig. 16. The results indicate an SCAS command signal bandwidth (frequency at -3 dB amplitude) of about 4 rad s^{-1} in pitch and roll, with a significantly lower command bandwidth for the thrust (vertical) axis. Actuator hardware response bandwidths should be 5–10 times the respective SCAS command bandwidths to avoid introducing significant phase lag in the control loops (Franklin *et al.* 1991). Similar analysis techniques were used by Blanken to determine the effect of control system design on changes in pilot control bandwidth (workload) and handling qualities. This study included an interesting comparison of pilot workload for the simulation and flight test environments (Blanken and Pausder 1994).

6. Development

At the development stage, flight control system hardware and software components and subsystems undergo bench testing to verify that the performance characteristics meet the design specifications. Sophisticated flight control development facilities (DF) or 'hot-benches' allow the test of prototype flight software and hardware integrated with the simulated aircraft dynamics. In helicopter development, model or full-scale rotors are dynamically tested in the wind tunnel and the responses are validated against design requirements and comprehensive analysis models. This section presents system identification techniques to support development-stage validation. Examples are drawn from the NASA VSRA project, helicopter actuator tests, and the Sikorsky Bearingless Main Rotor (SBMR) full-scale rotor wind tunnel tests.

An extensive development facility has been used in the NASA vertical/short takeoff and landing (V/STOL) systems research aircraft (VSRA) project, which equipped a YAV-8B Harrier aircraft with a fly-by-wire research flight control system (Foster *et al.* 1987) (Fig. 17). The overall flight control goals of the VSRA program are to assess critical technology elements for advanced short takeoff/ vertical landing (STOVL) aircraft, including: integrated flight/propulsion control, advanced control and display laws, and reaction-controlled bleed-flow requirements. The role of the DF has been for verification of control law flight software, system software and safety monitoring. Actual flight computers and flight hardware were included in the DF to validate flight systems during the final development stage. The aircraft dynamics are simulated by the VSRA math model, with inputs from a test console or a rudimentary pilot–cockpit station. CIFER® was exercised extensively to validate broken-loop and end-to-end closed-loop frequency responses of the DF flight systems against the design models and theoretical analyses. Signal processing and conditioning algorithms and digital timing were also verified during DF testing.

Actuator system dynamics are an important component of the overall high-frequency phase lag in modern flight control systems. Therefore, flight control system stability margins and overall closed-loop performance and handling qualities can be significantly degraded if the actuator dynamics do not meet the design specifications. System identification bench testing of aircraft actuators ensures that expected performance is achieved and that costly modifications can be avoided at the flight test stage. Frequency-response identification and transfer function modeling from a typical helicopter actuator test are shown in Fig. 18. Excellent coherence

Figure 17. NASA V/STOL system research aircraft (VSRA).

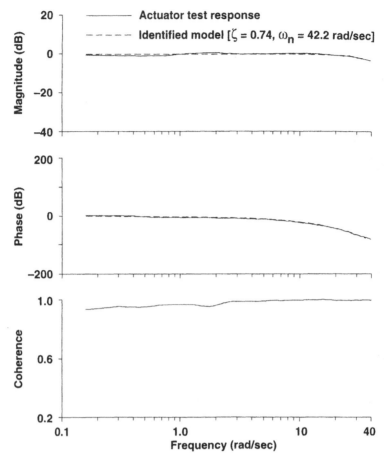

Figure 18. Helicopter actuator response identification and modeling.

is achieved over a broad frequency range (0·2–40 rad s^{-1}) using a computer-generated frequency-sweep excitation. The actuator dynamics are well characterized by the damped second-order response obtained by CIFER® (Fig. 18). These component system results are used to update simulation math models and to optimize flight control system gains prior to first flight.

Structural analysis programs such as NASA structural analysis (NASTRAN) are rarely able to predict accurately the flexible response beyond the first elastic modes of a new aircraft. Therefore, structurally scaled models or full-scale structural test vehicles are evaluated in special rigs to verify the elastic characteristics and make final adjustments to the structural compensation (e.g. notch filters) in the control system prior to first flight. Automated test/analysis facilities excite the individual structural modes of the aircraft with shakers and then use system identification methods to determine model characteristics. For modern rotorcraft development, system identification has also been effective in extracting dynamic response models of subscaled or full-scale rotor systems from dynamic wind tunnel test data. The control response dynamics of the SBMR were determined in a joint NASA/Sikorsky test in the Ames 40- by 80-Foot Subsonic Wind Tunnel (Fig. 19) (Tischler *et al.* 1994). Computer-generated frequency-sweep excitation signals to the

Figure 19. Sikorsky Bearingless Main Rotor (SBMR) test in NASA Ames 40- by 80-Foot
Subsonic Wind Tunnel.

SBMR swashplate actuator were carefully designed to ensure adequate identification within the limitations of the rotor and wind tunnel stand. Rotor blade and hub
moment frequency responses were then extracted using CIFER® and were compared with comprehensive simulation models of the SBMR. CIFER® was also used
to extract the rotor's physical parameters based on a linearized 14 DOF analytical
formulation of the SBMR dynamics (Tischler *et al.* 1994).

Figure 20 shows the identified on-axis roll moment response to a lateral stick
input. The simulation math model and 14 DOF identified model agree closely with
measured responses. The off-axis pitching moment to lateral stick input is shown in
Fig. 21. Here, the simulation model phase response deviates significantly from both
the measured response and identified parametric model, indicating a poor prediction of rotor cross-coupling.

The key identified physical parameters of the rotor system are compared with
the GENHEL simulation values in Table 3 (both are updated from the earlier
results of Tischler *et al.* (1994)). Many of the important rotor parameters such as
Lock number, blade inertia and effective hinge offset compare very favorably, and
reflect the good on-axis response prediction of the simulation model. The important
difference between the identified model and the GENHEL simulation is the control
phase angle. This parameter has a known geometric value of −14° in the wind
tunnel tests, but the identified value needed to capture the measured off-axis
response of Fig. 21 is −23.4°. This discrepancy in control phasing indicates a fundamental problem in the aerodynamic modeling of the rotor. Follow-on analyses of
these results have yielded a new approach for correcting the simulation math model
and improvements in the identification methods for free-flight results (Takahashi *et
al.* 1995). Accurate cross-coupling prediction is especially important for the design

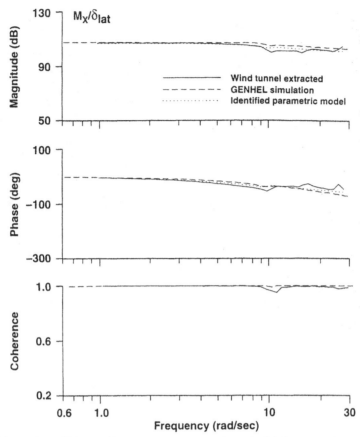

Figure 20.   SBMR roll moment response to lateral stick input (40 knots).

of decoupling compensators in modern rotorcraft flight control systems. Corrections to the flight control laws prior to final flight software installation and vehicle testing reduce development flight test costs and improve the final performance of the system.

7. Flight testing

The flight test program for flight control and handling-qualities validation and optimization has a significant impact on the overall development schedule and cost for modern fly-by-wire aircraft. System identification provides a critical technology for tracking aircraft dynamic response performance into flight, solving problems that arise in flight tests, and rapidly optimizing control system parameters. This section presents system identification methods for control system flight testing. Flight data results are presented for the VSRA and UH-60A Rotorcraft Aircraft Systems Concept Airborne Laboratory (RASCAL) projects.

Flight test verification of aeroservoelastic stability margins is an important concern for modern fly-by-wire aircraft, where dynamic coupling of the high-gain flight control system with lightweight structural dynamics can degrade flutter stability. Flutter margin verification using system identification has been adopted by British Aerospace in the development of a series of fly-by-wire high-performance

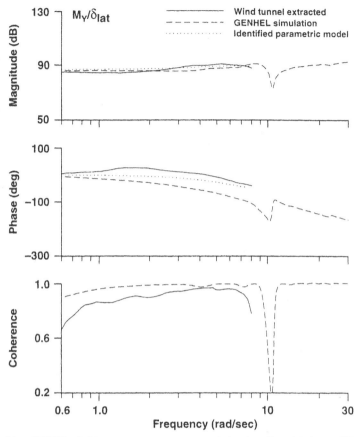

Figure 21.   SBMR pitching-moment response to lateral cyclic stick input (40 knots).

Rotor parameter	Symbol	Units	GENHEL value	Identified value
Lock number†	γ	ND	7·46	7·82
Lift-curve slope	a	rad^{-1}	5·73	5·33
Blade inertia†	I_b	slug-ft^2	552·81	489·8
Blade 1st mass moment†	S_b	slug-ft	38·76	48·78
Blade weight	$m_b g$	lbs	115	142·68
Flapping frequency	v_β	per rev	1·081	1·080
Effective hinge-offset	e	ND	0·097	0·095
Lag frequency	v_ζ	per rev	0·699	0·697
Lag damper	C_ζ	ft-lb-sec rad^{-1}	372	473·29
Collective-lag/shaft freq.	v_{ζ_0}	per rev	—	0·474
Collective-lag/shaft damping	C_{ζ_0}	ft-lb-sec rad^{-1}	—	1631·49
Trim coning angle	β_T	rad	0·0768	0·0654
Pitch-flap coupling	K_{P_β}	rad rad^{-1}	0	0
Pitch-lag coupling	K_{P_ζ}	rad rad^{-1}	−0·0225	−0·184
Control phase angle	Δ_{SP}	deg	−14·0	−23·4

† Mass moment parameters and Lock number are referenced to the hub center and not to the hinge-axis.

Table 3.   Comparison of SBMR identified parameters with GENHEL values.

aircraft, as described by Caldwell (1994). Near-real-time system identification was employed during the X-29 aircraft flight testing (Clarke *et al.* 1994) for on-line verification of stability margins in a highly efficient flight envelope expansion program. Piloted frequency sweeps were used to excite the vehicle structural modes at each test condition, and the telemetered data were then analyzed using high-speed array processing computers. Once the stability margins were verified, the pilot was cleared to proceed to the next flight condition, avoiding the normally time-consuming test technique of clearing one flutter test point per flight. In a similar application of near-real-time identification techniques, CIFER® was used to support flight tests of the 'Pathfinder', a large high-altitude solar unmanned air vehicle (UAV) (Dornheim 1995). Servo-loop stability margins were extracted based on telemetered data from computer-generated frequency-sweep tests, and then com-

YAV-8B Aeroelastic Identification (120 Kts)

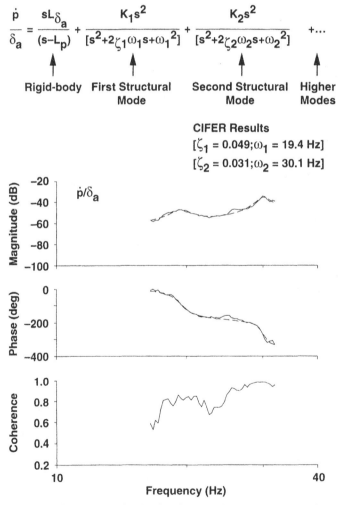

Figure 22.   YAV-8B aeroelastic wing-bending identification (120 knots).

pared with simulation predictions. When the CIFER® results indicated a reduction compared with predictions in stability margins at a high-altitude flight condition, the ground station pilot executed real-time switching commands to adjust the Pathfinder control law gains.

Theoretical analyses of the XV-15 tilt-rotor aircraft (Fig. 13) predicted that the reduction of whirl-mode flutter stability margins with increasing flight speed would limit the aircraft's usable flight envelope. An extensive flight test program was conducted to verify the expected margins. Early testing using the traditional dwell-delay method proved time consuming and resulted in considerable data scatter. Acree and Tischler (1993) conducted automated frequency-sweep tests using wing flaperon excitation and subsequently analyzed the data using the CIFER® identification tools. The frequency-domain test technique proved to be much more time efficient, and the results showed both a reduction in the scatter at specific conditions and an improvement in consistency across flight conditions.

Automated frequency-sweep flight testing was also conducted on the VSRA YAV-8B aircraft (Foster *et al.* 1987) (Fig. 17) to determine the locations of the (open-loop) first and second structural wing-bending modes, and to verify actuator and sensor processing dynamics. The parametric model shown in Fig. 22 was

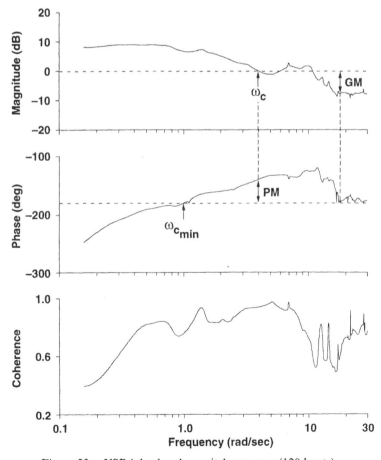

Figure 23.   VSRA broken-loop pitch response (120 knots).

obtained from CIFER®, and includes the rigid-body response and second-order representations of the two structural modes. Notch filters, included to avoid coupling of the flight control and aeroelastic dynamics, and control law gains were subsequently updated based on these identification results. Piloted frequency-sweep flight testing was also conducted in the VSRA program to document the final stability margins and closed-loop response for a number of flight conditions. The broken-loop pitch response for 120 knots as obtained from CIFER® is shown in Fig. 23. The figure shows that the dynamics are conditionally stable, with a minimum crossover frequency of 1 rad s^{-1} required for closed-loop vehicle stability. The nominal crossover frequency of 4 rad s^{-1} yields a phase margin of 40° (acceptable). A gain margin of about 8 dB is indicated over the broad frequency range 15–30 rad s^{-1} where the phase curve has a nearly constant value of about $-180°$.

The identified closed-loop response dynamics are shown in Fig. 24. In the frequency range of 0·3–5 rad s^{-1} the response is accurately modeled by a well-damped second-order system:

$$\frac{\theta}{\theta_{\text{com}}} = \frac{6·35e^{-0·048s}}{[0·953, 2·47]} \tag{3}$$

These dynamics closely match the design response of

$$\frac{\theta}{\theta_{\text{com}}} = \frac{4·0}{[1·0, 2·0]} \tag{4}$$

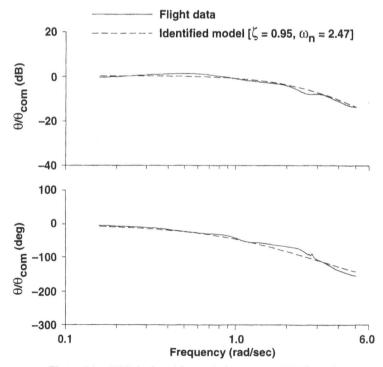

Figure 24.   VSRA closed-loop pitch response (120 knots).

The small equivalent time delay of $\tau = 48$ ms reflects the VSRA high-bandwidth fly-by-wire actuators and rapid digital calculations, and suggests no time-delay-related handling-qualities problems.

In some applications, simulation math models are not sufficiently accurate for control law design prior to first flight. For example, the current state of the art of rotorcraft flight dynamics simulation yields a fair prediction of the on-axis characteristics, but usually an inadequate prediction of the cross-coupling response as in the GENHEL simulation result of Fig. 21, often not even correct in sign (Curtiss 1992). Rotorcraft math models are still useful for initial simulation and control law development efforts, but are less satisfactory for the final determination of stability margins and decoupling controller gains. Initial flight tests with the SCAS-OFF or with reduced control system gains can be conducted to identify new aircraft dynamics or to update the simulation for final control law parameter selection. The DLR developed a high-bandwidth flight control system for the Bo-105 variable-stability aircraft (ATTHeS) based directly on bare-airframe state-space models extracted from flight data using frequency-domain system identification (von Grunhagen *et al.* 1994). This direct use of flight-identified state-space models for control law design represents the most sophisticated and demanding application of system identification tools.

An approach similar to that of the DLR has been adopted by the AFDD/NASA in the development of an advanced fly-by-wire flight control system for the RASCAL UH-60A helicopter (Takahashi *et al.* 1995), which uses the same airframe as the ADOCS demonstrator (Fig. 3). Extensive theoretical studies of combat rotorcraft control law concepts for application to RASCAL have been conducted by Takahashi (1994) and Cheng *et al.* (1995) based on UH-60A simulation math models. At the same time, Fletcher (1995) has conducted UH-60A flight tests and comprehensive frequency-domain identification studies to extract high-order state-space models of the aircraft for hover and cruise flight conditions. These efforts were brought together in the RASCAL control law study described in Takahashi *et al.* (1995). Figures 25 and 26 compare two flight mechanics simulation math models ('A' and 'B') used for the control law designs with the bare-airframe flight test data. The on-axis roll-response agreement between the math models and the flight test data is reasonable at mid-frequency (0.8–10 rad s^{-1}), but is inadequate beyond 10 rad s^{-1} owing to errors in the prediction of the in-plane rotor response. Large errors are also seen at low frequency. The simulation models show poor predictive capability for the cross-coupling response of roll rate to longitudinal stick input, with large phase errors in the critical frequency range of 1–10 rad s^{-1} (Fig. 26). While the simulation models were sufficient for the preliminary flight control and simulation studies, they are clearly inadequate for selecting final flight gains – especially for the response decoupling parameters.

The identified higher-order linear model is compared with the flight data and the simulation models in Figs 25 and 26. Significant improvement in the on-axis prediction is seen for both the high-frequency (rotor-response) and lower-frequency dynamics. The identified model also tracks the off-axis magnitude and phase very closely, showing clear improvement compared with the two simulation models. The excellent predictive capability of the identified model is also seen in the time-response comparison of Fig. 27.

The identified state-space model was then substituted into the model-following control system block diagram in place of the original simulation model ('A')

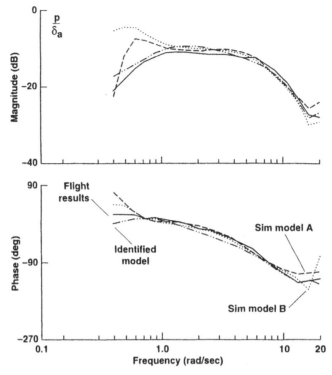

Figure 25.   UH-60A on-axis roll-rate response to lateral stick (hover); comparison of simulation and identified state-space model with flight data.

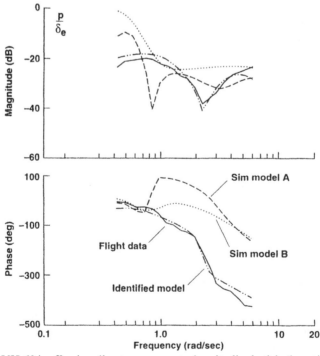

Figure 26.   UH-60A off-axis roll-rate response to longitudinal stick (hover); comparison of simulation and identified state-space model with flight data.

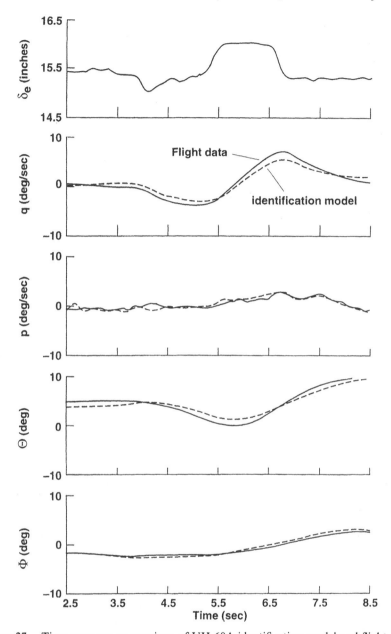

Figure 27. Time-response comparison of UH-60A identification model and flight data.

response to check the expected flight characteristics. Figure 28 shows that the design phase margin is significantly degraded when the identified model is incorporated. Further, the level of closed-loop cross-coupling (Fig. 29) increases by 20 dB (a factor of 10) in the critical handling-qualities frequency range of 1–10 rad s^{-1}. The control system design parameters were then retuned for the identified model response. Figure 28 shows that the original design crossover frequency, phase margin and gain margin are recovered. Also, the cross-coupling level for the

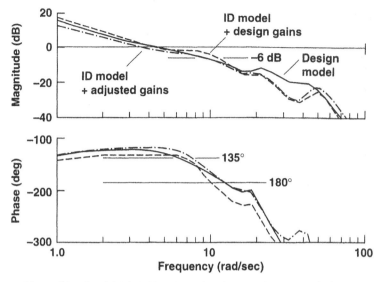

Figure 28. RASCAL UH-60A broken-loop roll response in hover.

retuned system closely tracks the coupling levels for the original control system design (Fig. 29).

The full exploitation of system identification tools early in the flight test development and control system optimization effort has been illustrated for the Bo-105 (ATTHeS) and UH-60A (RASCAL) programs. This approach will significantly reduce flight test development time for new aircraft, and will expedite the optimization of flight control system performance and handling qualities.

8. Concluding remarks

(*a*) System identification is a full life-cycle technology that supports aircraft flight control system development from design specification through flight test optimization. Significant reductions in development time and costs are realized by tracking open- and closed-loop dynamic response characteristics through the development process.

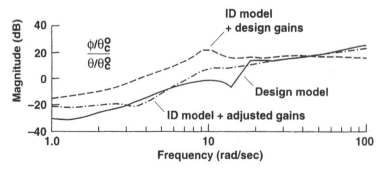

Figure 29. RASCAL UH-60A coupling response of roll rate due to longitudinal input for hover.

(*b*) Frequency-domain system identification methods are well suited to aircraft flight control development since many current design specifications, design and analysis techniques, and acceptance flight test techniques are based in the frequency domain.

(*c*) Reliable computational tools for system identification are available and have been successfully employed in many recent aircraft programs.

(*d*) System identification is especially effective in providing a transparent and integrated understanding of handling-qualities characteristics and system stability. Considerable improvements in system performance are facilitated by the rapid availability of accurate end-to-end and subsystem dynamic models.

REFERENCES

ACREE, C. W., Jr. and TISCHLER, M. B., 1993, Determining XV-15 aeroelastic modes from flight data with frequency-domain methods. NASA Technical Paper 3330, ATCOM Technical Report 93-A004.

ANON., 1954, Military specification, flying qualities of piloted airplanes. MIL-F-8785 (ASG).

ANON., 1961, Helicopter flying and ground handling qualities; general requirements for MIL-H-8501A.

ANON., 1975, Military specification, flight control systems – general specification for design, installation and test of piloted aircraft. MIL-F-9490D (USAF).

ANON., 1987, Military standard, flying qualities of piloted vehicles. MIL-STD-1797 (USAF).

ANON., 1989, Aeronautical design standard, handling qualities requirements for military rotorcraft. ADS-33C (AVSCOM).

ATENCIO, A., 1993, Fidelity assessment of a UH-60A simulation on the NASA Ames Vertical Motion Simulator. NASA TM-104016.

BALLIN, M. G. and DALANG-SECRETAN, M. A., 1991, Validation of the dynamic response of a blade-element UH-60A simulation model in hovering flight. *J. Am. Helicopter Soc.*, **36**(4), 77–88.

BLANKEN, C. L. and PAUSDER, H.-J., 1994, Investigation of the effects of bandwidth and time delay on helicopter roll-axis handling qualities. *J. Am. Helicopter Soc.*, **39**(3), 24–33.

BUCHHOLZ, J. J., BAUSCHAT, J. M., HAHN, K. U. and PAUSDER, H.-J., 1995, ATTAS & ATTHeS in-flight simulators: recent application experiences and future programs. AGARD Flight Vehicle Integration Panel Symposium 'Simulation: Where are the Challenges?', Braunschweig, Germany.

CALDWELL, B. D., 1994, The FCS-structural coupling problem and its solution. AGARD-CP-560, Paper 16.

CHENG, R. P., TISCHLER, M. B. and BIEZAD, D. J., 1995, Rotorcraft flight control design using quantitative feedback theory. Symposium on Quantitative Feedback Theory, 2–4 August, Purdue University, West Lafayette, IN.

CHURCHILL, G. B. and DUGAN, D. C., 1982, Simulation of the XV-15 tilt rotor research aircraft. NASA TM-84222.

CLARKE, R., BURKEN, J. J, BOSWORTH, J. T. and BAUER, J. E., 1994, X-29 flight control system: lessons learned. *Int. J. Control*, **59**(1), 199–219.

CURTISS, H. C., JR., 1992, On the calculation of the response of helicopter to control inputs. 18th European Rotorcraft Forum, Avignon, France.

DORNHEIM, MICHAEL A., 1995, Solar-powered aircraft exceeds 50000ft. *Aviat. Week Space Technol.*, Sept. 18, p. 67.

ENGELLAND, S. A., FRANKLIN, J. A. and MCNEILL, W. E., 1990, Simulation model of a mixed-flow remote-lift STOVL aircraft. NASA TM-102262.

FLETCHER J. W., 1995, Identification of UH-60A stability derivative models in hover from flight test data. *J. Am. Helicopter Soc.*, **40**(1).

FOSTER, J. D., MORALEZ, E. III, FRANKLIN, J. A. and SCHROEDER, J. A., 1987, Integrated control and display research for transition and vertical flight on the NASA V/STOL research aircraft (VSRA). NASA TM-100029.

FRANKLIN, J. A., STORTZ, M. W., ENGELLAND, S. A., HARDY, G. H. and MARTIN, J. L., 1991, Moving base simulation evaluation of control system concepts and design criteria for STOVL aircraft. NASA TM-103843.

GLUSMAN, S. I., DABUNDO, C. and LANDIS, K. H., 1987, Evaluation of ADOCS demonstrator handling qualities. Proceedings of the 43rd Annual National Forum of the American Helicopter Society, Washington, DC.

HAMEL, P. G. 1991, Rotorcraft system identification. AGARD AR 280.

HAMEL, P. G. and JATEGAONKAR, R. V., 1995, The evolution of flight vehicle system identification. AGARD Structures and Materials Panel Specialists' Meeting on Advanced Aeroservoelastic Testing and Data Analysis, Rotterdam, The Netherlands.

HOH, R. H. and ASHKENAS, I. L., 1979, Development of VTOL flying qualities criteria for low speed and hover. Systems Technology, Inc., Hawthorne, CA, TR-1116-1.

HOH, R. H., MITCHELL, D. G., ASHKENAS, I. L., KLEIN, R. H., HEFFLEY, R. K. and HODGKINSON, J., 1982, Proposed MIL standard and handbook – flying qualities of air vehicles. AFWAL-TR-82-3081, vol. 2.

HOH, R. H., MITCHELL, D. G., APONSO, B. L., KEY, D. L. and BLANKEN, C. L., 1988, *Proposed Specification for Handling Qualities of Military Rotorcraft. Volume 1 – Requirements.* US Army Aviation Systems Command, TR 87-A-4.

KALETKA, J. and FU, K.-H., 1993, Frequency-domain identification of unstable systems using X-31 aircraft flight test data. AIAA-93-3635, AIAA Atmospheric Flight Mechanics Conference, Monterey, CA.

KALETKA, J. and VON GRUNHAGEN, W., 1989, Identification of mathematical derivative models for the design of a model following control system. 45th Annual Forum of the American Helicopter Society, Boston, MA.

MATLAB®, 1992, High-performance numeric computation and visualization software. The Math Works, Inc., Reference Guide.

McFARLAND, R. E., 1988, Transport delay compensation for computer-generated imagery systems. NASA TM-100084.

MOORHOUSE, D. J. and CITURS, K. D., 1994, The control system design methodology of the STOL and maneuver technology demonstrator. *Int. J. Control*, **59**(1), 221–238.

TAKAHASHI, M. D., 1994, Rotor-state feedback in the design of flight control laws for a hovering helicopter. *J. Am. Helicopter Soc.*, **39**(1), 50–62.

TAKAHASHI, M. D., FLETCHER, J. W. and TISCHLER, M. B., 1995, Development of a model following control law for inflight simulation using analytical and identified models. American Helicopter Society, 51st Annual Forum, Fort Worth, TX.

TISCHLER, M. B., 1987, Frequency-response identification of XV-15 tilt-rotor aircraft dynamics. NASA TM-89428.

TISCHLER, M. B., 1990, System identification requirements for high-bandwidth rotorcraft flight control system design. *J. Guid., Control Dyn.*, **13**(5), 835–841.

TISCHLER, M. B. and CAUFFMAN, M. G., 1992, Frequency-response method for rotorcraft system identification: flight applications to BO 105 coupled rotor/fuselage dynamics. *J. Am. Helicopter Soc.*, **37**(3), 3–17.

TISCHLER, M. B., FLETCHER, J. W., MORRIS, P. M. and TUCKER, G. E., 1991, Flying quality analysis and flight evaluation of a highly augmented combat rotorcraft. *J. Guid., Control, Dyn.* **14**(5), 954–963.

TISCHLER, M. B., DRISCOLL, J. T., CAUFFMAN, M. G. and FREEDMAN, C. J., 1994, Study of bearingless main rotor dynamics from frequency-response wind tunnel test data. American Helicopter Society Aeromechanics Specialists Conference, San Francisco, CA.

VON GRUNHAGEN, W., BOUWER, G., PAUSDER, H.-J., HENSCHEL, F. and KALETKA, J., 1994, A high bandwidth control system for a helicopter in-flight simulator. *Int. J. Control* **59**(1), 239–261.

PART TWO

Rotorcraft and V/STOL

3

A high bandwidth control system for the helicopter in-flight simulator ATTHeS – modelling, performance and applications

WOLFGANG VON GRÜNHAGEN, GERD BOUWER,
HEINZ-JÜRGEN PAUSDER, FROHMUT HENSCHEL and
JÜRGEN KALETKA

1. Introduction

Manned simulation continues to be the important tool for flying qualities research as well as in systems design approach. The implementation of highly augmented flight control systems, new controller types and pilot information systems will give designers more flexibility to alter the response characteristics of the overall helicopter system and to tailor the desired flying qualities. This capability is also a two-edged possibility, in as far as it has to have adequate guidance from flying qualities specifications, concerning which types of helicopter dynamics are desirable for various piloting tasks. The demands placed upon the engineers continue to emphasize a strong need for realistic simulation. The pitfalls inherent in the exclusive use of ground-based pilot-in-the-loop simulation will become evident during the development of the next generation of rotorcraft. For this reason, an integrated approach, using both ground and in-flight simulators throughout the design process and the flying qualities research, is important.

The realistic visual and motion cues are the key feedback parameters for the pilot to adapt his control strategy to the vehicle and the flight task. The results obtained from ground simulation studies have to be verified in the 'real world'. This means that they have to be tested in flight. To avoid the appearance of problems late in a development phase, which can cause expensive delays and can increase costs, it is necessary to utilize a flying testbed. The demands of high fidelity, high flexibility, and high quality of pilot cues formulate the need to use flying simulators (Gmelin et al. 1986). The great advantages of an in-flight simulator, compared with a ground simulator, are the ability to fly the system in the real world with adequate visual and motion cues for the pilot and, compared with a technology demonstrator, the capacity to vary the system response characteristics. Hardware components, like new types of inceptors or information displays, can be installed.

Besides the complementary incorporation of an in-flight simulator in the development of a new helicopter system, and the integrated technologies of pilot vehicle interfaces, additional areas of use are more general and basic research

73

related:

(*a*) control law design research;

(*b*) investigation of the response system required by a flight task;

(*c*) establishment of credible data for the definition of generic evaluation criteria;

(*d*) investigation of the interference effects between the required overall system response; display format, and pilot's inceptor type;

(*e*) requirements for system mode blending characteristics; and

(*f*) investigations of allowable response degradations which occur after system failures.

In addition, the development of an in-flight simulator inherently includes the necessity to solve many problems of the development of a highly augmented rotorcraft. Consequently, many of the lessons learned and the technical solutions can be transferred directly to the development of an operational rotorcraft.

2. DLR in-flight simulator ATTHeS

Recognizing the requirements for a flying testbed with a broad simulation capability and flexibility, a helicopter in-flight simulator has been developed at the DLR Institute for Flight Mechanics. The ATTHeS (Advanced Technology Testing Helicopter System) is based on a BO 105 helicopter (Fig. 1).

The next generation of helicopters are required to fly with high pilot gain, high bandwidth, and high precision flight tasks, which also have to be performed in a helicopter airborne simulator. The capability to cover the required range of dynamic response behaviour is essentially limited by the characteristics of the helicopter being the host for the in-flight simulation. The high control power

Figure 1. In-flight simulator ATTHeS.

and the quick initial response to the pilot's inputs, in helicopters with a hingeless rotor, are an excellent precondition. Correspondingly, the BO 105 helicopter is well suited to be a host for an in-flight simulator. The high level of interaxis coupling and the high-order response induced by the rotor system complicate the design of a full-authority digital flight control system, which is required for a capable in-flight simulator.

The testbed is equipped with a full authority non-redundant fly-by-wire (FBW) control system for the main rotor and a fly-by-light (FBL) control system for the tail rotor. The test helicopter requires a two-men crew, consisting of a simulation pilot and a safety pilot. The safety pilot is provided with the standard mechanical link to the rotor controls, whereas the simulation pilot's controllers are linked electrically/optically to the rotor controls. The FBW/L actuator inputs, which are commanded by the simulation pilot and/or the flight control system, are mechanically fed back to the safety pilot's controllers. With this function, the safety pilot is enabled to monitor the rotor control inputs. This is an important safety aspect, because the safety pilot can evaluate whether the inputs are adequate to the flight task. The safety pilot can disengage the FBW/L control system by switching off the FBW/L system or by overriding the control actuators. In addition, an automatic safety system is installed, monitoring the hub and lag bending moments of the main rotor. The testbed can be flown in three modes:

(1) the FBW/L disengaged mode, where the safety pilot has the exclusive control;

(2) the 1:1 FBW/L mode, where the simulation pilot has the full authority to fly the basic helicopter; and

(3) the simulation mode, where the simulation pilot is flying a simulated helicopter system with full authority.

In the 1:1 FBW/L and the simulation mode, the flight envelope is restricted to 50 ft above ground in hover and 100 ft above ground in forward flight.

To incorporate the digital control system for in-flight simulation purposes, an onboard computer and a data acquisition system have been installed. In the specifications for the design, the following system conditions and requirements have been considered.

(i) Limited space is available in the helicopter.

(ii) Software modifications in the control system must be accomplished in a host computer on the ground.

(iii) A system simulation facility, which is compatible with the on-board system, is needed to check any software modifications before going into flight.

(iv) The on-board system tasks, control system and evaluation of the control system performance, have to be clearly separated.

(v) The flight tests have to be observed and managed from a ground station.

Figure 2 shows, in a block diagram, the onboard system. Two PDP11

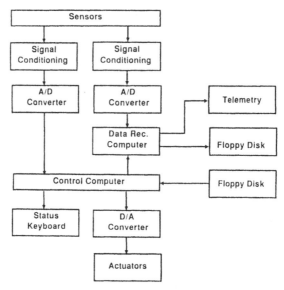

Figure 2.   Structure of ATTHeS on-board system.

computers, ruggedized for operation in the airborne environment, are installed. The data recording task and the control system task are assigned to the computers, which allow a largely autonomous treatment of the data streams needed for the control laws and required for the data recording for the control system performance evaluation.

The simulation pilot's inputs and the state variables, which are used in the control laws, are obtained directly from the preconditioned sensor signals with an installed 16 channel A/D converter. A sampling cycle of 25 Hz is realized. After the initialization, the control system is held in the trim position. The control system starts when the simulation pilot switches on the control status and the computer generates a subcycle of 1/5 of the frame time (8 ms). The subcycle allows the FBW/L actuator inputs to be refreshed in a lower time frame than the sampling frame. For the implemented explicit model-following control system, which is described in the next paragraph, this refreshing time is assessed with 16 ms. The overall computation time for the commanded model and the control laws is 13 ms. All data regarding cycle times and the measured test data included in this paper are related to a PDP 11/73 control computer. This control computer is now replaced by a PDP 11/93, which reduces the sampling and computational time to half the values.

The data recording computer is equipped with a 64 channel A/D converter. All sensor signals are sampled with a frequency of 100 Hz. A sampling frequency significantly higher than the control computer sampling frequency, has been specified to achieve a more precise assessment of the overall system performance. Both computers are linked by a dual port memory. The measured signals, which are used in the control computer, and the signals that are calculated in the control computer, are transmitted via the dual port memory for recording. The data are recorded onboard on a floppy disk. In addition, the data are transmitted by telemetry. The telemetry data are used only for quick-look purposes in the ground station. The ground station also contains a host computer compatible with the control system computer. Any modifications of the software code, including changes of the control laws and the command model parameters, are first performed in this host computer on the ground and then transferred to the on-board computers via a floppy disk.

For an examination of the real-time software, a system simulation tool has been developed. The basic helicopter and the actuating system are represented by a nonlinear model. The onboard computer system has been duplicated in the ground system simulation. This tool allows compatible hardware and software in the loop testing before implementation in the flying ATTHeS helicopter.

3. Model-following control system design

The most promising and also challenging method of a control system design for in-flight simulation is to force the basic helicopter to respond on the pilot's inputs as an explicitly calculated command model. Explicit model-following control is useful when a high flexibility is required to vary the command models. Figure 3 illustrates the principal structure of an explicit model-following control system and its relationship to system identification. The command model response on the pilot's inputs is calculated and is fed to the controllers. The feedforward controller, which has to compensate for the dynamics of the host helicopter, is calculated from a model of the host helicopter. A vehicle state feedback is implied to minimize the influence of noise and feedforward inaccuracies and to reduce the tendency of long-term drifts in the overall system response. The feedforward and feedback controllers are independent of the command model. This modern type of control system is used more and more for operational (Fogler and Keller 1993) and research (Snell *et al.* 1992) aircraft. The performance and accuracy of the model-following control system is highly dependent on the accuracy of the modelled dynamics of the host helicopter, but additional components contribute to the overall performance. The

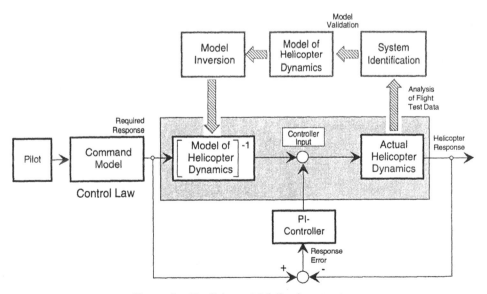

Figure 3.   Explicit model-following structure.

design has to consider, and has to be adapted to the pilot's controller dynamics, the pilot's input shaping, actuator dynamics, sensor dynamics and signal conditioning (Pausder *et al.* 1988, Pausder and Bouwer 1989).

As a consequence, an explicit model-following control system has been developed for use in the ATTHeS in-flight simulator. As mentioned above, this type of control system yields the required flexibility for the variation of the command model response behaviour. The presently designed MFCS is adapted for use in forward flight between 40 to 80 knots.

For the design of the control system, four main steps are to be performed:

(1) to define a model structure for the in-flight simulator host aircraft, including rotor states with high bandwidth capability;

(2) to determine the parameters of the above-defined model, either by system identification or simulation procedures;

(3) to determine the feedforward structure from the model of the host aircraft by formal inversion of the defined model;

(4) to define the feedback structure and to optimize the overall response by simulation and refinement in final flight tests.

3.1. *Modelling of the host*

For the layout of the control system, a special linear model of the host aircraft was developed and identified from flight test data. Usual six degree-of-freedom (DOF) flight mechanical models, neglecting the rotor states, have the disadvantage that they only describe the behaviour of the aircraft in the short term response if time delays are introduced. For the feedforward controller, the model of the host has to be inverted, leading to time 'lead', which cannot be realized. Therefore, an eight-DOF model of the BO 105 was formulated, minimizing the inherent time delays coming from the rotor degrees of freedom.

This can be explained by the roll equation of the aircraft. A similar approach is used for the pitch moment equation. For these derivations the following notation is used.

A_1, B_1 longitudinal and lateral blade control angle
a_1, b_1 longitudinal and lateral first-harmonic flapping coefficient
L roll moment coefficient
p roll rate
γ Lock number
τ_b time constant
Ω main rotor angular velocity

For a six-DOF linear model, the roll moment equation can be reduced to

$$\dot{p} = L_p p + L_{A_1} A_1 \tag{1}$$

An extended model structure, including first-order tip path plane dynamics for longitudinal and lateral flapping, was formulated as

$$\dot{p} = L_{b_1} b_1 \tag{2 a}$$

$$\dot{b}_1 = -\frac{1}{\tau_b} b_1 + \frac{1}{\tau_b} A_1 - p \tag{2 b}$$

$$\tau_b = \frac{\dfrac{\gamma^2}{16} + 16}{\Omega\gamma} \qquad (2\,c)$$

The validity of the assumption in (2 *a*) is demonstrated in Fig. 4. It clearly shows the high correlation between the lateral flapping b_1 and the roll acceleration $\dot{p}$. The figure shows results obtained from a nonlinear simulation programme with blade element modelling. In the lower part of the figure, the corresponding data obtained from flight test data are presented.

The second equation can be derived from linear rotor calculations (Chen 1979, Talbot *et al*. 1982). When small terms are neglected, the coupled differential equations for longitudinal flapping a_1 and lateral flapping b_1 are

$$\begin{bmatrix} \ddot{a}_1 \\ \ddot{b}_1 \end{bmatrix} + \Omega \begin{bmatrix} \dfrac{\gamma}{8} & 2 \\ -2 & \dfrac{\gamma}{8} \end{bmatrix} \begin{bmatrix} \dot{a}_1 \\ \dot{b}_1 \end{bmatrix} + \Omega^2 \begin{bmatrix} 0 & \dfrac{\gamma}{8} \\ -\dfrac{\gamma}{8} & 0 \end{bmatrix} \begin{bmatrix} a_1 \\ b_1 \end{bmatrix}$$

$$= \Omega^2 \begin{bmatrix} \dfrac{\gamma}{8} A_1 - 2\dfrac{p}{\Omega} - \dfrac{\gamma}{8}\dfrac{q}{\Omega} \\ \dfrac{\gamma}{8} B_1 + 2\dfrac{q}{\Omega} - \dfrac{\gamma}{8}\dfrac{p}{\Omega} \end{bmatrix} \qquad (3)$$

Making a first-order approximation by neglecting the flapping accelerations and

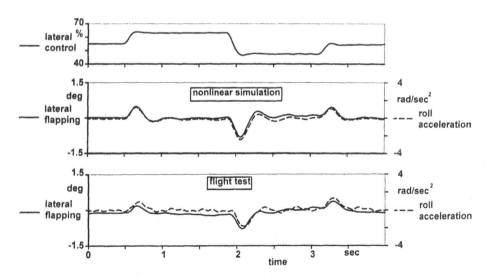

Figure 4.   Correlation between lateral flapping and roll acceleration.

solving for the flapping velocities leads to

$$
\begin{bmatrix} \dot{a}_1 \\ \dot{b}_1 \end{bmatrix} + \frac{\Omega}{\dfrac{\gamma^2}{64}+4} \begin{bmatrix} \dfrac{\gamma}{4} & \dfrac{\gamma^2}{64} \\ -\dfrac{\gamma^2}{64} & \dfrac{\gamma}{4} \end{bmatrix} \begin{bmatrix} a_1 \\ b_1 \end{bmatrix} = \frac{\Omega}{\dfrac{\gamma^2}{64}+4} \begin{bmatrix} \dfrac{\gamma^2}{64}A_1 - \dfrac{\gamma}{4}B_1 \\ \dfrac{\gamma}{4}A_1 + \dfrac{\gamma^2}{64}B_1 \end{bmatrix} - \begin{bmatrix} q \\ p \end{bmatrix}
$$

$$(4)$$

This approximation is possible due to the lack of stiffness parameters in the second matrix of (3).

The final structure of the roll equation is then derived by differentiating (2 *a*) and then inserting (2 *b*) into (2 *a*). It yields

$$ \ddot{p} = \tilde{L}_{\dot{p}}\dot{p} + \tilde{L}_p p + \tilde{L}_{A_1} A_1 \tag{5} $$

This is a second-order differential equation for the rotor/body motion. The main advantage of this approach is that the extended model explicitly represents the dynamics of the rotor degrees of freedom. For the identification, however, no blade flapping measurements are required. In practice, these data are often not available. In addition, for rigid rotors it is difficult to generate reliable data because of calibration and measurement problems. For the above-derived model, only rate accelerations are needed, which can accurately be obtained by differentiating the measured rates. The final model of the host is described by the linear differential equation system

$$ \dot{x} = Ax + Bu \tag{6a} $$

with the state vector

$$ x = (u, v, w, p, \dot{p}, q, \dot{q}, r, \Phi, \Theta)^{\mathrm{T}} \tag{6b} $$

and the control vector

$$ u = (\delta_x, \delta_y, \delta_0, \delta_p)^{\mathrm{T}} \tag{6c} $$

3.2. *System identification procedure*

The general approach used in aircraft system identification is shown in Fig. 5. In flight tests, specifically designed control input signals are executed to excite the aircraft modes of interest. The inputs and the corresponding aircraft responses are measured and recorded. A robust least-squares identification technique is initially applied to the measured data to check their internal compatibility. Data inconsistencies resulting from, for example, calibration errors, drifts or instrumentation failures are detected by comparing redundant measurements from independent sensors. This approach, which can be used on-line, helps to obtain the data quality required for system identification.

For this key step, the aircraft dynamics are modelled by a set of differential equations describing the external forces and moments in terms of accelerations, state and control variables, where the coefficients are the stability derivatives (Kaletka and Grünhagen 1989). The responses of the model and the aircraft resulting from the control inputs are then compared. The response differences are

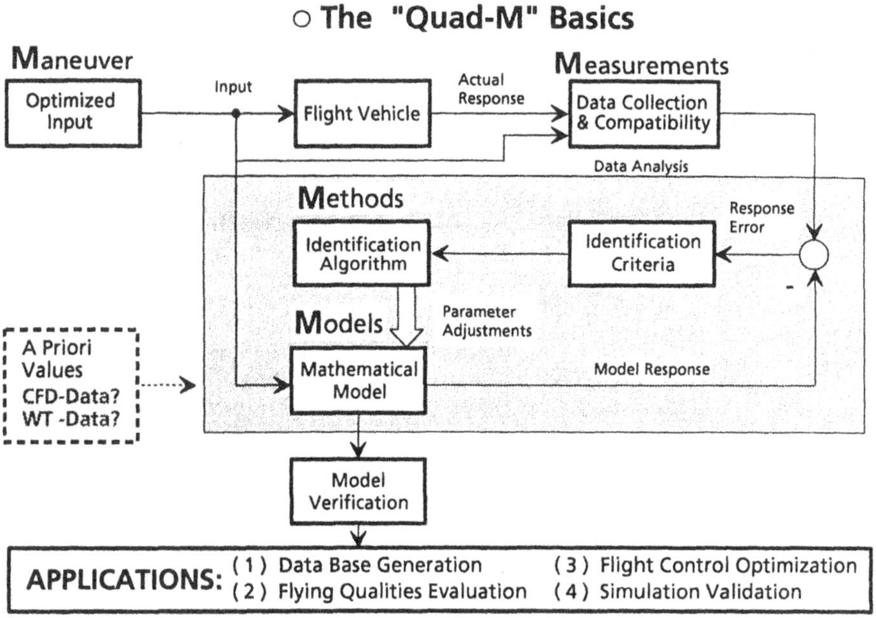

○ The "Quad-M" Basics

Figure 5. System identification approach.

minimized by the identification algorithm that iteratively adjusts the model coefficients. In this sense, aircraft system identification implies the extraction of physically defined aerodynamic and flight mechanics parameters from flight test measurements. Usually, it is an off-line procedure as some skill and iteration are needed to select appropriate data, develop a suitable model formulation, identify the coefficients and, finally, verify the results. Here, model formulation involves consideration of model structure, elimination of non-significant parameters, and inclusion of important nonlinearities. As the identified model is usually obtained from a small number of flight tests, a model verification step is mandatory to prove its validity and its suitability for different applications. An overview of the present status of rotorcraft system identification is given by Hamel and Kaletka (1994).

The nonlinear maximum likelihood identification technique was applied to determine the parameters of the extended eight-DOF model. Kinematic and gravity terms were formulated to be nonlinear.

As the model description with the two additional degrees of freedom only represents the rotor dynamics, some delay due to the hydraulic actuators must still be accepted. It was determined to be 30 to 40 ms for all controls. Time histories obtained from the identified extended model and the measured data are presented in Fig. 6.

3.3. *Feedforward design and calculation*

For the host aircraft, we have a linear system

$$\dot{x} = Ax + Bu \qquad (7\,a)$$

81

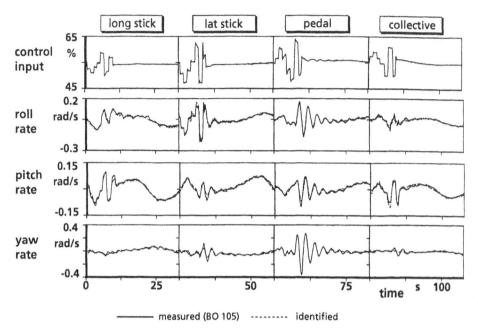

Figure 6. Comparison between flight test and identified model.

with

$$y = Cx + Du \tag{7 b}$$

y and u have the same dimensions. The system is assumed to have complete controllability and observability. The transfer function matrix for this system is

$$y = [C(sI - A)^{-1}B + D]u \tag{8}$$

From this equation a right prime representation

$$y = R(s)P^{-1}(s)u \tag{9}$$

is calculated. Right prime representation means $R(s)$ and $P(s)$ have no common polynomial matrix divisor, except for unimodular polynomial matrices (matrices with a constant determinant). From (9) the feedforward is calculated by formal inversion:

$$u = P(s)R^{-1}(s)y \tag{10 a}$$

Setting $y = y_m$ 'the model state' leads to the feedforward $u = u_f$

$$u_f = P(s)R^{-1}(s)y_m \tag{10 b}$$

The expression $P(s)R^{-1}(s)$ is a transfer function matrix with no strictly proper transfer functions. In a first step $P(s)$ is decomposed by polynomial matrix division to

$$P(s) = K(s)R(s) + H(s) \tag{11}$$

For the inversion of $R(s)$ a unimodular matrix $U(s)$ has to be calculated, so that the determinant of a matrix Γ_c is non-singular. Γ_c is the so-called column rank matrix, which is calculated from the coefficients of the highest powers of s for each column from $R*(s)$. The matrix

$$R*(s) = R(s)U(s) \tag{12}$$

is called column reduced and can be inverted numerically. Equation (11) has to be multiplied by the right-hand side by $U(s)$ to perform the inversion:

$$P(s)U(s) = [K(s)R(s) + H(s)]U(s) \tag{13}$$

By right-hand side multiplication with $R*^{-1}(s)$

$$P(s)R^{-1}(s) = K(s) + H(s)R^{-1}(s) \tag{14}$$

is calculated. In this equation the second term is a strictly proper transfer function. The expression $K(s) \cdot y_m$ is a product of a constant matrix $K*$ and a vector of the model state and its derivatives (see the table). Owing to unwanted couplings in turning flights, some of the matrix elements had to be set to zero. In particular, the activity of the collective control due to roll commands had to be cancelled in the matrix because of the opposite reactions in left and right turns. To consider the necessary increase of power in turning flights, a part which is proportional to the load factor is added to the collective control.

The second part of the feedforward is

$$H(s)U(s)R*^{-1}(s)y_m \tag{15}$$

which can be reformulated to

$$\frac{1}{r(s)} \hat{R}(s)y_m \tag{16}$$

$r(s)$ is the determinant of $R(s)$ and $\hat{R}(s)$ the adjugate matrix. By factorizing

$$r(s) = \prod_i (s - s_i) \tag{17}$$

one gets

$$\left[\sum_i \frac{1}{s - s_i} \bar{R}_i \right] y_m \tag{18}$$

Here, $\bar{R}_i$ are the residual matrices of the invariant zeros s_i. The functions

$$\frac{1}{s - s_i} y_m \tag{19}$$

are in the origin of the Laplace transformation convolutions

$$\int_0^t [e^{s_i I(t - \tau)} Y_m(\tau)] \, d\tau \tag{20}$$

They are solved 'on-line' by Simpson integration.

	φ	ϑ	$\int r$	p	q	r	w	$\dot{p}$	$\dot{q}$	$\dot{r}$	$\dot{w}$	$\ddot{p}$	$\ddot{q}$
η_x	X	X	X	X	X	X	X	X	X	X	X	X	X
η_y	X	X	X	X	X	X	X	X	X	X	X	X	X
η_0	O	X	X	O	X	X	X	O	X	X	X	O	O
η_p	O	X	X	O	X	X	X	X	X	X	X	O	X

Structure of feedforward.

83

3.4. Examples

3.4.1. *Lateral directional motion.* The equation of lateral directional motion is

$$\begin{bmatrix} \dot{v} \\ \dot{r} \end{bmatrix} = \begin{bmatrix} Y_v & (Y_r - u_0) \\ N_v & N_r \end{bmatrix} \begin{bmatrix} v \\ r \end{bmatrix} + \begin{bmatrix} Y_{\delta_p} \\ N_{\delta_p} \end{bmatrix} \delta_p \tag{21}$$

leading to the transfer function matrix representation

$$\begin{bmatrix} v \\ r \end{bmatrix} = \begin{bmatrix} \dfrac{s Y_{\delta_p} + (Y_r - u_0)N_{\delta_p} - N_r Y_{\delta_p}}{s^2 - s(Y_v + N_r) + Y_v N_r - (Y_r - u_0)N_v} \\[3mm] \dfrac{s N_{\delta_p} - Y_v N_{\delta_p} + N_v Y_{\delta_p}}{s^2 - s(Y_v + N_r) + Y_v N_r - (Y_r - u_0)N_v} \end{bmatrix} \delta_p \tag{22}$$

The observer equation is chosen

$$y = \begin{bmatrix} 0 & 1 \end{bmatrix} \begin{bmatrix} v \\ r \end{bmatrix} \tag{23}$$

Inversion leads to

$$\frac{\delta_p}{r} = \frac{1}{N_{\delta_p}} \frac{s^2 - s(Y_v + N_r) + Y_v N_r - (Y_r - u_0)N_v}{s - Y_v + N_v \dfrac{Y_{\delta_p}}{N_{\delta_p}}} \tag{24}$$

By polynomial division one gets the static part of the feedforward

$$\delta_p = \begin{bmatrix} \dfrac{1}{N_{\delta_p}} & -\dfrac{N_v Y_{\delta_p} + N_r N_{\delta_p}}{N_{\delta_p}^2} \end{bmatrix} \begin{bmatrix} \dot{r} \\ r \end{bmatrix} \tag{25}$$

The invariant zero is

$$s_1 = Y_v - N_v \frac{Y_{\delta_p}}{N_{\delta_p}} \tag{26}$$

3.4.2. *Roll motion.* The roll equation is

$$\begin{bmatrix} \dot{\Phi} \\ \dot{p} \\ \ddot{p} \end{bmatrix} = \begin{bmatrix} 0 & 1 & 0 \\ 0 & 0 & 1 \\ 0 & L_p & L_{\dot{p}} \end{bmatrix} \begin{bmatrix} \Phi \\ p \\ \dot{p} \end{bmatrix} + \begin{bmatrix} 0 \\ 0 \\ L_{\delta_y} \end{bmatrix} \delta_y \tag{27}$$

leading to the transfer function matrix representation

$$\begin{bmatrix} \Phi \\ p \\ \dot{p} \end{bmatrix} = \begin{bmatrix} \dfrac{L_{\delta_y}}{s^3 - s^2 L_{\dot{p}} - s L_p} \\[3mm] \dfrac{L_{\delta_y}}{s^2 - s L_{\dot{p}} - L_p} \\[3mm] \dfrac{s L_{\delta_y}}{s^2 - s L_{\dot{p}} - L_p} \end{bmatrix} \tag{28}$$

The observer equation is chosen to be

$$y = [1 \quad 0 \quad 0] \begin{bmatrix} \Phi \\ p \\ \dot{p} \end{bmatrix} \tag{29}$$

leading to the transfer function

$$\frac{\delta_y}{\Phi} = \frac{1}{L_{\delta_v}} [s^3 - s^2 L_{\dot{p}} - s L_p] \tag{30}$$

The feedforward can then be derived

$$\delta_y = \frac{1}{L_{\delta_v}} [1 \quad -L_{\dot{p}} \quad -L_p] \begin{bmatrix} \ddot{p} \\ \dot{p} \\ p \end{bmatrix} \tag{31}$$

an invariant zero does not exist.

3.5. *Feedback design*

Since modelling and identification inaccuracies, changes in nominal flight conditions and, mainly, gusts drive the base helicopter states x from the desired states x_m, feedback controllers are required to reduce these errors.

A decoupled minimum structure feedback system controls the pitch states with the longitudinal cyclic, the roll states with the lateral cyclic, the vertical velocity with the collective and the yaw states with the tail rotor control. The pitch, roll and yaw rates and the vertical velocity are controlled with proportional and integral (PI) controllers and the pitch and roll attitude and the sideslip angle with proportional feedback (Fig. 7).

The feedback gains were optimized in a non-real-time simulation and verified in flight with the same procedure. The helicopter was disturbed at 60 knots in level flight with a control input by the safety pilot in the collective for pitch and yaw optimization and with a longitudinal control input for roll and heave controller optimization. In the simulation, the quadratic error between the model states (trim values) and the helicopter states was minimized. At first, the

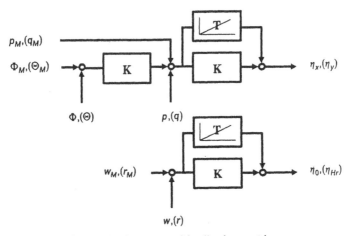

Figure 7.   Structure of feedback control.

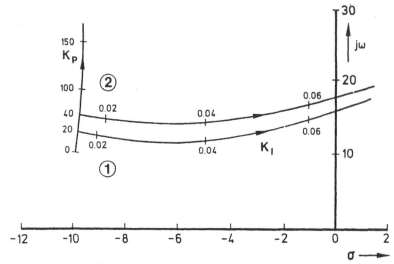

Figure 8.   Root loci of closed loop system.

proportional feedbacks were optimized, then the integral controllers, finally the attitude feedbacks.

   The influence of feedback loops on the system dynamics can be analysed with classical control theory methods. Figure 8 shows the roll root loci of an eight degree-of-freedom (DOF) linear, 60 knots level flight, model of the BO 105 dependent on linear and integral roll-rate feedback K_p and K_i. In the first layout step, K_p is increased, which increases the roll frequency, while the roll damping decreases. With a defined proportional roll-rate feedback, the integral roll-rate feedback is increased. The roll damping now decreases rapidly with

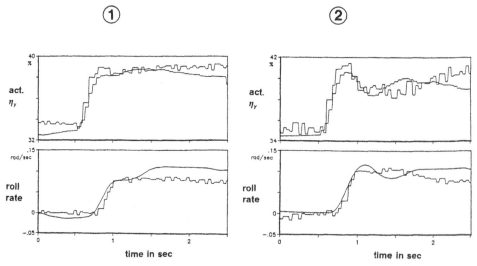

Figure 9.   Step responses at design points (1) and (2).

nearly constant roll frequency. Corresponding time histories of the design points

(1) $(K_p = (20\% \text{ s})/\text{rad}, \; K_i = (0.02\%)/\text{rad})$

(2) $(K_p = (40\% \text{ s})/\text{rad}, \; K_i = (0.02\%)/\text{rad})$

are shown in Fig. 9. The actuator and roll rate responses to pilot step inputs give good accord between linear simulation and flight test, in the design point (1); while in the design point (2) the real helicopter is more damped than the simulation result.

4. Model-following performance

A good initial response and high bandwidth capability of ATTHeS was emphasized in the MFCS design. This characteristic is essentially dependent on a well adapted feedforward controller. The use of the angular accelerations for pitch and roll in the BO 105 model, which is used for the feedforward calculation, improves the initial response behaviour of the ATTHeS simulator. The overall time delays are measured in flight tests with about 110 ms for the roll axis and 150 ms for the pitch axis. The increased value in the pitch axis results from the basic helicopter due to a higher moment of inertia (Pausder *et al.* 1991).

The quality of long-term following is demonstrated in Fig. 10. Both rate and attitude signals show satisfactory model matching. The ATTHeS system has been flown in roll manoeuvres like slalom, constant turns and figure eights without any long-term drift tendency. Corresponding to the assessed effective time delay and the time histories of the long-term behaviour, Fig. 11 shows the model-following performance for the roll attitude in the frequency domain. The left-hand side shows the roll axis response of the basic BO 105 identified from a flight test with a lateral sweep control input by the pilot. It includes lower frequency dutch roll dynamics at 2.5 rad s^{-1} and low damped lead–lag dynamics at 15.0 rad s^{-1} (Kaletka *et al.* 1991). The right-hand side compares the ATTHeS response with the analytical response of the designed rate command system. Here, the equivalent time delay of 110 ms was included. An acceptable model following up to 8 rad s^{-1} is achieved which allows roll bandwidth configurations to be covered up to 6 rad s^{-1}.

Figure 12 illustrates the achieved decoupling ratio of ATTHeS in comparison with the basic BO 105 in time histories and crossplots. In the upper time histories a comparison is made for disengaged and engaged MFCS modes. In both tests, the simulation pilot excited the helicopter with longitudinal control oscillations. With the disengaged MFCS, the strong coupling response of the basic BO 105 between the pitch and roll axes can be observed. With the MFCS engaged, the coupling to the roll axis is suppressed. The crossplots of the longitudinal and lateral actuator positions indicate the necessary control activity to compensate the coupling.

A criterion for the acceptance of the simulation by pilots has been derived by Anderson (1993) for fixed-wing aircraft. Here, the frequency responses of measured simulator rates are compared with the command model rates. Ideally, the frequency response should have a magnitude relationship of one and a phase angle of zero for a broad frequency range. For practical use, boundaries of so-called 'unnoticeable dynamics' were defined, where pilots will not notice simulation deficiencies. Using this criterion also as a quality assessment for the ATTHeS simulation, the frequency responses for measured roll rate to model roll rate are compared in Fig. 13 for two different cases: (1) an identified conventional six-DOF rigid-body model was used

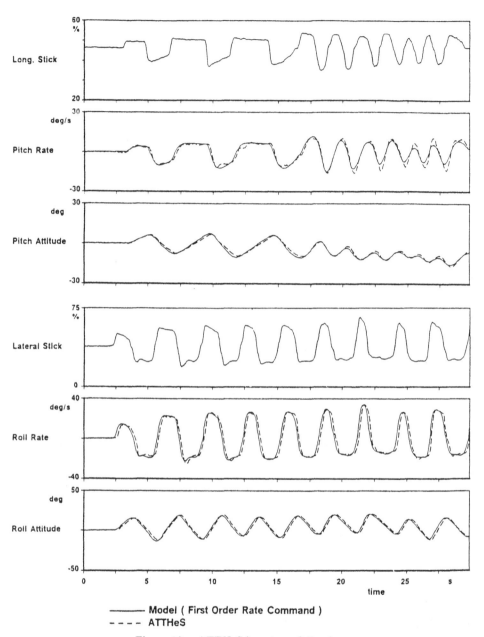

Figure 10. ATTHeS long term following.

to define the flight control system, and (2) an identified extended model with rotor DOF was applied for the flight control system design. It is clearly seen that only the ATTHeS frequency response based on the extended model lies within the boundaries. Since good pilot ratings for the simulation were also given, this confirms the definition of the boundaries and seems to indicate that they can be used as a more general quality criterion for high bandwidth flight control systems.

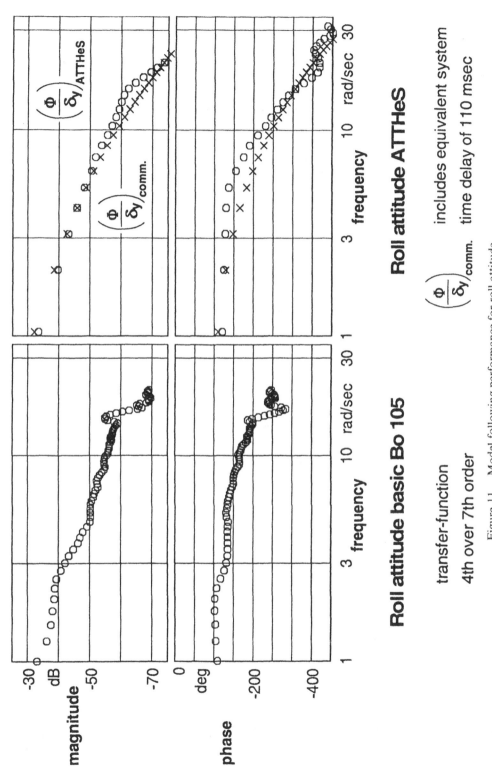

Figure 11. Model-following performance for roll attitude.

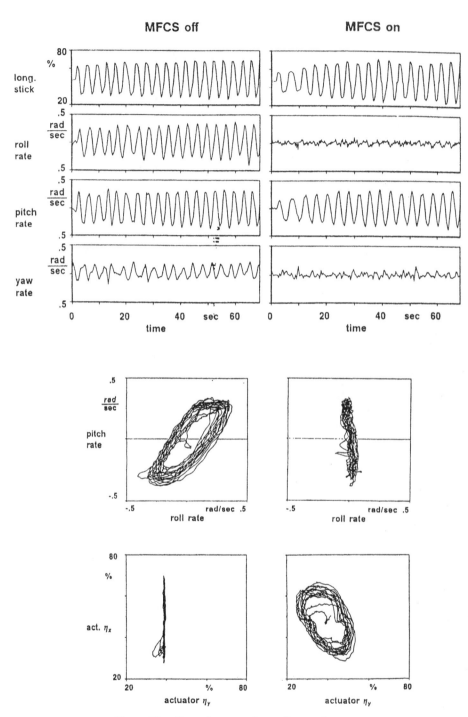

Figure 12.   Control system decoupling performance.

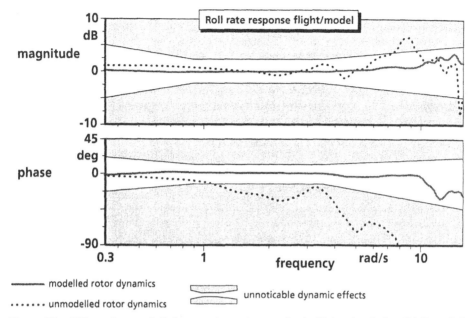

Figure 13. Effect of unmodelled rotor dynamics on the in-flight simulation fidelity of the fly-by-wire BO 105 ATTHeS.

5. ATTHeS utilization

ATTHeS is mainly utilized for handling-qualities research and helicopter in-flight simulation. Decoupled rate command/attitude hold and attitude command models were defined as transfer functions with selectable control sensitivities, dampings and time delays. The influence of these parameters on the handling quality level was investigated in extensive flight tests. For helicopter in-flight simulation purposes, a linear derivative simulation model with specific nonlinear extensions has been developed. Specific control systems of the helicopter to be simulated can be embedded. The high fidelity of the simulation was proven in flight tests. In addition, ATTHeS is used for test pilot training at various European test pilot schools.

5.1. Handling qualities research

A main objective of flight mechanics research is to generate credible data for the definition of handling qualities criteria. An updated military rotorcraft handling qualities specification (ADS-33) (Anon. 1989) has been published which is a mission oriented specification and considers the integration of modern cockpit and control technologies. Although the ADS-33C is a US specification at present, it is of international interest and international studies have contributed to the databases for the definition of requirements. In co-operation with the US Army, a roll axis bandwidth study was conducted using the ground-based vertical motion simulator at NASA Ames and the ATTHeS in-flight simulator of DLR in a complementary mode. The objective of this study was to verify and, if necessary, to refine the criteria boundaries specified in the ADS-33C. Based on previous slalom testing experience (Pausder 1985) a modified slalom task with precise tracking phases through a set of gates was built up. The modified slalom meets the demands of an appropriate small amplitude precision tracking task and takes into consideration the constraints of the test facilities.

For the investigations of bandwidth and phase delay influences on handling qualities, a special conceptual model was developed. This decoupled model generates pitch and roll-rate command/attitude hold (RCAH) or attitude command (AC) for longitudinal, and lateral cyclic, rate of climb for collective and sideslip command for the pedals. The model is extended with a co-ordinated-turn term defined in relation to the commanded roll attitude. Finally, the angular rates are Euler-transformed and integrated to the Euler angles.

The model structure (so-called conceptual models) is defined by transfer functions (Ockier and Pausder 1994)

$$\frac{p}{\delta_y} = \frac{L_{\delta_y}}{s + L_p} \tag{32}$$

For the implementation in the control system software, these transfer functions are transformed to the time domain, which leads to the differential equation

$$\dot{p} + L_p p = L_{\delta_y} \delta_y \tag{33}$$

To adapt this differential equation to the requirements of the second-order formulation of the feedforward design, it is reformulated by adding a time constant τ_p. The final formulation of the commanded model is then

$$\ddot{p} = \tau_p(-\dot{p} - L_p p + L_{\delta_y} \delta_y) \tag{34}$$

This is a differential equation of second order in the short-term response and first order for the demanded long-term response. The time constant is adapted to the BO 105 response due to the rotor fuselage coupling. The same formulation is used for the pitch axis. For the turn coordination the following equations are used

$$\dot{\psi} = \frac{g}{u} \tan(\Phi) \tag{35 a}$$

$$p_{\psi} = \dot{\psi} \sin(\Theta) \tag{35 b}$$

$$q_{\psi} = \dot{\psi} \cos(\Theta) \sin(\Phi) \tag{35 c}$$

$$r_{\psi} = \dot{\psi} \cos(\Theta) \cos(\Phi) \tag{35 d}$$

to allow the simulation pilot to fly coordinated turns with lateral cyclic control only.

The roll axis bandwidth has been varied between about 1·5 and 4·5 rad s^{-1} in the flight experiment. Additional configurations were defined with added time delays between 40 and 160 ms. The overall system transfer behaviour of ATTHeS showed an excellent matching around the bandwidth frequencies, especially compared with the transfer functions of the command models. The summarized data of the flight tests are shown in the evaluation diagram of Fig. 14. The results indicate the high performance of ATTHeS with respect to bandwidth capability, model fidelity and flexibility. The results of the handling qualities tests have been published in more detail by Pausder and Blanken (1992) and Ockier and Pausder (1994).

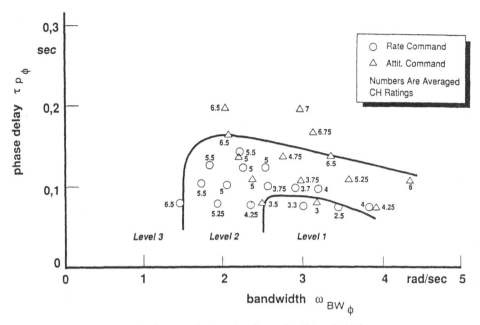

Figure 14. Summarized evaluations of roll bandwidth tests.

5.2. *Helicopter in-flight simulation*

For the simulation of helicopters in flight, a special model to be followed was developed (Bouwer and von Grünhagen 1992). Either from flight test data or from runs of a generic simulation program, the linear model

$$\dot{x}_\mathrm{M} = A_\mathrm{M} x_\mathrm{M} + B_\mathrm{M} u_\mathrm{p} \tag{36 a}$$

with the model state vector

$$x_\mathrm{M} = (u_\mathrm{M}, v_\mathrm{M}, w_\mathrm{M}, \dot{p}_\mathrm{M}, \dot{q}_\mathrm{M}, p_\mathrm{M}, q_\mathrm{M}, r_\mathrm{M}, \Phi_\mathrm{M}, \Theta_\mathrm{M})^\mathrm{T} \tag{36 b}$$

and the simulation pilot controls

$$u_\mathrm{P} = (\delta_x, \delta_y, \delta_0, \delta_p)^\mathrm{T} \tag{36 c}$$

are defined. The model dynamic matrix A_M and the model control matrix B_M are calculated with the identification procedure described above. The necessary nonlinear terms from coordinated turn, gravity, changes of flight path and Euler equations are programmed explicitly. In addition, the four-axis Stability Control Augmentation System (SCAS) of the helicopter to be simulated is programmed. The SCAS can be engaged/disengaged via software switches during the flights. For the investigation of SCAS failures, several failure situations are programmed.

The in-flight simulation of the Lynx helicopter shall serve as an example. This helicopter has some couplings opposite to the corresponding couplings of ATTHeS in its basic BO 105 mode. Figure 15 shows an airspeed change manoeuvre with constant altitude and heading, which is performed by pitching the helicopter up and down. It can be observed that the simulation pilot controls and the ATTHeS actuator positions (= MFCS output) in the left diagrams

93

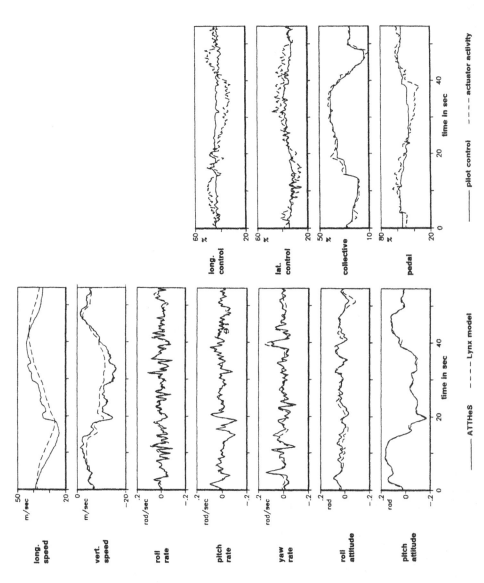

Figure 15. Acceleration/deceleration manoeuvre with Lynx model.

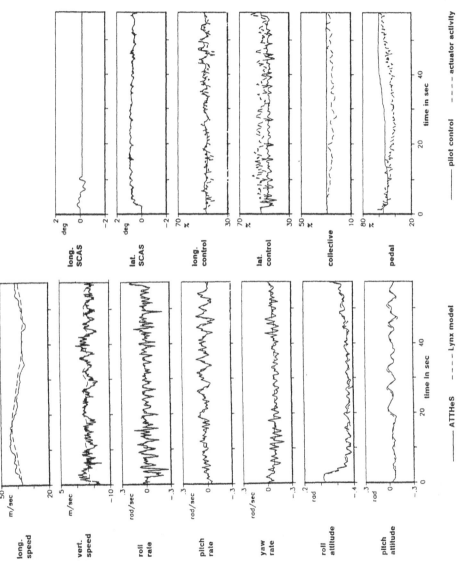

Figure 16. Longitudinal SCAS failure of Lynx model.

diverge, except for the collective control. This is a result of the opposite pitch rate due to collective control input coupling of the Lynx helicopter, which has a mechanical linkage between collective and longitudinal cyclic controls. All ATTHeS states match very well the commanded Lynx states. In general, the in-flight simulation was deemed to be representative for the Lynx helicopter (Howitt and Cheyne 1992).

Figure 16 documents a failure situation in the simulated helicopter. In a right turn at 10 s, the longitudinal SCAS quits. Now the Lynx helicopter is in its basic unstable mode for the pitch axis. At about 20 s, the simulation pilot's longitudinal stick control activity increases and the Lynx is oscillating in the pitch axis. Even this situation is matched very well by ATTHeS and was rated by the simulation pilot to be representative for the real flight case.

5.3. *Hover position hold*

The design of control systems for helicopters in hover and at low speed is a basic requirement for the extension of mission profiles and new mission demands. A special task for various applications is the position hold under windy and gusty conditions above a ground-fixed or moving target, like a shipboard reference, or a small vessel or lifeboat on a rescue mission.

ATTHeS was equipped with an innovative measurement system for the hover position above a target. A video camera in combination with a sophisticated computer for processing optical information was used as an integrated sensor system for the measurement of the relative position of the aircraft to a target. Based on the existing model-following control system of ATTHeS for the forward flight condition, this control system was modified and adapted to fulfil the special requirements of the position-hold task.

For a helicopter in hover, the longitudinal and lateral accelerations can be controlled either by changing the pitch and roll attitudes or, for short-term corrections, directly by the sideforce capability of the rotor system with only very small attitude changes. The relative position above a target is a second-order differential equation with a nonlinear function in the attitudes. In the existing MFCS for hover a special effort was made to stabilize the attitude loops in pitch and roll. For the position-hold task additional terms in these loops were formulated. These terms relate the commanded pitch and roll attitudes to the relative position between the helicopter and target and the corresponding velocities. The coefficients of these additional control loops were preoptimized in a non-real-time simulation.

Before it underwent any flight tests, the position-hold hardware and software had to be integrated in a real-time helicopter simulation (see above). To perform this integration, a special hardware configuration was built, where

- a visual system (Alvermann 1993) simulated the view of the downward-looking camera;
- the camera to be used in flight was mounted in front of the visual screen of this system;
- the computer system of the camera was connected to a duplicate of the on-board control computer.

With this configuration, the overall MFCS with the position-hold system was tested extensively under real-time conditions. In these simulations, the defined flight task was flown with additional manoeuvres, including hover position and hold turn in constant winds up to 20 m s^{-1}. The handling of the position hold was found to be adequate to start flight testing. The flight task for the evaluation pilot was defined in four steps:

(1) engage the fly-by-wire system in hover about 20 m behind the target;
(2) engage the basic MFCS (position hold off) with rate command/attitude hold in pitch, roll, yaw and heave;

Figure 17.   Position-hold flight test arrangement.

(3) fly above the target;

(4) when the camera has found the target within the defined range, engage the position hold.

The MFCS as well as the position hold could be engaged by pressing a button on the computer keyboard. The position hold was automatically disengaged when the pilot moved the cyclic stick.

The target was represented by a car on top of which was mounted a black square. The car crew were informed by radio to drive in a circle at a velocity of about 15 km h^{-1} and with a radius of about 40 m. In the position-hold mode the control system had to fly the helicopter above the target (Fig. 17) at a constant altitude and with a constant heading while the simulation pilot flew with hands off the controls. With the constant heading, the ground track of the car covers all the combinations for the forward (backward) and sideward airspeed of the helicopter.

Figure 18 shows the time histories of the longitudinal x_{POS} and lateral y_{POS} positions, the longitudinal d_x and lateral d_y cyclic stick inputs and the status of the control system from the continuously recorded telemetry data stream over 10 minutes The flight test data were gathered and recorded under 15 knots wind conditions with gusts up to 30 knots. At the beginning, the pilot flies the helicopter in the regular 1 : 1 fly-by-wire mode (status = 0). After engaging the basic MFCS (status = 1) the pilot flies the helicopter above the target until the tracking system detects it. After engaging the position hold (status = 3), the pilot flies hands off (d_x, d_y = constant). During the following 8 minutes, while three circles were completed by the car, the relative position error was about 5 m maximum, but with a non-moving target at $t = 2$ min and $t = 5$ min the error is about 1 m in the longitudinal

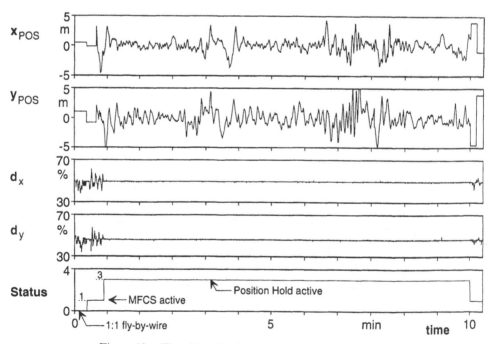

Figure 18.   Time histories from telemetered data recording.

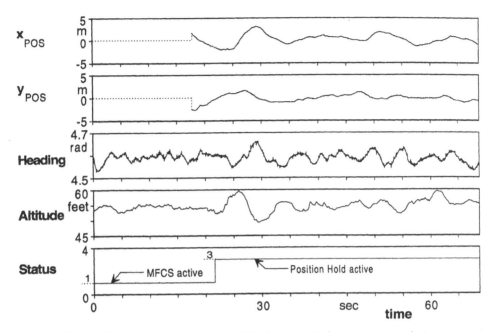

Figure 19. Position, heading and altitude states during autonomous hover.

and 1·5 m in the lateral position. The standard deviation during this flight test was 1·2 m in the longitudinal and 1·6 m in the lateral position.

Time histories of the more extensive on-board data recording system of the first 70 s of this flight test are shown in Fig. 19. The data recording was started with the MFCS active (status = 1). At $t = 18$ s, the camera computer system has found the target. At $t = 22$ s, the pilot engages the position-hold system (status = 3). Within 20 s, the position error is stabilized to less than 2 m. At $t = 50$ s, the target starts moving forward (indicated by increasing x_{POS}). During the complete run, the heading-hold error was in the range of 5° absolute and the altitude error was ±5 feet (1·5 m) maximum (Bouwer *et al.* 1995).

6. Concluding remarks and future aspects

The development of advanced helicopters with high authority control systems and sophisticated cockpit technologies requires the exclusive incorporation of pilot-in-the-loop simulation tools. In the process of flying qualities research and system development, the pragmatic complementary use of ground-based and in-flight simulation will provide the confidence in results and reduce the risks in rotorcraft design.

DLR has developed the ATTHeS in-flight simulation system which yields variable stability and control capacity with a high level of flexibility. A sophisticated explicit model-following control system was developed to fulfil the simulation performance demands. The measured performance of the implemented explicit model-following control system illustrates the potential of ATTHeS to cover the expected response characteristics of advanced helicopter systems. A good initial response, a high bandwidth capability, a high decoupling level and a good model-following fidelity have been demonstrated. The methods applied were transferred and successfully tested for the hover and low-speed flight condition.

Besides the continuing refinement of the simulation system, like the integration of advanced cockpit technology and innovative sensor systems, ATTHeS will be applied in extensive flying qualities studies to establish further data for bandwidth/ time delay requirements, coupling requirements and an evaluation of allowable degradation and blending characteristics when system failures occur. Additionally, ATTHeS will be involved in government and industry project support as well as in test pilot training courses.

REFERENCES

ALVERMANN, K., 1993, Visualization and view simulation based on transputers. *Journal of Aircraft*, **30**(4), 550–560.

ANDERSON, M. R., 1993, Robustness evaluation of a flexible aircraft control system. *AIAA Journal of Guidance, Control and Dynamics*, May–June.

ANON, 1989, Handling qualities requirements for military rotorcraft. ADS-33C.

BOUWER, G. and VON GRÜNHAGEN, W., 1992, LYNX helicopter in-flight simulation with ATTHeS, DLR Institute Report, IB 111-92/47.

BOUWER, G., OERTEL, H. and VON GRÜNHAGEN, W., 1995, Autonomous helicopter hover positioning by optical tracking. *Aeronautical Journal of the Royal Aeronautical Society*, May.

CHEN, R. T., 1979, A simplified rotor system mathematical model for piloted flight dynamics simulation. NASA Technical Memorandum 78575.

FOGLER, D. L. and KELLER, J. F., 1993, Design and pilot evaluation of the RAH-66 Comanche core AFCS. NASA/AHS Conference on Piloting Vertical Flight Aircraft, San Francisco, CA.

GMELIN, B., PAUSDER, H.-J. and HAMEL, P., 1986, Mission oriented flying qualities criteria for helicopter design via in-flight simulation. AGARD-CP-423, Paper No. 4.

HAMEL, P. and KALETKA, J., 1994, Rotorcraft system identification – an overview of AGARD FVP Working Group 18. AGARD FDP Conference on Aerodynamics and Aeroacoustics of Rotorcraft, Berlin, Germany.

HOWITT, J. and CHEYNE, S., 1992, Assessment of the DRA HELISIM Lynx math model on the DLR BO 105 ATTHeS in-flight simulator. Defence Research Agency (DRA) Working Paper FSB WP(92)046, Bedford, UK.

KALETKA, J. and VON GRÜNHAGEN, W., 1989, Identification of mathematical derivative models for the design of a model following control system. 45th Annual Forum of the American Helicopter Society, Boston, MA.

KALETKA, J., TISCHLER, M. B., VON GRÜNHAGEN, W. and FLETCHER, J. W., 1991, Identification of mathematical derivative models for the design of a model following control system. *Journal of the American Helicopter Society*, **36**(4), 25–38.

OCKIER, C. J. and PAUSDER, H.-J., 1994, Experiences with ADS-33 helicopter specification testing and contributions to refinement research. AGARD FMP-Symposium on Active Control Technology: Applications & Lessons Learned, Turin, Italy.

PAUSDER, H.-J., 1985, A study of roll response required in a low altitude slalom task. 11th European Rotorcraft Forum, London, UK.

PAUSDER, H.-J. and BLANKEN, C. L., 1992, Investigation of the effects of bandwidth and time delay on helicopter roll-axis handling qualities. 18th European Rotorcraft Forum, Avignon, France.

PAUSDER, H.-J. and BOUWER, G., 1989, Recent results of in-flight simulation for helicopter ACT research. 15th European Rotorcraft Forum, Amsterdam, The Netherlands.

PAUSDER, H.-J., BOUWER, G. and VON GRÜNHAGEN, W., 1988, A highly maneuverable helicopter in-flight simulator – aspects of realization. 14th European Rotorcraft Forum, Milan, Italy.

PAUSDER, H.-J., BOUWER, G. and VON GRÜNHAGEN, W., 1991, ATTHeS in-flight simulator for flying qualities research. International Symposium In-Flight Simulation for the 90s, DGLR-91-05.

SNELL, S. A., ENNS, D. F. and GARRARD, W. L., 1992, Nonlinear inversion flight control for a supermaneuverable aircraft. *Journal of Guidance, Control and Dynamics*, **15**(4), 976–984.

TALBOT, P. D., TINLING, B. E., DECKER, W. A. and CHEN, R. T., 1982, A mathematical model of a single main rotor helicopter for piloted simulation. NASA Technical Memorandum 84281.

4

Advanced flight control research and development at Boeing Helicopters

KENNETH H. LANDIS, JAMES M. DAVIS, CHARLES DABUNDO and
JAMES F. KELLER

1. Introduction

Boeing Helicopter's role in the technology advancement of full-authority digital fly-by-wire systems was spurred by requirements to improve handling qualities and mission effectiveness, and reduce weight and maintenance activities. The transition of Boeing research programs into production readiness is illustrated in Fig. 1. Development efforts on the Tactical Aircraft Guidance System (TAGS), Heavy Lift Helicopter (HLH) and Advanced Digital/Optical Control System (ADOCS) demonstrator programs form the basis for design of the V-22 and RAH-66. Currently, the V-22 is progressing through a 5 year engineering manufacturing development (EMD) contract for further refinement and demonstration of a production design. The aircraft is being co-developed by Boeing and Bell Helicopter Textron. Lessons learned from the V-22 flight test program are influencing design of the RAH-66 prototype. The first flight occurred in January 1996. The Comanche is being co-developed in a teaming arrangement between Boeing Helicopters and Sikorsky Aircraft Division of United Technologies Corporation. Boeing has lead responsibility for the flight control system development and mission equipment integration on both programs.

Safety is the primary objective of the flight control system development activities at Boeing. System design concepts are guided by a 'safety first' design philosophy, wherein a highly reliable control path is always maintained, thereby minimizing and constraining the effects of system failures. Production-type safety requirements are imposed at the outset of the design process, and although backup systems have been available on some programs, the design has never relied on them for flight safety.

Requirements for flight control systems integration at Boeing Helicopters have broadened and cover a wide range of technology disciplines. With higher levels of integration, advanced flight control systems have taken on a centralized role and have been renamed vehicle management systems. Significant emphasis has been directed toward achieving the following:

- *Improved flight safety and reliability* through the implementation of an architecture which minimizes the effects of system failures through intelligent fault detection algorithms.

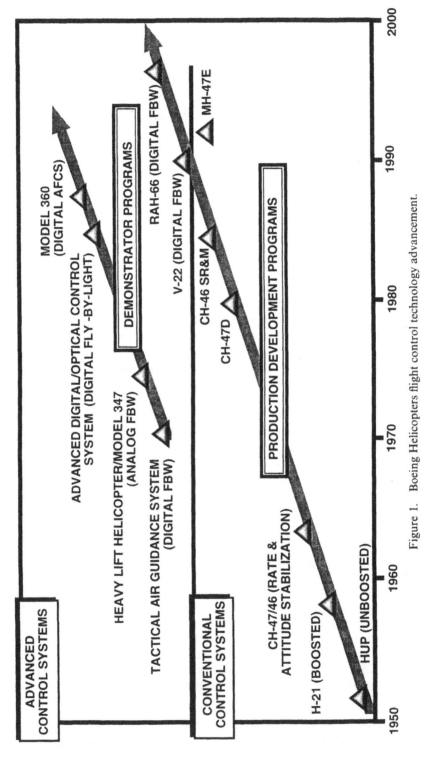

Figure 1.   Boeing Helicopters flight control technology advancement.

- *Advanced control law techniques* using model-following control applications for tailoring of the aircraft response to achieve favorable handling qualities and structural loads characteristics.

- *Increased integration of mission-required functions* within the control system including engine, navigation and fire control systems, along with integration of advanced cockpit controllers.

- *Enhanced mission effectiveness* through selectable control law moding which allows optimization of control characteristics for each task, and automatic modes, such as programmed evasive maneuvers, to counteract known threats.

The purpose of this chapter is to highlight the advances in flight control design which will provide mission-capable rotorcraft for the future. Important developmental programs conducted at Boeing over the last several decades are summarized. A detailed discussion of vehicle management system design approaches provides insight into the control law architecture/implementation used in Boeing rotorcraft. Integration of subsystems to enhance efficiency and optimization of control law moding are highlighted as features required for a mission-capable aircraft. The control law design process employed to develop these features is provided for reference. The culmination of this experience as related to the RAH-66 Comanche and the V-22 Osprey programs is discussed in detail.

2. Development program overview

The broad experience accumulated at Boeing Helicopters is summarized in Fig. 2. The experience base established is backed up with extensive flight testing to verify system design concepts. On all programs, Boeing developed and refined an approach to control law architecture and redundancy management which took advantage of the benefits of model-following control laws. Significant experience with the integration of sidestick controllers was obtained, leading to a production-ready design. The flight control system has taken on a more centralized vehicle management system role through its integration with other subsystem functions, including navigation, engine, and fire control. This section provides an overview of the major flight control research and development programs conducted at Boeing Helicopters. Technology achievements of each program are discussed.

2.1. *Tactical Aircraft Guidance System*

Fly-by-wire efforts at Boeing began in the late 1960s with the Tactical Aircraft Guidance System (TAGS) research program. TAGS incorporated a triplex full-authority digital system using a three-axis sidearm controller. Advanced control concepts developed on TAGS included ground-referenced velocity vector control, command modeling and sensor complementary filtering. These concepts formed the basis of many flight control features included in the RAH-66 and V-22.

2.2. *Heavy Lift Helicopter*

Advancements in flight control technology were demonstrated from 1971 to 1974 as part of the US Army Heavy Lift Helicopter program (Davis *et al.* 1977). A dual-fail operational, full-authority, triplex fly-by-wire flight control system was

VMS Technology	PROGRAM FLIGHT HOURS				
	TAGS 139	HLH 316	ADOCS 551	V-22 1000+	RAH-66 FIRST FLIGHT JAN 1996
Explicit Model Following	●	●	●	●	●
Multi-mode Control Laws	●	●	●	●	●
Sidestick Controllers	●	●	●		●
Integrated FCS / Navigation			●	●	●
Integrated FCS / Engine				●	●
Integrated Fire & Flight Control					●
Advanced Redundancy Management	●	●	●	●	●
Integrated Diagnostics				●	●

Figure 2.   Boeing Helicopters vehicle management system experience.

Figure 3. Advanced flight control research and development programs at Boeing.

designed and flight tested on the Boeing Model 347 HLH demonstrator helicopter shown in Fig. 3. It was the first fly-by-wire helicopter with no mechanical backup control system. The aircraft had a retractable, rear-facing, load-controlling crew station equipped with a four-axis sidearm controller. As described in Landis *et al.* (1975), reconfigurable automatic flight control system (AFCS) modes provided Level 1 handling qualities along with a fully automatic approach to hover capability that enabled the pilot to perform rapid and precise external load operations in all weather conditions. With the hover-hold mode engaged, the aircraft demonstrated position hold to within 7·0 inches (18 cm) in steady winds of 12 knots, gusting to 24 knots. The load-controlling crew member, using the four-axis sidearm controller and operating through the hover-hold mode, demonstrated an impressive capability to control the aircraft precisely for rapid hookup and precision placement of external loads. An automatic load stabilization system provided a significant increase in load damping, both in hover and forward flight, to decrease pilot workload and improve performance during external load operations.

2.3. *Advanced Digital/Optical Control System (ADOCS) demonstrator*

Under a US Army contract during the 1980s, Boeing designed and flight tested an Advanced Digital/Optical Control System called ADOCS. Over 500 hours of testing with the digital fly-by-optics control system were accomplished on a modified UH-60A Black Hawk (Fig. 3). The ADOCS program was sponsored by the Army Aviation Applied Technology Directorate (AATD), Fort Eustis, Virginia, to demonstrate a battlefield-compatible helicopter with a flight control system that would enhance aircraft mission capabilities through decreased pilot workload and improved handling qualities. An increasing hostile ballistic environment, along with consideration of the effects of lightning and electromagnetic interference on the battlefield of the future, led to the selection of fibre optic technology for the ADOCS flight control system. The application of multimode, mission-tailored control laws to improve scout/attack helicopter handling qualities, and the development of sidestick controller technologies, were advanced to a state of readiness for the RAH-66 Comanche.

The early stage of ADOCS development included a comprehensive piloted simulation study (Landis and Glusman 1984) of sidestick controller configurations and levels of system command/stabilization for use under both day and night-time conditions. Various sidestick controller designs (Fig. 4), ranging from a right-hand four-axis controller to a conventional (2 + 1 + 1) configuration, were evaluated for numerous attack helicopter flight control tasks. The test matrix also evaluated controller compliance characteristics ranging from rigid force controllers to compliant small-displacement controllers. As a result of this simulation study, rigid force controller configurations were eliminated prior to flight test. For the flight demonstrator program, design improvements were made to the cockpit controllers, culminating with the flight demonstration of a small-displacement four-axis sidestick configuration and a (3 + 1) controller configuration. Figure 5 shows the sidestick installed in the Black Hawk cockpit, as modified for the ADOCS demonstrator program. The (3 + 1) configuration, comprising a three-axis small-displacement sidestick for pitch, roll and yaw control, and a left-hand medium-displacement collective controller with 4·0 inches (10 cm) of travel, was identified as the best configuration, leading to a similar application in the RAH-66.

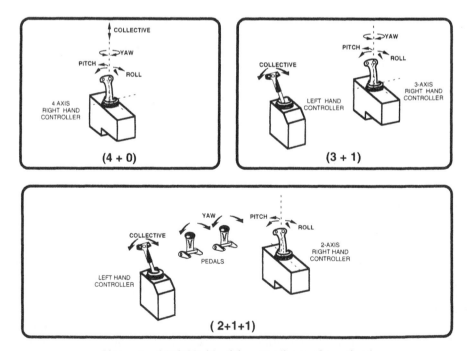

Figure 4.   ADOCS sidestick controller configurations.

Figure 5. Advanced Digital/Optical Control demonstrator sidestick controller.

The ADOCS demonstrator aircraft provided an excellent test bed for the demonstration of major advancements in multimode explicit model-following control laws to enhance scout/attack helicopter mission performance. The reconfigurable digital fly-by-light control system provided the capability to evaluate a wide variety of selectable control laws. Overall success of the ADOCS control law design in providing excellent scout/attack helicopter mission performance and handling qualities was demonstrated in 1987 during nap-of-the-earth (NOE) flight testing conducted by the Army Aviation Engineering Flight Activity (AEFA) at the Army Fort Indiantown Gap test facility. Figure 6 compares pilot ratings for the ADOCS with the preferred (3 + 1) sidestick configuration and the UH-60A with conventional helicopter controls for a number of low-speed evaluation tasks. The flight test data shows the performance of the ADOCS with a sidestick controller configuration to be better than the conventional controls for a wide variety of tasks. Documentation of the ADOCS demonstrator design and flight test results can be found in Glusman *et al.* (1987, 1990).

2.4. *RAH-66 Comanche*

The ADOCS program preceded two pivotal events in the helicopter handling-qualities community, namely the initiation of the RAH-66 scout/attack helicopter program and the definition of a new handling-qualities specification (Aeronautical Design Standard ADS-33C, August 1989). The RAH-66 Comanche as shown in Fig. 3 will be the first helicopter to be procured under the new ADS-33 handling-qualities specification. Designed to be the next-generation scout/attack helicopter,

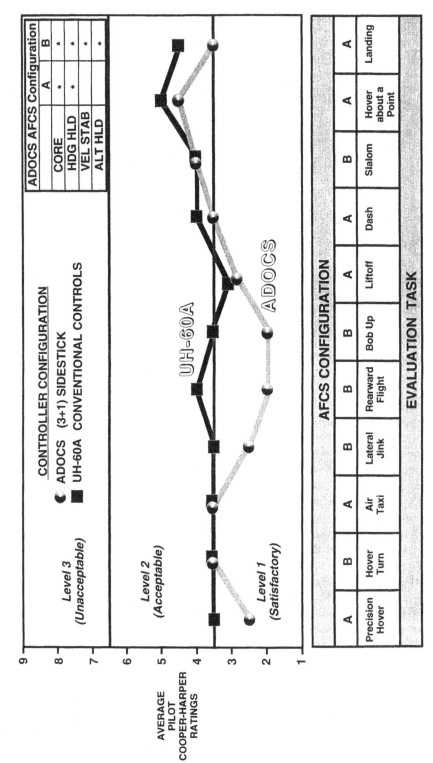

Figure 6.   ADOCS demonstrator low-speed handling-qualities evaluation.

the Comanche incorporates many advanced technology features, including a FANTAIL anti-torque system and a bearingless main rotor with high equivalent flap hinge offset. It will also be the first US Army helicopter with a fully digital fly-by-wire control system and a three-axis sidestick controller. The flight control system technology base developed during the ADOCS program was well suited for direct extension to the Comanche design.

The Comanche flight control system utilizes a multimode control law design that enables the pilot to select flight control system characteristics according to the handling-qualities demands of each mission. In satisfying the requirements of ADS-33, the default (normal-mode) control laws are designed to achieve Level 1 handling-qualities ratings in day/VFR conditions (usable cue environment (UCE) = 1 in the nomenclature of ADS-33). Selectable modes and vision aids allow the Comanche to remain Level 1 under degraded visual conditions (UCE = 2 and 3), and provide further handling-qualities enhancements for automated navigation and targeting. In addition, adequate (Level 2) handling qualities are provided for safe return-to-base capability in the unlikely event of multiple flight control system (FCS) failures.

The Comanche flight control system capitalizes on many advanced design technologies. To meet stringent weight and cockpit ergonomic specifications while providing the desired handling qualities, the primary pilot controller for the longitudinal, lateral and directional axes is a small-displacement sidestick controller, a direct outgrowth of the ADOCS program. Full-time vertical control is provided by a left-hand medium-displacement collective lever having a total travel of 6·0 inches (15 cm). When the altitude-hold mode is engaged, a limited small-displacement vertical control axis provided on the right-hand sidestick frees up the pilot's left hand for cockpit management tasks. The Comanche also uses a binocular helmet-mounted display (HMD) as its primary instrument display to allow the pilot to keep 'eyes out of the cockpit' at critical times. A visual and aural cueing system reduces workload and allows the pilot to access full aircraft capability throughout the flight envelope while not exceeding limits.

2.5. *V-22 Osprey*

The V-22 Osprey (Fig. 3) is an advanced technology tilt-rotor aircraft developed for multiservice use in a wide variety of missions. It is capable of airspeeds ranging from 45 knots rearward in VTOL mode to 345 knots forward in airplane mode. The aircraft, described in detail in Rosenstein and Clark (1986), is characterized by a high forward-swept wing and H-tail empennage. The tilting nacelle at each wingtip is comprised of an engine, transmission and three-bladed rotor and rotor swashplate controls. The rotors are mechanically interconnected for continuous power transmission during single-engine operation. The V-22 flight control system is a full authority, two-fail operative, triplex digital fly-by-wire system. Control of the V-22 in the hover/low-speed regime is achieved through cyclic and collective inputs to the swashplate and power demand signals to the Allison T406 engines. Thrust vectoring is achieved by variation of the nacelle incidence between 0° in airplane mode and 97·5° (7·5° aft of vertical) in helicopter mode. Conventional airplane control surfaces (elevator, rudder, flaperons) are active throughout the flight envelope, while helicopter swashplate controls are phased out in forward flight as nacelle incidence is

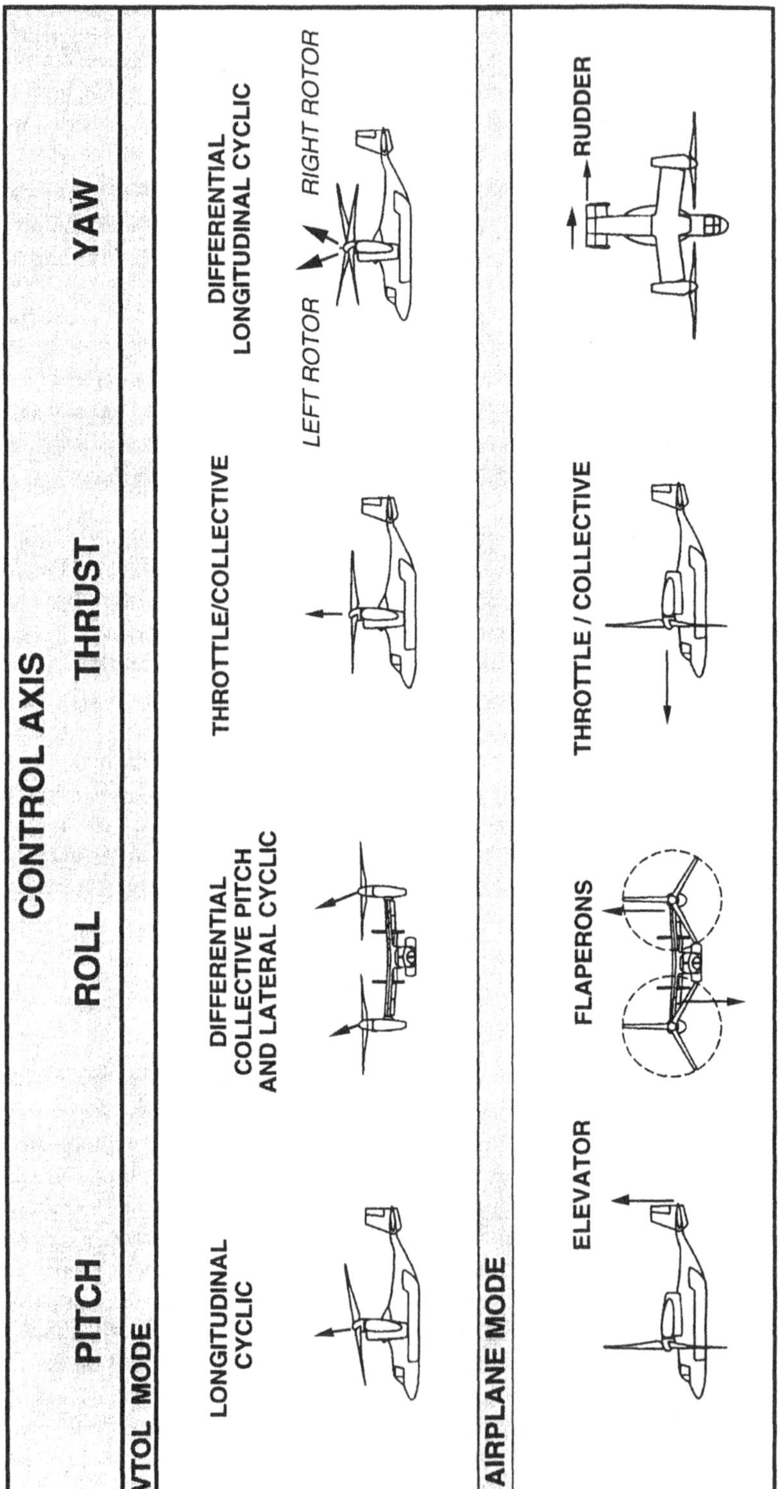

Figure 7. V-22 flight control configuration.

decreased. Figure 7 summarizes the V-22 control mechanisms used in the helicopter and airplane flight modes.

The cockpit controls of the V-22 include a conventional displacement center-stick, directional pedals and a small-displacement thrust control lever (TCL) having a total travel of 4 inches (10 cm). A programmable force–feel system provides airspeed-scheduled force–feel characteristics throughout the flight envelope. Early in the full-scale development program, a design study was conducted to evaluate conventional cockpit controls versus a $2 + 1 + 1$ sidestick controller configuration. In this study, handling qualities were judged to be similar with either control configuration design. Although weight and control system complexity are reduced with a sidestick configuration, conventional controls were selected for the V-22 aircraft to reduce program risk and provide for an easier transition of pilots from other VTOL aircraft.

Development of the V-22 has involved a balance between handling qualities, structural and aeroservoelastic stability requirements (Dabundo *et al.* 1991). The stringent V-22 design requirements of Military Specification SD-572 mandate a maneuver envelope to $4 \cdot 0g$ at speeds up to 345 knots. These requirements, along with the wide range of flight conditions within the operational flight envelope, present design challenges not seen in conventional helicopters or fixed-wing airplanes. The aircraft components must provide the performance to hover at 47 500 lb (21 205 kg) gross weight, the agility to attain Level 1 handling qualities for aggressive operational mission tasks, and the strength to withstand maneuvering at maximum aerodynamic capability as limited by the FCS. Further challenges are imposed by aeroservoelastic stability characteristics of the V-22 due to the proximity of structural and rigid-body mode frequencies. The flexibility offered by digital flight control systems has enabled the achievement of an effective design for the V-22 which satisfies all requirements.

3. Vehicle management system design

The integration of systems and subsystems is a key element of the vehicle management system (VMS) design process. Figure 8 illustrates the major technologies involved in the development of an integrated VMS. There is an increased demand for improving aircraft effectiveness through functional integration of flight control, engine control and fire control systems. In addition, simultaneous tailoring of handling qualities, structural design and aeroservoelastic (ASE) stability characteristics is required to meet design specifications. Some of the system architecture and control law features developed at Boeing which have proven to be successful include:

- Explicit model-following control incorporated within an architecture designed to maximize safety.

- Manual/automatic reconfiguration to tailor the handling characteristics and/or control functions to the task at hand.

- Control law implementation to integrate conventional or sidestick cockpit controllers in a fly-by-wire control system.

- Forward-loop control shaping to quicken the aircraft response to pilot inputs.

- Airspeed/nacelle gain scheduling to tailor aircraft stability and control characteristics throughout the flight envelope. '

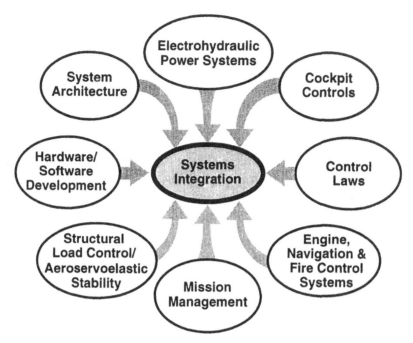

Figure 8. Elements of flight control system integration.

- Control command filtering to eliminate pilot biomechanical feedback for acceptable closed-loop structural mode damping.
- Structural load control algorithms embedded in the flight control laws to provide acceptable maneuvering flight loads throughout the operational flight envelope of the aircraft.

The following sections highlight the design approach used to integrate these features while satisfying overall air vehicle design requirements. Examples of the integration of system control functions in the design of the V-22 and RAH-66 flight control laws are also provided. In addition, the importance of ASE modeling and analysis to achieve a satisfactory total aircraft design is discussed.

3.1. *Control law architecture*

3.1.1. *PFCS/AFCS architecture.* Through the development programs described earlier, Boeing has designed an architecture which maximizes safety and reliability by partitioning flight-critical and mission-critical control laws. As shown in Fig. 9, the control laws are functionally separated into a primary flight control system (PFCS) and an automatic flight control system (AFCS). Each level is also physically separated using its own processor. The PFCS provides flight-critical control laws and possesses a higher level of reliability than the mission-aiding AFCS. The PFCS provides the minimum required flight capability in the event of multiple system failures, while the AFCS provides enhanced handling qualities for mission performance under all conditions. The required level of PFCS reliability is achieved through the use of redundant hardware and software processing, minimization of sensor inputs and reduced control law complexity.

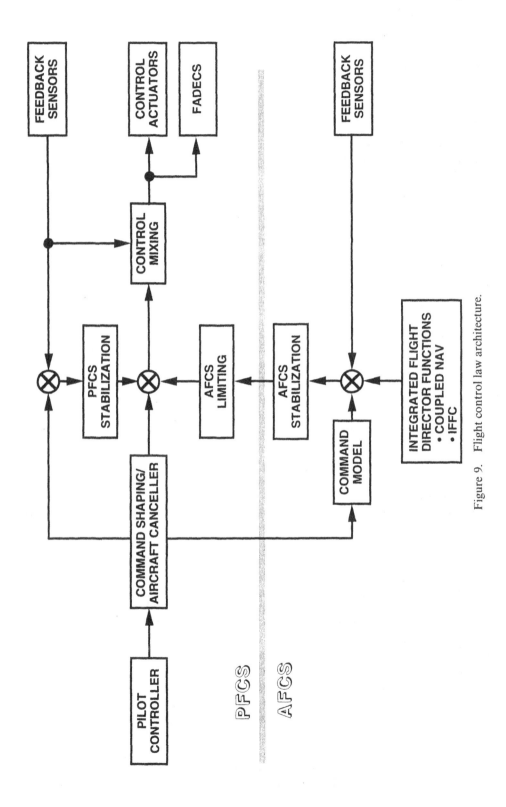

Figure 9. Flight control law architecture.

The AFCS processor interfaces to the PFCS through a limit and switching function denoted as *AFCS port limiting*. This function is designed to provide adequate control authority for the operation of model-following control laws, while isolating the AFCS in the unlikely event of multiple system faults.

3.1.2. *Control mixing/scheduling.* Mixing of the cockpit controllers and the AFCS commands to generate the required rotor, control surface, and engine commands is performed in the PFCS. The capability to vary mechanical mixing and rotor control gearing gains of current production helicopters, e.g. the CH-47 Chinook and UH-60 Black Hawk, is limited, thereby restricting the optimization of control response and coupling characteristics.

Alternatively, the Comanche features airspeed-scheduled control mixing with a fully populated mixing matrix to formulate rotor control commands in the aircraft control axes. In this manner, the highly coupled high hinge offset rotor is decoupled as a function of airspeed, providing desirable control responses in all axes over a wide range of flight conditions. V-22 mixing is programmed to provide optimal use of both rotor and control surface commands. The mixing gains are scheduled with both airspeed and nacelle angle to provide consistent handling characteristics throughout the helicopter, conversion and airplane flight envelope.

3.1.3. *Explicit model following.* Explicit model-following control laws (Fig. 10) are well suited for implementation in fly-by-wire digital control systems which provide the flexibility for tailoring of control law characteristics. The model-following design approach has been refined at Boeing to provide a control law implementation within the PFCS/AFCS architecture which is both safe and effective. In the design of model-following control laws, a cancellation of the inherent aircraft response is accomplished in the PFCS by the command-shaping function. The command model, located within the AFCS, is used to generate the desired state response of the aircraft. The desired response is then compared to the sensed states to form state errors for stability and command response augmentation. Primary attributes of the model-following control approach are: (1) aircraft response consistency over the complete range of ambient conditions and aircraft configurations, (2) independently designed stability and command response characteristics since sensor feedback is used only to augment the trajectory with respect to the commanded aircraft state, and (3) robust full-time feedback stabilization by minimizing errors between the desired and actual control response. Since the feedback signal in each axis is an error signal based on commanded minus actual response, the reduced control sensitivity typical of a conventional SAS with high rate feedback is overcome. Furthermore, synchronization and loss of attitude stabilization while maneuvering is avoided, and mode-switching transients associated with a conventional synchronized attitude feedback system are eliminated.

As shown in Fig. 10, the AFCS error signal is passed through a port-limiting/switching algorithm which fails safely in the event of multiple system failures. The port is designed to provide adequate control authority for operation of the model-following control laws. The ADOCS design implemented all feedback augmentation within the AFCS, resulting in AFCS port saturation for large-amplitude maneuvers. To remedy this situation, the RAH-66 design incorporates the feedforward angular rate response command models and the associated feedback loops in the PFCS as illustrated in Fig. 9. The rate feedback loops are only active when the AFCS is

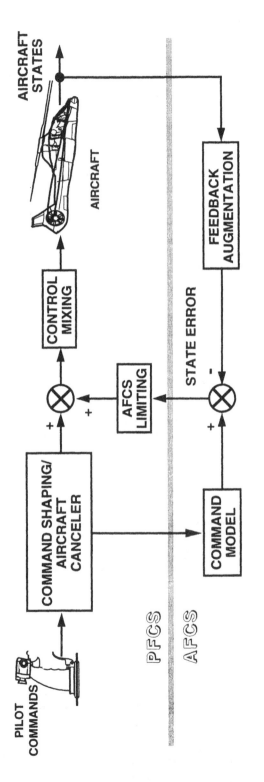

MODEL FOLLOWING ATTRIBUTES

- CONSISTENT AIRCRAFT RESPONSE OVER COMPLETE RANGE OF FLIGHT CONDITIONS AND CONFIGURATIONS

- STABILITY AND COMMAND RESPONSE CHARACTERISTICS DESIGNED INDEPENDENTLY

- FULL TIME FEEDBACK STABILIZATION AND GUST REJECTION DURING MANEUVERS

- NO CONTROL LAW SWITCHING TRANSIENTS

Figure 10.   Model-following control law architecture.

operational but, by being resident in the PFCS, they are not subject to port limiting. By providing control response fidelity with respect to the rate-command model in the PFCS, desired levels of control augmentation (i.e. attitude-command/hold functions) can be implemented in the AFCS with a minimum port authority. Improved redundancy management algorithms to detect and eliminate failure modes of the PFCS provide the required degree of system safety.

3.1.4. Attitude-command model following. As discussed above, the model-following control law approach allows command response characteristics of the vehicle to be tailored based on mission requirements. Figure 11 describes the command/ stabilization characteristics implemented on the RAH-66 Comanche. Since the Comanche is designed for aggressive air combat maneuver scenarios, its control laws must operate to extreme aircraft orientations. Although the angular rate sensors are based in the aircraft body axes, the attitude references need to be inertially based for large-amplitude maneuvers. Thus, the attitude signals must be accurately transformed for precision attitude control during extreme maneuvers. To address this requirement, the Comanche design performs Euler angle transformations on the attitude signals within the model-following control law structure. The attitude errors are then inversely transformed back into the body axis, where the aircraft control is referenced. Since the Euler angle reference system inherently possesses singularities at $+90°$ and $-90°$ of pitch attitude, the control laws are synchronized within a narrow cone about these orientations to eliminate divergence of the attitude signal.

3.2. Cockpit controller integration

Control law functionality for landing/takeoff and control trim operation is based on the selected cockpit controller configuration. Implementation strategies for both conventional large-displacement centerstick controls and unique-trim sidestick controllers are discussed below.

	RAH-66 COMMAND RESPONSE / STABILIZATION				
	CORE AFCS		SELECTABLE AFCS (VELOCITY STAB / ALT HOLD)		
AXIS	HOVER & LOW-SPEED	HIGH SPEED FORWARD FLIGHT	HOVER	LOW-SPEED	HIGH-SPEED FORWARD FLIGHT
LONGITUDINAL	RATE / ATTITUDE		VELOCITY COMMAND/ POSITION HOLD	ATTITUDE COMMAND/ ATTITUDE HOLD	ATTITUDE COMMAND/ AIRSPEED HOLD
LATERAL	RATE / ATTITUDE		VELOCITY COMMAND/ POSITION HOLD	ATTITUDE COMMAND/ ATTITUDE HOLD	RATE COMMAND/ ATTITUDE HOLD
DIRECTIONAL	RATE / HEADING	RATE / HEADING WITH AUTOMATIC TURN COORDINATION	RATE COMMAND/ HEADING HOLD	RATE COMMAND/ HEADING HOLD WITH AUTO-TURN COORDINATION BASED ON GROUNDSPEED	RATE COMMAND/ HEADING HOLD WITH AUTO-TURN COORDINATION
VERTICAL	RATE OF CLIMB		RATE OF CLIMB	RATE OF CLIMB /ALTITUDE	

Figure 11.   RAH-66 Comanche command response/stabilization characteristics.

3.2.1. *Automatic trim transfer.* Regardless of the type of controller selected for the aircraft, the explicit model-following approach may result in control trim building up in the AFCS. Therefore, control laws are required to automatically transfer control trim offsets from the limited-authority AFCS to the PFCS.

The conventional displacement controller used on the V-22 has a relatively large range of control motion. Stick positions are sensed electronically by LVDTs and passed to a digital flight computer which processes rotor and control surface commands. Pitching moment variation due to changes to longitudinal center-of-gravity and nacelle position creates a wide range of possible trim elevator positions at a given airspeed. As illustrated in Fig. 12, parallel backdriving of the longitudinal stick is utilized to avoid storing trim within the limited-authority AFCS, thereby reducing the potential for port saturation. The low-frequency component of the AFCS port signal is fed to a backdrive actuator which smoothly drives the stick to reduce the AFCS offset to zero. In a trim condition, the AFCS port is centered, and the cockpit stick trim position provides the pilot with a tactile cue of the actual aircraft rotor and control surface position.

Apart from the obvious physical differences with respect to conventional displacement controls, sidestick controllers are unique in the manner in which aircraft trim is represented. The unique-trim sidestick controller, because of its size and mechanical simplicity, cannot provide sufficient motion to reflect both trim and maneuvering commands. Therefore, the sidestick uses a common, or unique, centered reference position to represent trim for all flight conditions. The actual trim control surface position must be stored within the control laws. The low-frequency component of trim resident in the AFCS output is continually transferred to the PFCS so all trim resides at a single point in the PFCS regardless of augmentation level. In this manner, the transients associated with deselection or loss of the AFCS are minimized.

3.2.2. *Landing and takeoff/ground handling.* Since displacement cockpit controllers provide a direct relationship between cockpit stick position and the resultant control surface position, the pilot is provided with tactile cues to establish a control strategy for landing and takeoff. Pilots rely on experience to anticipate the magnitude of control displacement required for a smooth transition between a ground and fly state. In contrast, unique-trim sidestick controllers do not provide such cues. With a sidestick controller the pilot can no longer judge control surface position based on stick deflection since the trim is stored in the FCS and not at the pilot controller. This sidestick limitation is of no consequence for hover or forward-flight operations where the pilot adjusts the control input based purely on aircraft response. In addition, pilot workload with both controller configurations is increased during landing or takeoff since the pilot must compensate for a change in control responsiveness which occurs upon transition from a ground- to fly-state condition.

Sidestick controller handling qualities during landing and takeoff maneuvers were identified as an area of risk during ADOCS demonstrator flight testing. A detailed design study, including piloted simulation along with flight testing on the SHADOW aircraft, was conducted for the Comanche to resolve key control law issues regarding the landing and takeoff tasks. The study, discussed in Bauer (1993), produced new control law features for the Comanche to improve pilot kinesthetic cues, resulting in reduced workload and improved task performance for landing and

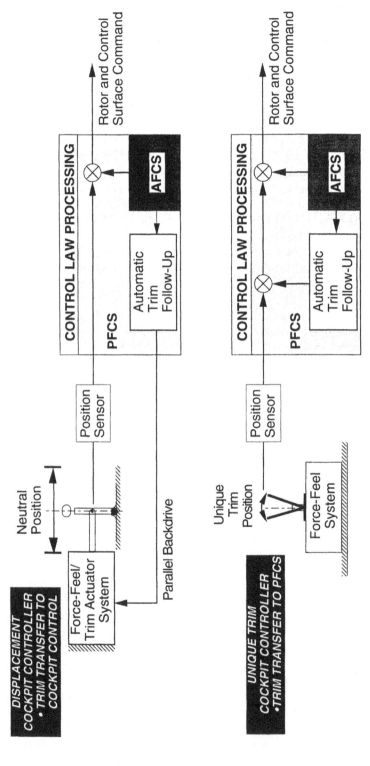

Figure 12. Automatic trim transfer implementation.

takeoff maneuvers. To accomplish this, in-flight control laws are progressively altered to a ground-state mode, based on the combinations of landing gear constraint imposed by ground contact. Weight-on-wheel switches are installed in each landing gear strut to sense ground and flight states. The sensed weight-on-wheel state is filtered to prevent rapid toggling between logic states during a light-on-gear condition.

The combinations of gear contact, listed sequentially for a landing maneuver, include:

- initial contact of any landing gear;
- individual aircraft axis (pitch, roll or yaw) constrained by the ground;
- initial contact of all gear with the ground plane;
- aircraft heavy on all gear.

These combinations, illustrated in Fig. 13, are used to transition control law functions for feedforward shaping, rate stabilization, attitude stabilization, automatic trim follow-up and rotor reference position with respect to cockpit control position. For example, upon contact of any gear with the ground, the feedforward dynamic shaping functions required to achieve desired fly-mode control response characteristics are changed to proportional control for ground operations. Additionally, upon any gear contact, the trim follow-up function in each axis is held so that the constant-gain forward path allows the pilot to judge accurately control requirements based on stick position. All AFCS stabilization, i.e. attitude, velocity or position, is also eliminated at this time through transient-free switching to prevent degradation of aircraft controllability resulting from false command model errors which may arise due to ground contact.

To enhance handling qualities during the transition between the ground and in-flight states, rate stabilization is retained in each control axis if the associated rotational degree of freedom is not fully constrained by gear contact. When an individual axis becomes constrained, the corresponding rate feedback channel is faded to zero. The elimination of rate stabilization when constrained is essential for

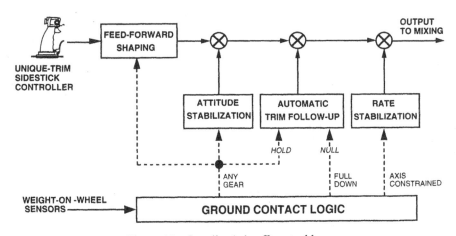

Figure 13. Landing/takeoff control laws.

121

shipboard operations where deck motions must not be transferred to the rotor while the aircraft is on the deck.

When the aircraft is fully down on the ground (heavy on all gear), the rotor control rigging position is referenced directly to the sidestick controller neutral position. The reference positioning process is initiated when the collective stick has been lowered 5% below the point at which all gear initiated ground contact. Trim follow-up in all axes is also nulled as the rotor reference position for ground operation is acquired. The setting of all stored trim values to zero restores an attribute inherent with conventional large-displacement controls, i.e. a fixed relationship of cockpit controller position and rotor control position while on the ground.

3.3. *Control law moding*

The experience base with multimode control laws accrued from the TAGS, HLH, ADOCS and V-22 programs was transferred to control law moding design requirements for the RAH-66 Comanche scout/attack mission. The PFCS modes are designed to minimize handling-qualities degradation when multiple failures result in loss of the AFCS. AFCS control laws are designed to allow the pilot to tailor handling qualities to mission requirements with minimal cognitive effort. AFCS moding enhances handling qualities based on the usable cueing environment (UCE) available to the pilot. Thus, a wide range of enhancing control modes are made available with minimum pilot interaction.

The pilot interface to the multimode control laws is through the FCS panel located in the center of the cockpit console. Figure 14 illustrates the FCS panel and levels of augmentation available to the pilot ranked from highest to lowest. The system powers up in the core AFCS configuration, shown as the center level of control augmentation. Higher levels of stability augmentation, i.e. velocity stabilization/hover hold and altitude hold, are selected by the pilot. The flight director can be selected to provide a maximum level of augmentation and to automate many facets of navigation and weapons delivery to reduce pilot workload and improve mission effectiveness.

3.3.1. *Core/mission PFCS.* During the definition of design criteria for degraded control mode operation, fully attended nap-of-the-earth (NOE) flight was identified as a primary design driver for the PFCS control laws. Based on simulation studies, yaw rate stabilization, as a minimum, was required to provide Level 2 handling-qualities ratings for NOE flight on PFCS (Fig. 15). Thus, the *mission PFCS* design employs yaw rate feedback layered on top of a minimal set of flight-critical control laws and allows the pilot to return to base using an NOE flight profile. As illustrated in Fig. 14, the mission PFCS is engaged manually by pilot deselect of the AFCS, or automatically by loss of the AFCS due to a sensed second failure. The *core PFCS* provides the minimum control laws required for Level 2 handling qualities for up-and-away flight, and landings on level, sloped and shipboard sites. Reversion to this mode will occur only in event of failure of three yaw rate sensors in the PFCS. Using four yaw rate sensors for redundancy, a two-fail operative capability is provided to meet PFCS system-level reliability requirements.

3.3.2. *Core/selectable mode AFCS.* The *core AFCS* is the normal flight control mode for the Comanche. It is designed with rate-command/attitude-hold functions

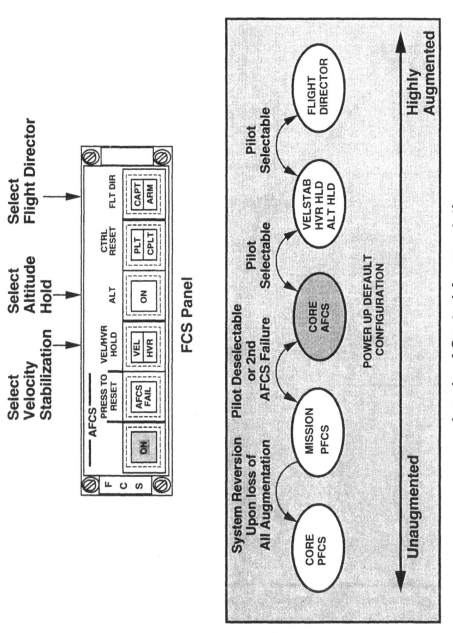

Figure 14. Pilot interface with RAH-66 flight control system.

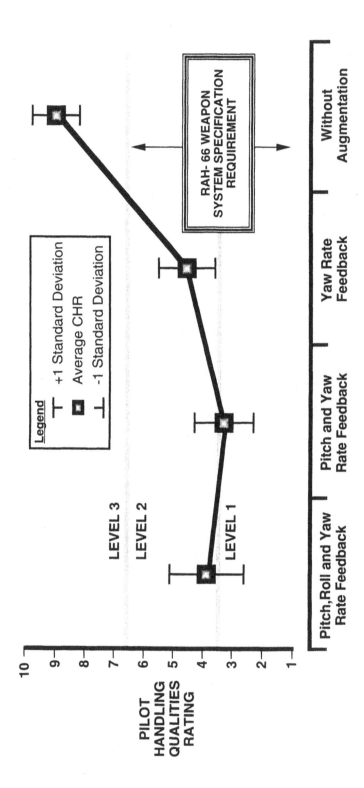

Figure 15.   Effect of stability augmentation on handling qualities for nap-of-earth flight.

in pitch, roll and yaw to allow the pilot to maximize maneuverability and agility of the Comanche. Automatic turn coordination is provided at airspeeds above 60 knots to improve handling qualities through reduced workload. Figure 11 provided a description of the command response and stabilization characteristics of the core AFCS as a function of airspeed. Additional details of the design are given in Fogler and Keller (1992).

Two *selectable AFCS mode* sets (altitude hold and velocity stabilization) are available to the pilot from the FCS panel for mission enhancement. *Altitude hold* provides a blended sensor hold function that uses a radar sensor reference while the aircraft is at low altitude and a barometric sensor reference for high-altitude flight. When the pilot desires operation with a specific sensor reference, the blended altitude reference function may be overridden through the MEP. The *velocity stabilization* mode provides the highest level of platform stabilization available to the pilot. Figure 11 depicts the features of these control laws, which vary according to groundspeed and airspeed. Hover hold is a submode of velocity stabilization which automatically engages when the aircraft is brought to hover to provide a hands-off position-hold capability. As seen in Fig. 14, the velocity stabilization button features a split-lamp indicator to annunciate when the hover-hold submode is active. Altitude hold is automatically enabled when the hover-hold submode is engaged. Additional information on the design of the Comanche selectable modes can be obtained from Dryfoos and Gold (1993).

3.4. *Integrated flight control law design*

3.4.1. *Flight control system/engine integration.* Traditionally, helicopters use engine governing, where the pilot sets blade collective and the control system adjusts engine power to maintain rotor rpm. This control approach provides a predictable pilot control response in hover since thrust response to collective is of relatively high bandwidth. Airplanes, on the other hand, generally use 'beta' governing where the pilot sets engine power and the governor adjusts propellor angle of attack to maintain rpm. This control approach is generally acceptable for airplanes since the pilot operates the thrust controller at relatively low frequencies.

The design of a rotor speed governor was an interesting challenge on the V-22. Precise height control for hover operations and shipboard landings is required in helicopter mode, while tight airspeed control and gust rejection are important in airplane mode. Precise power control and rotor speed governing are required in all flight modes. The throttle-governing approach typical of helicopters was not practical for the V-22 owing to the large sensitivity of mast torque to collective pitch at high inflow ratios in airplane mode.

Unique engine control requirements for the V-22 (Schaeffer *et al.* 1991) are met through incorporation of the thrust power management system (TPMS), which utilizes a collective ('beta') governor in all modes of flight. Acceptable hover characteristics are achieved using TCL quickening in the control laws, with high-frequency thrust control response actuated through collective blade pitch and low-frequency thrust commands provided through the engine response. Figure 16 provides an overview of the integrated V-22 engine/rotor control system, i.e. the TPMS. Pilot commands are input through the engine control levers (ECLs) during startup operations, and through the thrust control lever (TCL) during flight operations. The TPMS, which is integrated within the primary flight control system, provides

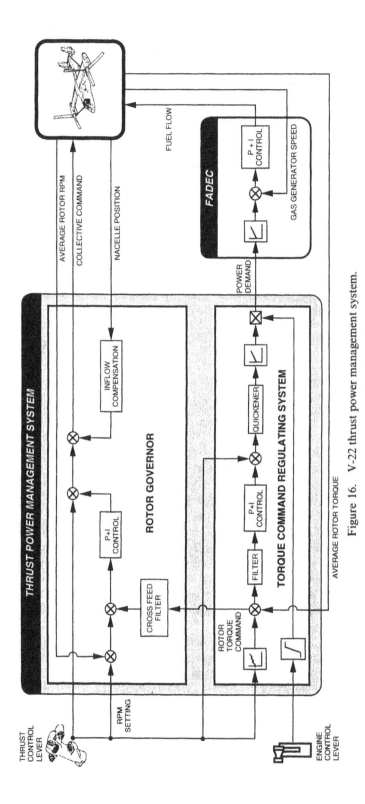

Figure 16. V-22 thrust power management system.

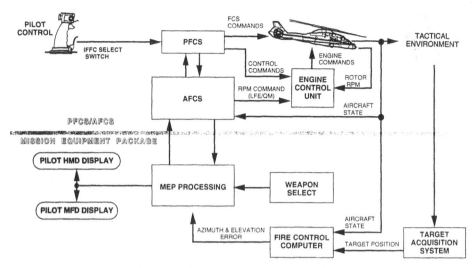

Figure 17. Flight/engine/fire control system integration.

command quickening, rotor governing, and torque regulation during all modes of flight. Key TPMS parameters are scheduled with nacelle angle and airspeed to provide desired characteristics throughout the flight envelope. Inflow compensation is provided to offload rpm governing requirements as the nacelle is varied. The primary outputs of the VMS are the collective command to the rotor for rpm governing, and the power demand to the FADEC (Full-Authority Digital Engine Control) to regulate rotor torque. The FADEC modulates fuel flow to the engine based on power demand and a proportional-plus-integral (P + I) regulator which governs gas generator speed. Extensive flight testing of the TPMS has demonstrated satisfactory performance and good handling qualities throughout the flight envelope.

Collective, lateral, and directional control commands typically couple with rotor torque demand to degrade targeting precision for single-rotor helicopters. Precise regulation of rotor rpm during targeting operations has been shown to be critical during adaptive fuel control flight testing (Sweet 1989). The Comanche is powered by twin LHTEC T800 engines controlled by a FADEC. To minimize rpm excursions during targeting maneuvers, Comanche flight control commands are directly coupled to the FADEC rpm governing control laws (Fig. 17) to provide engine response quickening and compensation. This alleviates undesired coupling and enhances handling qualities.

Integration of the flight and engine controls on the Comanche also includes two selectable modes that allow the pilot to change engine set speed. The *quiet mode* (QM) reduces main rotor rpm to 95% nominal to reduce the Comanche's acoustic signature. The *load factor enhancement mode* (LFE-mode) automatically increases the main rotor rpm to 107% nominal to enhance the normal load factor envelope and to reduce control loads at high speeds. These additional rotor speed options expand the mission capability of the RAH-66.

3.4.2. Integrated fire/flight control (IFFC). Fire control systems typically operate autonomously from other aircraft systems. Wide-regard weapons such as seeker

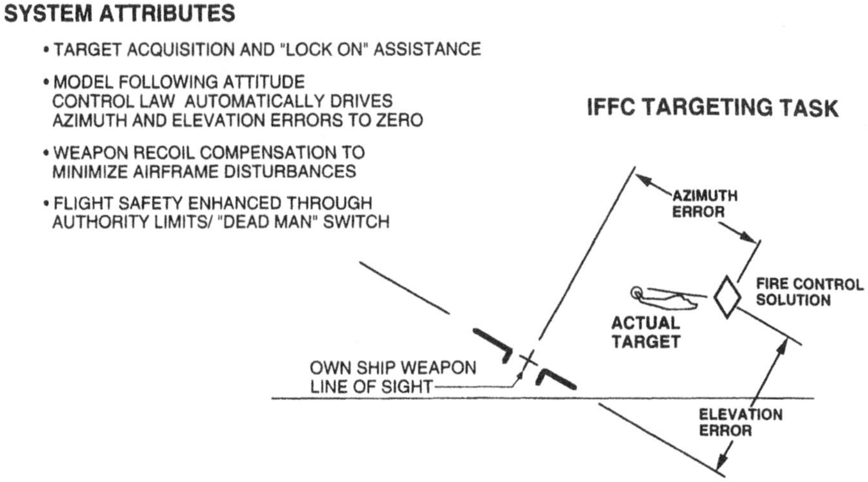

SYSTEM ATTRIBUTES

- TARGET ACQUISITION AND "LOCK ON" ASSISTANCE

- MODEL FOLLOWING ATTITUDE
  CONTROL LAW AUTOMATICALLY DRIVES
  AZIMUTH AND ELEVATION ERRORS TO ZERO

- WEAPON RECOIL COMPENSATION TO
  MINIMIZE AIRFRAME DISTURBANCES

- FLIGHT SAFETY ENHANCED THROUGH
  AUTHORITY LIMITS/ "DEAD MAN" SWITCH

IFFC TARGETING TASK

AZIMUTH ERROR

FIRE CONTROL SOLUTION

ACTUAL TARGET

OWN SHIP WEAPON LINE OF SIGHT

ELEVATION ERROR

Figure 18.   RAH-66 IFFC functionality.

guided missiles, e.g. HELLFIRE, do not require significant aircraft maneuvering, since the fire control system automatically compensates for aircraft motion. With unguided limited-regard weapons, the success rate is directly affected by the crew's ability to coordinate the aircraft trajectory within target sight constraints. Therefore, handling qualities and hence the aircraft control laws directly influence the successful accomplishment of the targeting task. The launch precision accuracy of unguided limited-regard weapons is affected foremost by crew targeting errors, and secondarily by the accuracy of the fire control solution.

The Comanche fly-by-wire digital system architecture (Fig. 17) is well suited for integration of the weapons' fire and flight control systems to reduce pilot workload associated with weapons aiming. When the IFFC system is engaged, the fire control solution is presented as a maneuvering command on the helmet-mounted display (HMD). The IFFC maneuvering command is also coupled to the flight control system by referencing the rate and attitude state variables to the fire control solution. The authority of the IFFC control is highest when the aircraft is closely aligned with the target. As the aircraft attitude is moved outside of the engagement window, the IFFC authority is reduced to zero. In this manner, the pilot is provided with a 'sticky pipper' to improve basic targeting accuracy. The IFFC system is designed to allow the aircraft to be stabilized within $\pm 0.1°$ of the fire control solution. The pilot command path remains active, so that the pilot can override the system at all times. The IFFC system is easily disengaged by release of a grip switch which must be depressed at all times when using IFFC. Functionality and attributes of the Comanche IFFC system are illustrated in Fig. 18. The design and development of IFFC control laws is discussed in Fowler *et al.* (1992).

3.5. *Structural loads limiting*

The unique requirements of the V-22 mandate flight operations to the maximum aircraft aerodynamic capability is limited by the FCS. The V-22 design requires a maneuver envelope to $4.0g$ at speeds up to 345 knots (Fig. 19). However, the aerodynamic capability is greater than $4.0g$ above 260 knots. Since the strength of struc-

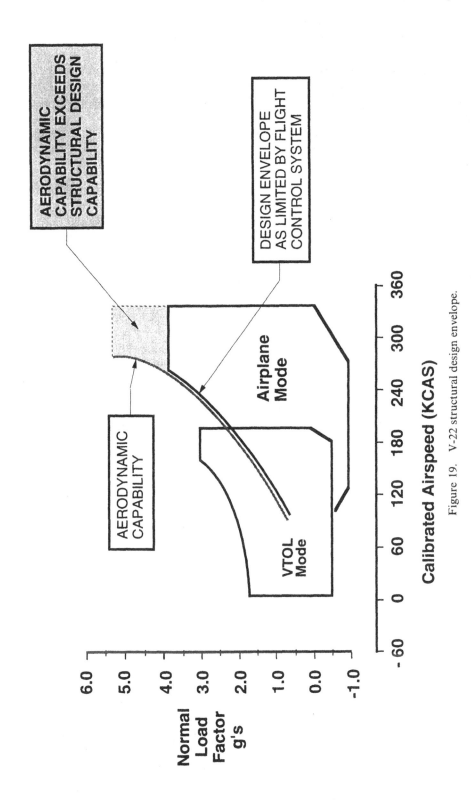

Figure 19.   V-22 structural design envelope.

tural components is constrained by weight, size and cost considerations, the control laws were optimized to alleviate trim/maneuver loads as described in King *et al.* (1993). The V-22 system design ensures that structural component limit and fatigue loads are not exceeded for the full range of aircraft maneuvers, while retaining Level 1 handling qualities. The digital flight control system provides a cost/weight effective means of meeting both structural loads and handling-qualities requirements throughout the flight envelope.

Two examples of structural load-limiting control laws are presented, namely (1) the use of roll-rate feedback applied in airplane mode and (2) the implementation of a longitudinal flapping limiter in the VTOL/conversion flight mode.

(1) *Roll-rate feedback.* The V-22 rotor system is comprised of counter-rotating rotors mounted at the wingtip, connected by an interconnect drive shaft which maintains equal rotor rpm at both rotors. During roll maneuvers in airplane mode, aircraft roll rate induces an effective increase in the rate of rotation of the downward rolling rotor and an effective decrease in the rate of rotation of the upward rolling rotor. This variation in the effective rate of rotation on the left and right rotors causes a torque split which is transmitted through the interconnect drive shaft during maneuvers. These maneuver-induced loads are a design concern with respect to the fatigue and limit-loading characteristics of the shaft.

To minimize weight and cost of the drive system, roll rate to differential collective pitch feedback is provided in airplane mode to alleviate cross-shaft loads during roll maneuvers. By sensing roll rate and making the appropriate differential collective pitch change, the torque split between the rotors, and thus the loads carried through the interconnect shaft during maneuvers, can be minimized. Figure 20 presents simulation and flight test

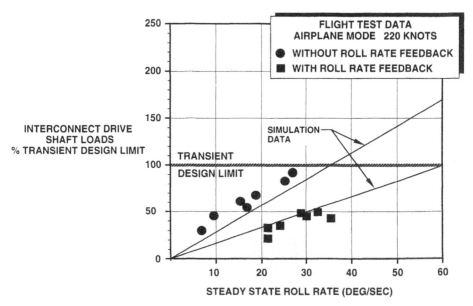

Figure 20.   Effect of roll-rate feedback on roll maneuver capability.

data showing the benefit of roll-rate feedback in reducing interconnect drive shaft torque for roll maneuvers. With structural load limiting, i.e. roll-rate feedback, the maneuvering capability of the V-22 as limited by transient design limit loads is increased from 30° to 60° per second.

(2) *Flapping limiter.* The relatively small disk area of the V-22 rotor results in operation at relatively high thrust coefficients in VTOL/conversion-mode forward flight. Regions of blade stall can develop at moderate speeds which induce increased blade flapping and rotor yoke chord bending loads. Furthermore, the aircraft operates over a large range of nacelle/airspeed and c.g. conditions. At the envelope extremes, trim flapping increases, thereby presenting fatigue issues in trim and limit flapping issues during maneuvering flight.

A control law was developed to feed rotor blade flapping to the elevator in order to reduce trim flapping of the rotor in both VTOL and conversion flight modes. A frequency splitter was also implemented in the longitudinal stick command path to alleviate high-frequency longitudinal cyclic inputs which can lead to excessive flapping during aggressive pitch maneuvers. The frequency splitter, which modifies the command split between the rotor and elevator, was set to maintain required control power throughout the conversion envelope. Flight test results indicate that this control law was successful in reducing trim and maneuver flapping throughout the V-22 envelope, while maintaining acceptable handling characteristics.

3.6. *Aeroservoelastic stability*

Aeroservoelastic (ASE) stability characteristics have become a significant consideration in the design of modern aircraft as the requirement to minimize weight and maximize maneuverability becomes more critical. Analysis, design and testing of ASE stability characteristics are key tasks in the development of air vehicle characteristics. Both the V-22 and RAH-66 programs have shown that the flight control system provides a powerful tool to meet weight and maneuverability design requirements.

The large roll/yaw inertia of the V-22 coupled with the relatively soft wing/fuselage interface (induced by the wing fold mechanism) creates a situation where the frequencies of the aircraft's structural modes are relatively low (less than 5·0 hertz) and in proximity of the rigid-body modes (Fig. 21). As described by Parham *et al.* (1991), flight testing of the V-22 identified three conditions where the pilot/control system tended to couple with the aircraft, resulting in a pilot-assisted oscillation (PAO). Each case of PAO involved a destabilizing acceleration feedback through the pilot/control stick at a rigid-body or structural mode frequency shown in Fig. 21. Initial ASE analyses for the V-22 did not include a model of biodynamic feedback and, therefore, did not predict the PAOs. Subsequently, generalized pilot/control models were established using measured data taken from ground-shake tests and in-flight excitations of the aircraft/pilot. These models were correlated to measured stability data and then used to design system enhancements as described below.

The initial PAO experienced during V-22 testing occurred while light on the gear and is attributed to coupling of lateral pilot motion with a 1·4 hertz rigid-body mode, known as the high focus mode. This coupling, illustrated in Fig. 22, involves

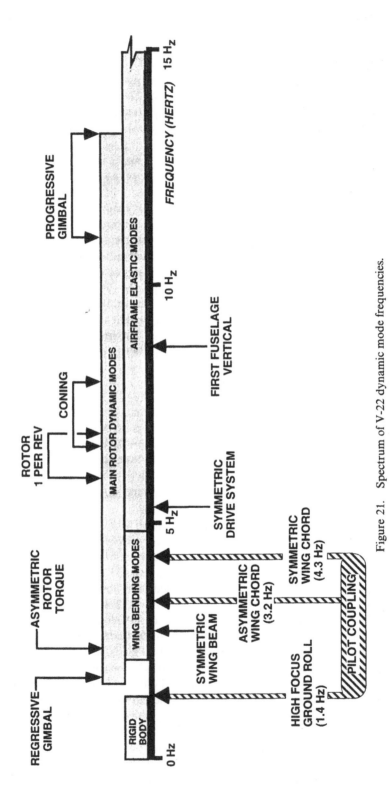

Figure 21.   Spectrum of V-22 dynamic mode frequencies.

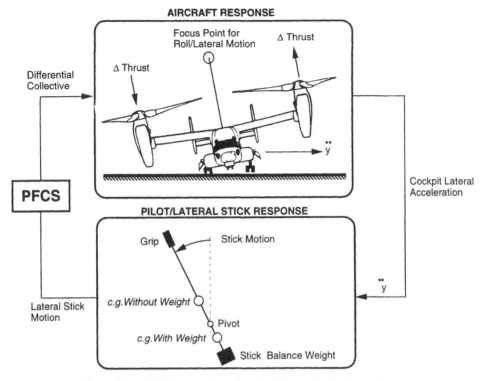

Figure 22.   High focus roll mode/pilot biomechanical coupling.

both lateral and roll motion about a focus point above the aircraft. When the air-craft oscillates, lateral control motions are introduced by the pilot inertial reaction to cockpit lateral acceleration. The gain of this loop with the pilot gripping the stick tightly was large enough to destabilize the high focus roll mode. Mass balancing of the cockpit lateral control (a large-displacement centerstick), i.e. lowering the stick c.g. below the pivot point, reduced the gain from lateral acceleration to pilot/stick input and effectively eliminated this mode.

Two in-flight occurrences of PAO were encountered which involved a destabi-lizing pilot/control system feedback loop at a structural wing-bending mode fre-quency. These PAOs were present in airplane mode and involved: (1) a coupled pilot/control system response to lateral acceleration at the asymmetric wing chord (AWC) mode frequency (3·4 hertz) at speeds greater than 250 knots, and (2) a coupled pilot/control system response to longitudinal acceleration at the symmetric wing chord (SWC) mode frequency (4·2 hertz) at speeds greater than 300 knots. Both in-flight PAO cases were addressed by incorporating digital 'notch' filters in the feedforward command path to effectively break the feedback loop at the critical structural mode frequency without inducing unacceptable phase lag in the handling-qualities frequency range.

The dynamics associated with the 4·2 hertz SWC mode are illustrated in Fig. 23. The thrust control lever (TCL) commands increased thrust, which causes elastic deformation of the wing in the chord axis. This motion results in fore and aft longi-tudinal acceleration at the cockpit, thereby closing the biodynamic feedback loop

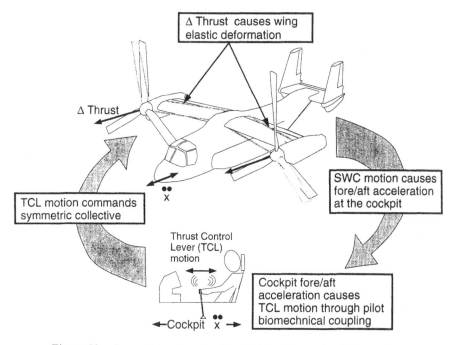

Figure 23.   Symmetric wing chord mode/pilot biomechanical coupling.

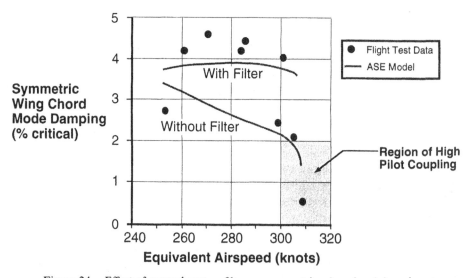

Figure 24.   Effect of control system filter on symmetric wing chord damping.

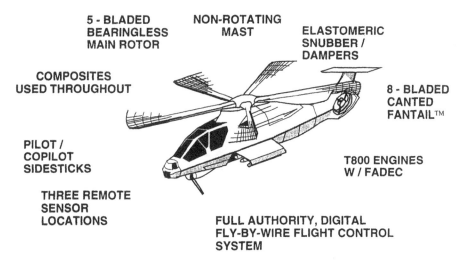

5 - BLADED
BEARINGLESS
MAIN ROTOR

NON-ROTATING
MAST

ELASTOMERIC
SNUBBER /
DAMPERS

COMPOSITES
USED THROUGHOUT

8 - BLADED
CANTED
FANTAIL™

PILOT /
COPILOT
SIDESTICKS

T800 ENGINES
W / FADEC

THREE REMOTE
SENSOR
LOCATIONS

FULL AUTHORITY, DIGITAL
FLY-BY-WIRE FLIGHT CONTROL
SYSTEM

Figure 25.   Key sources of RAH-66 aeroservoelastic coupling.

through the pilot. As seen in Fig. 24, this loop closure has a large effect on the SWC mode damping above 300 knots. A model update to include this loop was incorporated and used to design a notch filter that eliminated the destabilizing coupling through the pilot without adversely affecting handling qualities. The effect on SWC damping as shown in Fig. 24 illustrates the effectiveness of this filter.

Extensive use of composites throughout the Comanche to reduce weight has led to the lowest fuselage structural modes being in close proximity to the rotor 1/Rev frequency. The analysis of the complex interactions between the various sources of aeroelastic forces was performed at an early stage of the Comanche program due, in part, to lessons learned on the V-22 program. Figure 25 summarizes elements of the Comanche design which drive the requirement for a comprehensive ASE analysis. Pilot inputs into the control system come from a small-displacement sidestick and medium-displacement collective controller. Accurate empirical models of pilot biodynamic feedback with the RAH-66 pilot controller configuration were developed using motion-base simulation. A comprehensive ASE analysis tool, including the modeling of pilot biodynamic feedback, was developed and utilized throughout design of the Comanche VMS to meet all design requirements while minimizing cost and weight of the aircraft structure.

4. Control law design process

The control law design process employed at Boeing Helicopters has evolved from experience with major flight control research and development programs. Advanced computer analysis, design and simulation tools have been developed over the course of these programs, thereby increasing overall quality and productivity of the design process. The flight control law design process, highlighted in Fig. 26, includes four phases: namely, preliminary design, detailed design, implementation/ verification/validation and acceptance testing.

The *preliminary design* process begins with the development of design criteria based on air vehicle specification requirements. An iterative process is applied in

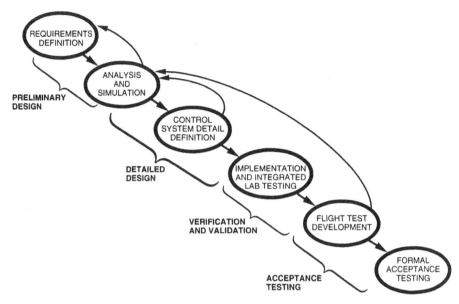

Figure 26.   Vehicle management system design process.

which the combined vehicle/system characteristics are analyzed, simulated and assessed against the air vehicle specification requirements for handling qualities, structural loads and aeroservoelastic stability. This phase of development establishes the control law architecture, sensor requirements and system component bandwidth requirements. For example, the handling-qualities specification may quantify the minimum required level of dutch roll damping. An assessment of the vehicle's inherent stability and control characteristics may indicate that artificial stabilization is required to increase yaw damping to meet the established design criteria. The need for stabilization will then flow down a set of requirements for the system architecture, sensor characteristics and actuator performance.

Once the preliminary control law design process is complete, a *detailed design* phase is conducted. Again, an iterative process is applied which utilizes extensive analysis and piloted simulation evaluations to ensure acceptability of the control laws for a wide range of maneuvers. Mission effectiveness of the aircraft is assessed for specific mission task elements based on the handling-qualities requirements defined during the preliminary design phase. The detailed control law development process ensures that all handling-qualities, system and structural requirements are simultaneously met.

Comprehensive models are developed during detailed design to assess rigid-body and elastic stability characteristics, along with their interaction with the coupled pilot/flight control system (Fig. 27). Aeroelastic stability analyses are conducted to ensure that no ground or air resonance conditions occur in the coupled rotor/ fuselage/drive/landing gear system. These analyses are considered to be 'open loop' in the sense that the control surfaces remain fixed. Aero*servo*elastic (ASE) stability analyses, however, examine the stability of the 'closed-loop' system where the control surfaces respond to control law commands, elastic structural deformations at the hub or sensors, and any undesired pilot inputs created by the biodynamic

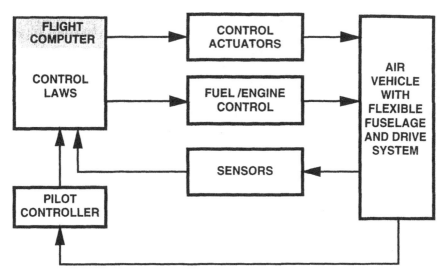

Figure 27. Rotorcraft analytical model overview.

transmission of pilot seat accelerations to the control sticks (also called 'pilot-assisted oscillations').

Figure 27 shows the primary components of the rotorcraft linear model used to analyze stability and control, ASE, and structural loads characteristics. Using this model, the ASE stability of the closed-loop system including pilot dynamics is determined and compared to the gain and phase margin guidelines established in Military Specification MIL-F-9490D. Components of the analytical model are discussed below.

The *air vehicle model* is a linear representation of the vehicle dynamics of interest including fully coupled rigid-body dynamics, rotor dynamics, landing gear dynamics and flexible fuselage and drive systems. This model is based on the same full-force, non-linear aircraft model used for real-time piloted simulation. The air vehicle model provides outputs to represent the linear accelerations felt at the crew stations, and sensor feedback signals referenced to sensor package locations in the flexible fuselage. Structural loads models are also included for real-time assessment of loads during batch and piloted simulation of control response characteristics.

The *sensor models* receive the feedback signals from the air vehicle model and calculate the dynamic response characteristics of each individual sensor package. Location of the sensor packages and the design of their mounting structures are important variables that influence the ASE stability characteristics.

The *pilot models* are linear transfer functions representing the pilot and copilot biodynamic transmission of seat accelerations through the body to the control sticks. Ideally this model should account for accelerations in all three orthogonal axes and produce control motions in all axes simultaneously, but single-axis models have proven adequate. These models are developed empirically from flight- or ground-based shake test data. Although defined as 'pilot' models, this element of the model represents the dynamics of the combined pilot/seat/controller system.

The *control law model* provides a detailed representation of the flight control system. In addition to the feedback stabilization loops, all feedforward control paths are included since the biodynamically closed acceleration feedback paths are of primary importance. Since total loop computational delay has a significant effect on levels of stability, all input, processing and output delays must be represented accurately.

The *fuel control and engine model* couples the flight control laws with the flexible drive system and rotor dynamics. Accurate modeling of fuel control and engine dynamics is essential to insure a high-fidelity model.

The end product of this phase of the detailed design process is the definition of a complete set of integrated control laws. A formal control law definition process ensures that a rigorous and disciplined review of the system is accomplished and that a valid control law definition results prior to coding flight-critical control system software. Over the past 20 years Boeing Helicopters has developed numerous computer design tools to streamline and shorten the time required for detailed design. These tools are packaged in a workstation environment to maximize efficiency of the process.

Verification and validation of the total flight control system, including the control laws, is performed to insure that system integrity is maintained. Verification of the control laws insures proper coding to the requirements by testing at the unit, module and function levels. Implementation errors are effectively eliminated by using the exact flight control element algorithms in the design phase. Validation of the flight control software involves an assessment at the system level. Non-piloted, open-loop test conditions are run in the Flight Control System Integration Rig (FCSIR) to evaluate the overall system response. These runs are compared to expected results generated by the FCS simulation. After successful open-loop testing

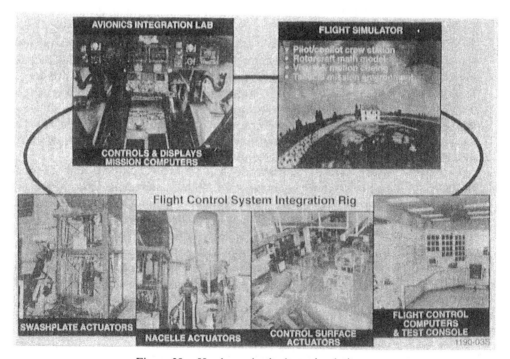

Figure 28.   Hardware-in-the-loop simulation.

is accomplished, the simulator is interfaced with the FCSIR and Avionics Integration Laboratory (Fig. 28) to conduct a formal final pilot acceptance test of the total system as described for the V-22 in Robinson *et al.* (1989). A wide range of maneuvers and flight conditions are evaluated. System failures are introduced to evaluate failure-mode effects and degraded mode handling qualities with hardware-in-the-loop. The validated software is then installed on the aircraft, and a flight assessment of the aircraft is made to demonstrate that handling-qualities, system and structural requirements are met prior to delivery of the aircraft to the customer for acceptance testing.

5. Summary

Advanced flight control system technology at Boeing Helicopters has matured over two decades of developmental programs, culminating with integration of advanced technologies in the RAH-66 Comanche and V-22 Osprey. An integrated design is achieved by applying a balanced approach using state-of-the-art analysis tools and real-time piloted hardware-in-the-loop simulation.

The PFCS/AFCS system architecture provides a safe approach to control partitioning for explicit model-following control laws. This approach has proven to be adaptable to a wide range of aircraft configurations, providing the flexibility to address new directions in handling-qualities requirements. Enhanced mission effectiveness is achieved through integration of stability and control augmentation and structural load alleviation control law functions. Integration of these control law functions, along with advanced cockpit displays and controls, engine control and fire control systems have proven to be key in providing a VMS capable of meeting future military requirements.

6. Future development trends

Research into advanced flight control techniques continues at Boeing Helicopters. Future control system designs will be optimized for multiple missions and configurations, and provide increased levels of stability and command augmentation throughout the flight regime, with intelligent system reconfiguration to compensate for failures. System bandwidth will be increased through control system algorithms that provide active stabilization of fundamental airframe and rotor modes. Control algorithms based on rotor-state feedback and individual blade sensing will provide means of extending usable bandwidth and improving gust rejection and stability characteristics. Integrated design approaches will simultaneously optimize handling qualities, maneuver control and aeroservoelastic stability while minimizing airframe and rotor structural loads.

Advances in computing speed are enabling adaptive compensation for changing aircraft configuration, flight condition and level of aggressiveness through the application of real-time system identification techniques. Increasingly sophisticated analytical redundancy techniques (including real-time system identification methods) will be used for failure prognosis detection and isolation. These analytical redundancy techniques will also provide improved failure tolerance with less physical hardware. Computer speed advances will also make digital control algorithms feasible for active vibration suppression, higher harmonic and individual blade control, and active modal damping enhancement. Parallel on-line computing will allow the implementation of high-order controller structures using finite impulse or neural techniques.

Integration of cockpit displays, sensors, stability augmentation and flight director functions will allow the pilots of the future to survive an increasing array of threats while improving mission effectiveness. Automatic control system technologies will be applied to on-line mission planning and flight director functions based on more accurate navigational, threat and obstacle avoidance information provided from phased-array radars, forward-looking infrared (FLIR), video and global positioning sensor (GPS) technologies.

REFERENCES

AERONAUTICAL DESIGN STANDARD ADS-33C, 1989, Handling qualities requirements for military rotorcraft. United States Army Aviation Systems Command, St. Louis, Missouri.

BAUER, C. J., 1993, A landing and takeoff control law for unique-trim, fly-by-wire rotorcraft flight control systems. Presented at the 49th Annual National Forum of the American Helicopter Society, St. Louis, Missouri.

DABUNDO, C., WHITE, J. and JOGLEKAR, M., 1991, Flying qualities evaluation for the V-22 tiltrotor. Presented at the 47th Annual National Forum of the American Helicopter Society, Phoenix, Arizona.

DAVIS, J., GARNETT, T. and GAUL, J., 1977, *Heavy Lift Helicopter Flight Control System, Volume III – Automatic Flight Control System Development and Feasibility Demonstration*, USAAMRDL TR-77-40C.

DRYFOOS, J. B. and GOLD, P. J., 1993, Design and pilot evaluation of the RAH-66 Comanche selectable control modes. Presented at the AHS Specialist Meeting, Monterey, California.

FOGLER, D. L. and KELLER, J. F., 1992, Design and pilot evaluation of the RAH-66 Comanche core AFCS. Presented at the 48th Annual National Forum of the American Helicopter Society, St. Louis, Missouri.

FOWLER, D. W., LAPPOS, N. D., DRYFOOS, J. B. and KELLER, J. F., 1992, RAH-66 Comanche integrated fire and flight control development and tests. Presented at the 48th Annual National Forum of the American Helicopter Society, Washington, DC.

GLUSMAN, S. I., DABUNDO, C. and LANDIS, K. H., 1987, Advanced flight control system for nap-of-the earth flight. NATO-AGARD Symposium Paper No. 28, Stuttgart, West Germany.

GLUSMAN, S. I., CONOVER, H. W. and BLACK, T. M., 1990, *Advanced Digital Optical Control System (ADOCS), Volume III – Handling Qualities*, USAASC TR 90-D-11C.

KING, D. W., DABUNDO, C., KISOR, R. L. and AGNIHOTRI, A., 1993, V-22 load limiting control law development. Presented at the 49th Annual National Forum of the American Helicopter Society, St. Louis, Missouri.

LANDIS, K. H. and GLUSMAN, S. I., 1984, *Development of ADOCS Controllers and Control Laws, Volume 3 – Simulation Results and Recommendations*, NASA Technical Report NAS2-10880.

LANDIS, K. H., DAVIS, J. M. and LEET, J. R., 1975, Development of heavy lift helicopter handling qualities for precision cargo operations. Presented at the 31st Annual National Forum of the American Helicopter Society.

MILITARY SPECIFICATION MIL-F-9490D, 1975, Flight control systems – general specification for design, installation, and test of piloted aircraft. United States Air Force.

MILITARY SPECIFICATION SD-572, 1993, Detailed design specification for V-22 engineering manufacturing development (EMD). Naval Air Systems Command, Washington, DC.

PARHAM, T., POPELKA, D., MILLER, D. G. and FROEBEL, A. T., 1991, V-22 pilot-in-the-loop aeroelastic stability analysis. Presented at the 47th Annual National Forum of the American Helicopter Society, Phoenix, Arizona.

ROBINSON, C., DABUNDO, C. and WHITE, J., 1989, Hardware-in-the-loop testing of the V-22 flight control system using piloted simulation. AIAA Flight Simulation Technologies Conference, Boston, MA.

ROSENSTEIN, H. and CLARK, R., 1986, Aerodynamic development of the V-22 tiltrotor. AIAA/AHS/ASEE Aircraft Systems Design and Technology Conference, Dayton, Ohio.

SCHAEFFER, J. M., ALWANG, J. R. and JOGLEKAR, M. M., 1991, Thrust/power management control law development. Presented at the 47th Annual National Forum of the American Helicopter Society, Phoenix, Arizona.

SWEET, D., 1989, S-76 adaptive fuel control ground and flight tests. USAAVSCOM TR-89-D-8.

5

Application of nonlinear inverse methods to the control of powered-lift aircraft over the low-speed flight envelope

JAMES A. FRANKLIN

Nomenclature

AX_{CMD}, $\dot{U}_c$	axial acceleration command (ft s^{-2})
AZ_{CMD}	normal acceleration command (ft s^{-2})
a_z	normal acceleration (ft s^{-2})
C_D	drag coefficient
C_L	lift coefficient
C_T	thrust coefficient
g	acceleration due to gravity (ft s^{-2})
h	altitude (ft)
$\dot{h}$, $\dot{h}_c$	vertical velocity; vertical velocity command (ft s^{-1})
i_w	wing incidence (deg)
m	aircraft mass (slug)
$\dot{m}$	engine mass flow rate (slug s^{-1})
q, $\bar{q}$	dynamic pressure (lb ft^{-2})
S	wing area (ft^2)
s	Laplace operator
T, T_G, T_{TOT}	gross thrust (lb)
U	flightpath velocity (knots, ft s^{-1})
u	generalized control
u_B, w_B	longitudinal and vertical body axis velocities (ft s^{-1})
V, V_x	longitudinal groundspeed (knots, ft s^{-1})
V_C, V_{CF}	filtered calibrated airspeed (knots, ft s^{-1})
V_S, V_{SF}	airspeed command (knots, ft s^{-1})
V_T	true airspeed (ft s^{-1})
V_e	jet velocity ratio, $\sqrt{q/q_{jet}}$
W, GW	gross weight (lb)
X, Z	axial and normal force (lb)
x	generalized state variable
α	angle-of-attack (deg)
α_c	commanded angle-of-attack $\alpha_c = \theta - \gamma_c + i_w$ (deg)

γ flightpath angle (deg)
γ_c flightpath angle command (deg)
ΔD incremental drag due to direct thrust (lb)
ΔL incremental lift due to direct thrust (lb)
$\Delta L/T$ normalized jet-induced aerodynamic ground effect
δ_{LN} lift nozzle deflection (deg)
δ_{SP} spoiler deflection (deg)
δ_{USB} upper surface blown flap deflection (deg)
θ pitch attitude (deg)
θ_c commanded aircraft attitude (deg)
ϕ bank angle (deg)
τ time constant (s)
ω spoiler washout frequency ($\text{rad}\,\text{s}^{-1}$)

1. Introduction

For several years, NASA has been engaged in research on control of powered-lift aircraft during low-speed flight. Aircraft types have included short take-off and landing (STOL) transports, vertical and short take-off and landing (V/STOL) transports, and short take-off and vertical landing (STOVL) fighters. The need for this research has been motivated by the poor stability and control characteristics of the basic aircraft associated with low aerodynamic stability and damping, control cross-coupling, the strong influence of jet-induced aerodynamics due to propulsion systems flows, and the number and redundancy of control effectors that must be employed in this flight regime. Furthermore, the desire to operate these aircraft in adverse weather conditions, including poor visibility, heavy winds and turbulence, to austere sites and aboard ship, imposes demands for precision of control and good flying qualities that generally exceed those demanded of conventional take-off and landing aircraft. In order to achieve the necessary precision of control of these aircraft, while coping with poor characteristics inherent in the basic airframe, it is necessary to provide stability and command augmentation systems (SCAS) that improve control of aircraft pitch, roll and yaw attitudes, and to provide the pilot with direct command of the aircraft's states associated with following a commanded trajectory from cruise flight to the final landing condition.

Trajectory control involves control of the aircraft's velocity vector, which, in turn, requires the continuous control of axial and normal force. Either axial or normal force is functionally dependent on several aircraft states and control variables, such as airspeed, angle-of-attack, thrust magnitude, thrust deflection, leading and trailing edge flap settings and flow spoilers, and typically is described by nonlinear functional relationships over the governing range of these independent variables. Examples of lift and drag characteristics for a STOL aircraft (Stevens 1981) are presented in Fig. 1 to illustrate the extent to which their non-dimensional coefficients vary with states such as angle-of-attack, thrust coefficient, and flap setting for pertinent ranges of these variables. A comparable illustration is shown in Fig. 2 for a STOVL fighter configuration to indicate the effects of engine thrust and airspeed on lift in ground proximity. In

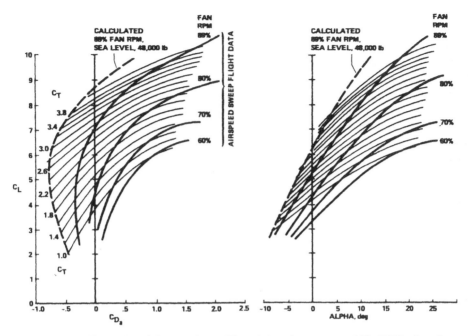

Figure 1. Examples of thrust-induced lift and drag for a powered-lift STOL aircraft.

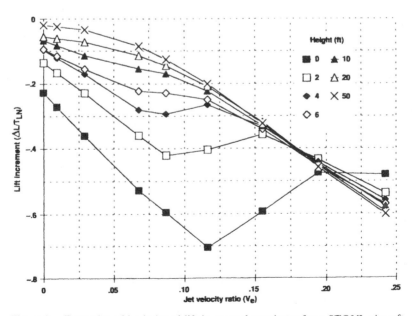

Figure 2. Examples of jet-induced lift in ground proximity for a STOVL aircraft.

principle, the design of a control system that manipulates axial and normal force to control the velocity vector must account for variation in each of these states and controls, throughout the low-speed flight regime. The influences of the aircraft design on the control system would be realized in feedback and feedforward control commands which are gain scheduled to accommodate variation in flight condition and aircraft configuration within the flight envelope, thus presenting considerable design complexity and difficulty for the designer.

During the past several years, this author had the responsibility to select and implement system concepts for velocity control for a flight experiment to be conducted on a STOL transport, and simulation experiments to be carried out on conceptual STOVL fighter designs. Familiarity with earlier work by the author's colleagues (Meyer and Cicolani 1981) in the use of nonlinear inverse methods for an autopilot design performed for a powered-lift STOL transport formed the basis for the control designs reported in this article. Alternatively described as feedback linearization (Slotine 1991), this method employs the knowledge of the aircraft's nonlinear aerodynamic and propulsion system characteristics at the current state, compares these characteristics with the commanded total force required for trajectory tracking, and commands the additional control inputs to produce the additional forces necessary to match those required to follow the desired trajectory. It is a very methodical and highly intuitive process and, when employed properly, keeps the designer intimately familiar with the physics of the aircraft aerodynamic and propulsion system forces and their influence on the aircraft's control.

For a system defined by $\dot{x} = f(x, u)$, the inverse method involves the solution of $\dot{x}_c = f(x, u)$ to determine the control u that satisfies the commanded acceleration $\dot{x}_c$. It should be noted that $f(x, u)$ represents the total force imposed on the vehicle, not just those forces associated with perturbations in states about an operating condition. Since the contribution of the current states x to the function $f(x, u)$ can be defined, the value of the control u to achieve $\dot{x}_c$ can be determined, and is the control required to trim as well as manoeuvre the aircraft. The process is illustrated in Fig. 3. The command $\dot{x}_c$ is composed of the pilot's control input and errors between the command and aircraft state feedback, shaped by the command generator and regulator to achieve the desired dynamic response for satisfactory control and flying qualities. This command passes to the nonlinear inverse of the system that computes u to force $f(x, u)$ to follow the desired commands. It should be appreciated that the front end of the system, comprising the pilot's command and regulator, is based on the demands of the pilot's task and is independent of the aircraft's characteristics

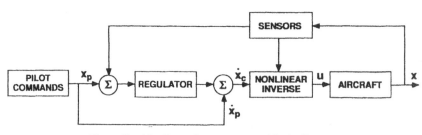

Figure 3.   Nonlinear inverse system block diagram.

or flight condition. Hence, no scheduling of command or feedback gains is necessary to achieve the desired dynamic response and, in fact, the relation of these gains to the desired dynamics is typically a simple, linear, time-invariant, first or second-order transfer function.

The complexity of variation in aircraft characteristics with flight condition and configuration is encompassed entirely in the function $f(x, u)$, and its manipulation to achieve the control outputs to satisfy the commanded acceleration $\dot{x}_c$ accounts for the variation in characteristics. It should be appreciated that alteration in the aircraft's design is accommodated by appropriate changes in $f(x, u)$ without the need for iteration of the command or feedback system structure or gains. The benefits of this approach to streamlining the control design process are substantial.

Applications of this method to control design for the STOL and STOVL aircraft have been performed (Franklin *et al*. 1986 and Franklin and Engelland 1991) and are summarized in the discussion that follows.

2. STOL aircraft control design and flight evaluation

Operations of STOL aircraft typically involve conversion from conventional to powered-lift flight at airspeeds on the order of 150 knots, deceleration on landing approach speeds ranging from 60 to 90 knots, a precision final approach segment, and are concluded with a spot landing on a runway 1500 to 2000 ft in length. The NASA Quiet Short Haul Research Aircraft (QSRA), shown in Fig. 4, was used as a test facility around which to design and evaluate control systems for such operations with this class of vehicle. It is a four-engine jet transport that employs an upper-surface-blown flap to generate propulsive lift. Control effectors, noted in Fig. 4, are the elevator for pitch, ailerons and spoilers for roll, rudder for yaw, and flaps in the engine exhaust wake, spoilers, and engine thrust for lift and drag control. The latter three effectors were those used for control of the velocity vector in the vertical plane. It was this application to which nonlinear inverse methods were applied and, due to the brevity of the discussion, on which this section of the paper will focus. A more thorough description of the complete control system design, which deals with the pitch, roll and yaw axes can be found in the original paper (Franklin *et al*. 1986).

In this application, velocity vector control can be more appropriately considered in terms of the flightpath angle in the vertical plane and the aircraft's velocity along the flightpath. Forces associated with control of these two states are referenced to axes aligned with the velocity vector, and are defined by the equations for the components along and normal to the velocity vector, i.e. for axial force

$$m\dot{U} = X - W \sin \gamma$$

and for normal force

$$mU\dot{\gamma} = -Z - W \cos \gamma$$

These equations are used to determine the total forces imposed by aerodynamics and the engine thrust required to satisfy commanded accelerations for control of

Figure 4.   Quiet Short Haul Research Aircraft.

the velocity vector, $\dot{U}_c$ and $U\dot{\gamma}_c$. Solving for the total forces gives

$$X = m\dot{U}_c + W \sin \gamma_c \quad \text{and} \quad Z = -mU\dot{\gamma}_c - W \cos \gamma_c$$

For powered-lift aircraft whose propulsion system flows are intended to augment the basic aerodynamics of the wing, it is customary to represent the total axial and normal forces in terms of their respective non-dimensional aerodynamic coefficients, dynamic pressure, and reference wing area, where

$$X = -C_D qS \quad \text{and} \quad Z = -C_L qS \cos \Phi$$

The total drag coefficient is composed of two components

$$C_D = C_{D_{FA}} + C_{D_{RAM}}$$

the first of which is associated with airframe and thrust-induced forces; the second with engine inlet flow momentum, referred to as ram-drag. Ram-drag is,

148

in turn, defined by

$$C_{D_{RAM}} = \dot{m}U/qS$$

The next step is to solve for the two coefficients in terms of the control commands. The required airframe drag coefficient, including thrust-induced effects is defined by

$$C_{D_{FA}} = -(W/qS)(\dot{U}_c/g + \sin \gamma_c) - \dot{m}U/qS$$

and the total airframe lift coefficient, including thrust-induced lift is

$$C_L = (W/qS \cos \Phi)(U\dot{\gamma}_c/g + \cos \gamma_c)$$

It now remains to determine the appropriate controls that produce the desired drag and lift coefficients C_D and C_L. For this aircraft, these coefficients have the functional relationships

$$C_{D_{FA}} = f(\alpha, C_T, \delta_{USB}, \delta_{SP})$$

$$C_L = g(\alpha, C_T, \delta_{USB}, \delta_{SP})$$

to angle-of-attack, thrust coefficient, blown flap and spoiler deflection. The redundancy in the relationship of dependent to independent variables is resolved by specifying angle-of-attack and spoiler deflection separately, where angle-of-attack is defined by the commanded pitch attitude and flightpath angles ($\alpha_c = \theta - \gamma_c + i_w$) and spoilers are deployed to a nominal bias position to permit them to be used to quicken lift response. Now, drag and lift coefficients may be determined uniquely in terms of thrust coefficient and blown flap deflection and, given these functional relationships, it is possible, in turn, to solve for the thrust coefficient and flap deflection required to produce the desired drag and lift coefficients. Thrust commands to the engines are then defined by $T = C_T qS$. Control of the lift coefficient typically requires a dynamic range that exceeds the rate of response of which the engines are capable. Thus, the spoilers are controlled in the short term to produce the initial lift increment, then are restored to the bias position by washing out the lift command, e.g.

$$\delta_{SP} = f(C_L)[s/(s + \omega_{WO})]$$

where $f(C_L) = g[C_L - C_L(\alpha)]$ and is the inverse function relating spoiler position to the desired lift coefficient. The washout corner frequency is chosen to complement the engine response time constant.

The velocity vector control commands $\dot{U}_c$ and $U\dot{\gamma}_c$ are produced by the pilot's commands and the regulator. Specifically, for the longitudinal axis, the pilot commands a selected airspeed to hold or a rate of change of airspeed if it is desired to accelerate or decelerate to a different reference condition. The regulator provides proportional-plus-integral control on speed error and generates the acceleration command $\dot{U}_c$, based on this speed error and the pilot's acceleration command. The regulator dynamics are designed to a bandwidth of 0.2 rad s^{-1} to reject turbulence inputs while retaining authority to suppress lower frequency shear disturbances. For the vertical axis, the pilot commands flightpath angle and angular rate, and the regulator provides proportional control on errors to these commands. Commanded flightpath is also used to provide orientation of the gravity component in the total force computation. Flightpath regulation was designed to achieve a bandwidth of $1\,\mathrm{rad\,s^{-1}}$. The bandwidths of control for airspeed and flightpath are separated sufficiently from the pitch attitude control bandwidth ($3\,\mathrm{rad\,s^{-1}}$) so that attitude and velocity vector control can be dealt with independently.

A block diagram of the control system structure, as implemented in the aircraft's flight computer, is presented in Fig. 5. The pilot's commands, sensor feedback of aircraft states, and regulator outputs appear at the left of the diagram. It should be noted that the gains associated with the command inputs and the regulator were constant over the powered-lift flight envelope for which this system was designed. The gains were selected to produce the dynamic response desired for speed and flightpath control noted above. Effects of variations in flight condition, aircraft loading and configuration are accommodated by the nonlinear inverse calculations that convert these commands to commanded aerodynamics and then to the associated thrust coefficient and flap setting. That computation of thrust coefficient and flap setting was accomplished through a three-variable table look-up, which is illustrated in Fig. 5 by the plots of C_T and δ_{USB} as a function of $C_{D_{FA}}$, C_L and α. The data in these tables were obtained from large-scale wind-tunnel model tests of the aircraft.

Examples of performance of the system based on flight data, in terms of transient response to the pilot's flightpath commands at a representative flight condition and to a command for deceleration to a final approach airspeed, are presented in Fig. 6. At the left of the figure, the flightpath responds quickly and precisely to the pilot's column force command, in accord with the criteria set forth for the system design. Airspeed remains constant at 70 knots as intended. Engine thrust, blown flap and spoiler response derived by the inverse calculations to meet the pilot's commands, are indicated and represent reasonable behaviour without excessive activity or high frequency content. The spoiler washout to the bias position is evident. During the decelerating transition shown on the right side of Fig. 6, the command to decelerate (V_s) is initiated shortly after the start of the run and ramps steadily from 98 to 70 knots. Airspeed follows the commanded deceleration and smoothly captures the final approach speed. The flightpath holds essentially constant. The inverse calculation reduces engine thrust to initiate the deceleration, then smoothly increases thrust and flap deflection to sustain lift and modulate drag to maintain the flightpath as speed is reduced.

Evaluations of system performance were obtained from flight experiments that involved precision approaches and landings on a STOL runway. The approaches were conducted along curved trajectories on a 6° descending path, beginning at flaps-up cruise airspeeds and decelerating to typical final approach speeds from 65 to 70 knots. Instrument flight conditions were simulated down to a minimum of 100 ft altitude, below which the final segment of the approach and the landing were performed visually. Winds varied from calm to strong crosswinds with moderate turbulence. Three NASA pilots who flew the aircraft to assess the system reported achieving fully satisfactory flying qualities for these operations. Specifically, they were able to obtain the precision of performance desired for an instrument approach and landing and to do so without objectionable effort on their part to compensate, mentally or physically, for deficiencies in the aircraft's behaviour. These results were insensitive to variations in wind and turbulence.

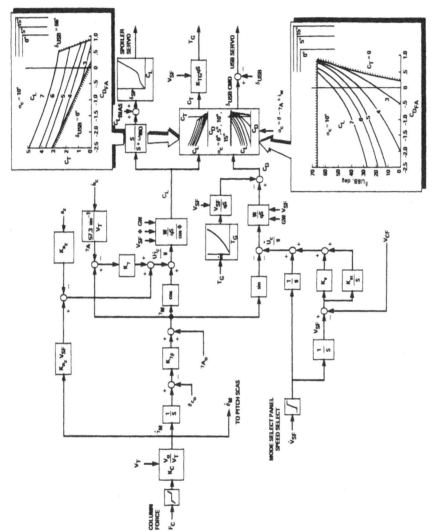

Figure 5.   System block diagram for flightpath–airspeed command and augmentation system.

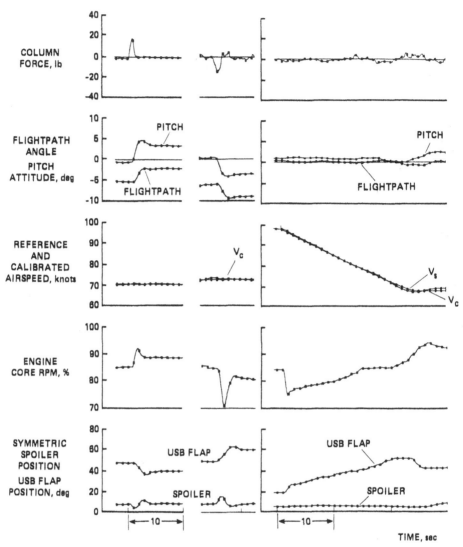

Figure 6. Representative time histories of transient response and transition between flight conditions.

3. STOVL fighter control design and moving-base simulation evaluation

Research on STOVL fighter flight controls proceeded from the base of experience gained from work with similar systems on STOL aircraft. However, with STOVL aircraft, it is necessary to cater for full deflection of thrust to the vertical to sustain the aircraft in hover, compared with the partial thrust deflection used for STOL aircraft. Like those of their STOL counterparts, STOVL operations typically commence around 150 knots, and additionally include a deceleration to hovering flight and a vertical landing. The aircraft on which the nonlinear inverse control concept was investigated was a conceptual design of a single engine supersonic STOVL fighter/attack aircraft shown in Fig. 7. Its powered-lift system uses a turbofan engine whose fan and core engine

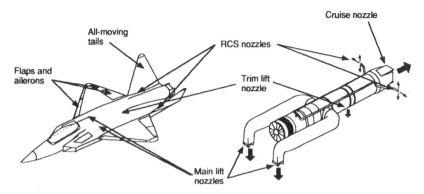

Figure 7.   STOVL aircraft and propulsion system.

flows are mixed aft of the turbine and are directed aft to the cruise nozzle for conventional flight or progressively diverted forward to the lift nozzles to support the aircraft in low-speed and vertical flight. Control effectors noted in Fig. 7 are symmetric empennage deflection, reaction controls, cruise nozzle deflection and ventral nozzle thrust for pitch; ailerons and thrust transfer between the lift nozzles for roll; differential empennage deflection, reaction controls and cruise nozzle deflection for yaw; and thrust transfer between the lift and cruise nozzles, deflection of lift nozzle thrust and total engine thrust for axial and vertical force control. Control of the velocity vector in the vertical plane was accomplished with the latter three controls. The system is described in further detail by Franklin and Engelland (1991).

The structure of a nonlinear inverse control for a STOVL aircraft bears a good deal of similarity to that used for the STOL example. However, for operations that include vertical flight, direct control of the velocity components is appropriate. Axial and normal forces used for control are referenced to axes aligned with the velocity vector in forward flight down to speeds for which conventional aerodynamic forces no longer predominate and where control of the cartesian components of velocity rather than the angle and magnitude of the total velocity vector is pertinent. The force equations are now slightly modified from their earlier form for axial force to

$$mAX_{CMD} = X - W \sin \gamma_c$$

and for normal force to

$$mAZ_{CMD} = Z + W \cos \gamma_c$$

and the flightpath command, γ_c, becomes a scaled vertical velocity command below 60 knots. For aircraft whose propulsion systems directly produce forces to sustain the aircraft in low speed and hovering flight, the total axial and normal forces are typically partitioned into components that include aerodynamic contributions in the absence of thrust effects, increments to the aerodynamic forces that account for jet-induced effects, and direct thrust contributions. Thus, the axial force contains the following terms

$$X = -C_D qS - (\Delta D/T)T - \Delta D_{prop}$$

153

Similarly, the normal force is

$$Z = -[C_L qS + (\Delta L/T)T + \Delta L_{prop}] \cos \Phi$$

Since all but the direct propulsion forces are known based on the aircraft's current state, these equations can be solved along with the command equations above to determine the axial and normal force increments due to the propulsion system required to meet the pilot's commands. That is

$$\Delta D_{prop} = -[C_D qS + (\Delta D/T)T + W \sin \gamma_c + mAX_{CMD}]$$

$$\Delta L_{prop} = -C_L qS - (\Delta L/T)T + (W \cos \gamma_c - mAZ_{CMD})/\cos \Phi$$

These increments required of the propulsion system are then resolved into aircraft body axes from which inlet flow momentum contributions are extracted to leave the direct axial and normal propulsion system thrust components, ΔX and ΔZ, that must be produced to achieve the pilot's commands. These direct thrust components can also be expressed in terms of total thrust magnitude and thrust deflection.

$$T = (\Delta X^2 + \Delta Z^2)^{1/2} \quad \theta_N = \cos^{-1} \Delta X/(\Delta X^2 + \Delta Z^2)^{1/2}$$

The thrust magnitudes from the lift and cruise nozzles and deflection of the lift nozzle thrust to produce the total thrust and thrust deflection can then be determined.

The velocity vector control commands AX_{CMD} and AZ_{CMD} are outputs of the pilot's control commands and the regulator. For the longitudinal axis, the pilot commands acceleration or deceleration, with a reference speed being established when the acceleration command is nulled. The regulator provides proportional-plus-integral control on speed error and generates the acceleration command AX_{CMD} based on this speed error and the pilot's acceleration command. The regulator dynamics are designed to a bandwidth of $1 \, \mathrm{rad \, s^{-1}}$. For the vertical axis, the pilot commands a flightpath angle for speeds above 60 knots, and vertical velocity at lower speeds and in the hover. Proportional-plus-integral control is provided by the regulator. Flightpath regulation was also designed to achieve a bandwidth of $1 \, \mathrm{rad \, s^{-1}}$.

The overall control system layout is shown in the diagram in Fig. 8. The pilot's commands, sensor feedback of aircraft states, and regulator outputs are shown at the top of the figure. Similar to the STOL aircraft example, the gains associated with the command inputs and the regulator were constant over the powered-lift flight envelope. Computations of the individual nozzle thrust and lift nozzle deflection described in the equations above were performed as shown at the bottom of the figure. Variations in flight condition, aircraft loading and configuration are contained in the aerodynamic tables and in the explicit inputs for aircraft weight, airspeed and dynamic pressure. For this conceptual aircraft design, the tabular data were obtained from small-scale wind-tunnel model tests and analytical predictions.

This application of the nonlinear inverse system was evaluated in a ground-based simulator experiment conducted on Ames Research Center's Vertical Motion Simulator. This is a six degree-of-freedom facility with large motion capability in the vertical and longitudinal axes and acceleration bandwidths in all axes that encompass the frequency range important for the pilot's motion

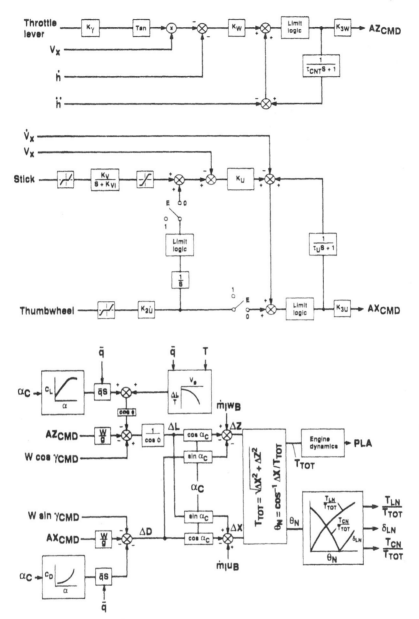

Figure 8. STOVL velocity command and augmentation system.

sensing in low-speed flight. Visual displays are produced by a three-window computer-generated imaging system. Four pilots conducted evaluations of curved, decelerating approaches under instrument flight conditions, concluding with a vertical landing in clear visibility. Visibility minima consisted of a 100 ft ceiling. Landings were performed on a 100×200 ft pad. Operations were conducted with calm air and in moderate winds and turbulence. An example of a decelerating transition is presented in the time history plot of Fig. 9, based on data taken from the simulator. Initially, the flightpath angle is reduced to

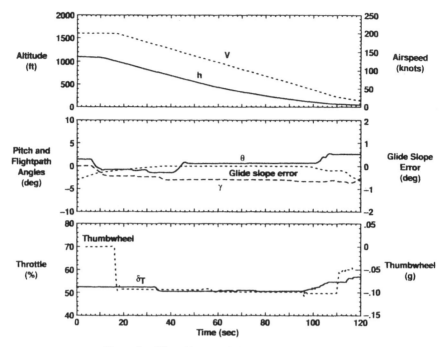

Figure 9. Time history of decelerating transition.

acquire the descending approach path, whereupon the deceleration is commenced at 200 knots. The deceleration proceeds smoothly to a near hover condition and the glideslope is tracked precisely, with little effort on the part of the pilot evident in the activity of either the deceleration command thumbwheel or the throttle for flightpath control. The NASA pilots conducting the evaluations found it easy to control the aircraft's velocity vector with the precision desired for both the decelerating approach to the hover and for the vertical landing, and consistently rated flying qualities as fully satisfactory. Their assessment was not altered by the effects of external disturbances from varying winds and turbulence.

4. Design experience

Both the STOL and STOVL control system designs were carried through their piloted evaluations without the need to iterate pilot command or feedback gains to obtain the desired control response characteristics. Furthermore, changes were made in the STOVL aircraft aerodynamic characteristics during the course of the control system design, without the need to adjust control system gains. The initial gains were calculated, literally 'back of the envelope' based, on second-order transfer function models of the axial and vertical velocity responses to the pilot's controls and had easily understood effects on the transient velocity response.

The control system designer who uses this nonlinear control concept must be aware of the physical characteristics of the system under control, in this case, the aerodynamic force characteristics of lift and drag. These characteristics are used in the control design in a form consistent with data extracted from wind tunnel or flight tests of the aircraft. The designer cannot be isolated from the physical properties of the system being controlled.

Flight-identified lift and drag characteristics of the QSRA were found to differ by 10–20% from those predicted based on the large-scale wind tunnel tests. While the aerodynamic data tables in the QSRA nonlinear inverse control used the wind tunnel data, control system performance in flight was not degraded due to differences with the actual aircraft lift and drag characteristics.

Use of spoilers to complement thrust to extend the bandwidth of lift control proved to be satisfactory. Dynamics of the engine, which could not be directly controlled by the nonlinear inverse, were evaluated by linearizing the nonlinear inverse process at an approach operating point and carrying out a classical frequency response analysis of the engine thrust-spoiler complement.

The effects of one compute cycle delay in computing the contributions of ram-drag or jet-induced lift as functions of engine thrust were not apparent for either the STOL or STOVL cases. Compute cycle times of 30–50 ms are sufficiently faster than the thrust transient response times so that no appreciable change in thrust occurs within the compute frame.

5. Conclusions

Nonlinear inverse control systems have been developed and evaluated experimentally for control of the velocity vector over the low-speed envelope of STOL and STOVL aircraft. A design method has been employed which directly utilizes the aircraft's aerodynamic and propulsion system characteristics, as extracted from model or flight tests or analytical predictions in the system structure. Command and feedback gains in the control system can be established solely on the dynamic response to achieve the desired flying qualities and are independent of the aircraft's characteristics. The nonlinear inverse, which relates the forces to achieve the pilot's commands to the required control positions, encompasses all the characteristics associated with variations in flight condition, aircraft loading and configuration. Based on flight and simulation experiments, conducted for example on STOL and STOVL aircraft, the nonlinear inverse control produces satisfactory flying qualities over the low-speed flight envelope of these aircraft. The system was also successful in accommodating errors in modelling the aerodynamic and propulsion system characteristics of these aircraft.

REFERENCES

FRANKLIN, J. A., and ENGELLAND, S. A., 1991, Design and piloted simulation evaluation of integrated flight/propulsion controls for STOVL aircraft. AIAA Conference Paper 91-3108, *Proceedings of the Aircraft Design Systems and Operations Meeting*, Baltimore, Maryland, U.S.A.

FRANKLIN, J. A., HYNES, C. S., HARDY, G. H., MARTIN, J. L., and INNIS, R. C., 1986, Flight evaluation of augmented controls for approach and landing of powered-lift aircraft. *AIAA Journal of Guidance, Control, and Dynamics*, **9**, 555–565.

MEYER, G., and CICOLANI, L., 1981, Application of nonlinear systems inverses to automatic flight control design—systems concepts and flight evaluations. *AGARD Journal*, No. 251.

SLOTINE, J.-J. E., 1991, *Applied Nonlinear Control* (Englewood Cliffs, NJ: Prentice Hall).

STEVENS, V. C., 1981, A technique for determining powered-lift STOL aircraft performance at sea level from flight data at altitude. *Proceedings of the AIAA Conference*, Paper 81-2480.

6

Flight control and handling research with the VAAC Harrier aircraft

G. T. SHANKS, S. L. GALE, C. FIELDING and D. V. GRIFFITH

1. Introduction

The Harrier has, for many years, been the Western world's only operational fixed-wing fighter aircraft with a vertical/short takeoff and landing (V/STOL) capability. From the pilot's perspective, much of its success can be attributed to the airframe/powerplant geometry and the chosen method of thrust vectoring. These features have resulted in an aircraft that can be flown manually, i.e. without a fly-by-wire system, throughout its flight envelope. Notwithstanding this basic capability, the earlier production aircraft (Harrier I) were fitted with limited-authority auto-stabilization systems (working in pitch, roll and yaw) to give an acceptable pilot workload during low-speed flight. The handling of the latest-generation aircraft (Harrier II) has been further enhanced by the provision of increased-authority stability augmentation and a pilot-selectable pitch- and roll-attitude hold capability.

Although the Harrier II has very good handling qualities in low-speed flight, a future short takeoff/vertical landing (STOVL) aircraft replacing the Harrier will certainly require a significantly more complex airframe/powerplant configuration to satisfy the more demanding operational requirements (e.g. supercruise, high AoA capability, stealth and carefree handling throughout the flight envelope). Such an aircraft will undoubtedly be statically unstable and, hence, will not be controllable by the pilot without assistance from an integrated flight and powerplant control system (IFPCS) using active control technology (ACT). Apart from providing basic stabilization and integrated management of the flight and powerplant control systems, the introduction of ACT will also provide the opportunity to add control features to reduce the pilot workload further during low-speed flight.

In order to investigate the handling, control and display requirements for future STOVL aircraft, the UK Ministry of Defence (MOD) is sponsoring a research programme called Vectored thrust Aircraft Advanced flight Control (VAAC). VAAC is exploring the low-speed regime from conventional aerodynamic wing-borne flight down to the hover (fully jet-borne flight). The research is extending the database used to derive the handling-qualities requirements for conventional takeoff and landing configurations (MOD Def Stan 00-970 and Mil-Std-1797A) to future STOVL aircraft, assisted by Mil-F-83300 and appropriate sections of ADS-33C.

The programme is supported by the experimental, fly-by-wire, VAAC Harrier aircraft, which provides a flexible and unique V/STOL flight test facility. The VAAC Harrier enables the rapid development and evaluation of different control strategies and response types that design, analysis and manned simulation have shown may offer a reduction in pilot workload. The facility's safety strategy, of safety pilot aided by software protection, has been flight certified and allows rapid flight test development without the overhead of meeting formal certification requirements for each control law software update. Moreover, the safety strategy allows any pilot, regardless of their particular flying experience (and even non-pilots), to assess the experimental control strategies without additional risk.

2. Standard Harrier aircraft control

The Harrier has five primary motivators (or effectors): tailplane, aileron, rudder, thrust magnitude (gross thrust) and thrust direction (nozzle angle). Each motivator is linked to an independent pilot control, referred to as an 'inceptor', via a mechanical control run. During jet-borne flight, rotational control in pitch, roll and yaw is provided by a three-axis, jet reaction control system (fed by high-pressure engine bleed air) which is mechanically linked to the relevant aerodynamic motivators. In addition to these primary motivators are the usual secondary motivators: flaps and airbrake.

In the cockpit of a standard Harrier, the pitch inceptors are arranged such that the centrestick controls the tailplane, and two separate, left-hand levers determine the thrust magnitude and direction. Cockpit switches control the flaps and the airbrake. Large variations of forces and moments occur between wing-borne and jet-borne conditions. Cross-axes coupling effects are experienced from inputs to these controls.

The Harrier's unique, two-inceptor left-hand arrangement (throttle *and* nozzle lever) prevents simultaneous modulation of the thrust magnitude and direction. Instead, the pilot is required to use his or her experience to judge the left-hand control combination required for the task; specifically to apportion the total thrust vector into the appropriate longitudinal and normal components. For example, during the final approach to the hover, the decelerating transition is usually performed by selecting increasing nozzle angle in a series of discrete steps, each at a suitable speed or range from the landing point. During the final stages of the deceleration, the pilot maintains a nominally constant pitch attitude, using the centrestick, while regulating the aircraft's flight path angle with the throttle. As the aerodynamic lift reduces, the pilot is required progressively to increase engine thrust to maintain the necessary total normal force. Additionally, the pilot must also make fine nozzle angle and pitch attitude adjustments to control the aircraft's speed and longitudinal position, allowing for local wind conditions, during the final landing sequence. Some further reduction in workload, specific to the Sea Harrier variant, is achieved with a 'speed trim' system. This system, controlled by the pilot using a switch located on the throttle, allows a fixed, limited-authority movement of the nozzles independent of the nozzle lever. With this basic, open-loop control system, the pilot has the ability to accelerate and decelerate at constant pitch attitude.

3. VAAC Harrier description

The VAAC research aircraft, a Harrier I two-cockpit trainer, is shown in Fig. 1. A detailed description of this aircraft is given by Shanks (1991). The front cockpit

Figure 1.   The VAAC Harrier.

has a near-standard production fit and is occupied by the safety pilot. The rear cockpit has been extensively modified to implement an experimental, digital, fly-by-wire flight control system (FCS) and ACT.

The rear cockpit's pitch inceptors have been disconnected from the basic aircraft's mechanical circuits. Their positions are now measured by electrical transducers and used as inputs to the experimental FCS, which has full-authority control of the tailplane, pitch reaction control valves, flaps, throttle and nozzle channels; it also controls roll and yaw to the auto-stabilizer authority limits. The FCS back drives the front cockpit's pitch inceptors (pitch stick, throttle and nozzle) through clutch mechanisms to the standard mechanical circuits. This feature assists the pilot in the front cockpit, in his or her role as safety pilot, when the experimental FCS is engaged. The rear cockpit's lateral stick and rudder pedals are connected, conventionally, to the front cockpit and aerodynamic motivators. Control reverts to the safety pilot through several forms of force override and discrete disconnect mechanisms; any disconnect action is accompanied by audio and visual indications to both pilots. The FCS installation was designed to allow experimental flight control laws to be installed and assessed, in flight, without significant clearance and flight safety implications.

To aid the overall safety protection of the aircraft, the FCS architecture has a mixture of duplex- and simplex-monitored systems. The control law software, which determines the aircraft's modified handling and control characteristics, is simplex monitored by what is referred to as the independent monitor (IM). The IM is a separate piece of software (stored in segregated memory and executed on a separate processor within the FCC) which detects both FCS failures and control law demands which could result in a potentially dangerous situation. If such a problem were detected by the IM, it would automatically disconnect the FCS and leave the safety pilot in control (within the cleared flight envelope of the Harrier) to effect a safe recovery. The design requirement for the IM is determined by the stringent safety needs during the transition from wing-borne to jet-borne flight. The IM is independent of the range of control laws to be evaluated. Because of the protection afforded by the safety pilot aided by the IM, the control law software can be declared as non-safety critical.

A fully programmable head-up display (HUD) system serves both cockpits. Rapid changes can be made to the rear cockpit HUD format, as required by the task objectives. The HUD computer and flight control computer (FCC) can exchange digital information via dedicated data-links. This flexible capability allows the maximum advantage to be gained in researching the requirements for HUD symbology, harmonized with the specific response types and flight control law functions.

4. The VAAC programme

It was anticipated that, for future advanced STOVL aircraft, the major handling and control questions would be associated with the longitudinal/pitch axes (because of the introduction of additional moments from a variety of sources, including the powerplant). Therefore, the VAAC programme has initially addressed the longitudinal plane, where the integrated management of the thrust vector and aerodynamic forces, to provide decoupled aircraft responses, was predicted to offer significant potential to reduce pilot workload during low-speed flight. Experience has also shown that, to achieve this objective, the flight control laws must also compensate for any limitations imposed by either the aerodynamics or the powerplant. A potential advanced STOVL configuration, with a typically complex motivator arrangement, is shown in Fig. 2 (Fielding 1994). The application of ACT to STOVL aircraft offers a design solution, but the multiple motivators raise some questions: how many inceptors are needed to control the forces and moments, and how should they be distributed between the pilot's left and right hands? These issues are referred to as the control strategy.

A reduction in pilot workload can be achieved by employing a control strategy that has response types appropriate to the operational task; the necessary stability and control should be obtained with due regard to current handling-qualities guidelines. Good handling qualities can only be achieved by implementing suitable flight control laws, which provide the link between the pilot's inceptors and the resulting aircraft response types. Therefore, fundamental to meeting the primary programme objective is the determination of the best aircraft control strategy and the response type. Various approaches can be taken, particularly where there are redundant motivators, including the circumstances where the motivator action changes between operational tasks. The use of discrete mode changing, either manually or automatically, needs to be considered. An automatic mode change option presents a complex system with implications on pilot awareness and acceptability; however, a manual mode change may present a more complex management task.

ACT offers the designer several alternative control strategies which must be linked with the appropriate response type. For example:

(a) Standard Harrier-like, with the retention of the three primary inceptors controlling pitch rate, thrust magnitude and thrust direction. The appropriate response type is achieved with feedback control added to one or more channels to provide the required handling or stability augmentation. Both manual or automatic mode changes, depending on the flight regime, could be incorporated.

(b) Three inceptors with additional feedbacks, controlling pitch rate, the normal thrust component and the longitudinal thrust component.

(c) Two inceptors, e.g. centrestick and one left-hand inceptor, controlling the total normal force and the total longitudinal force, independent of airspeed, with the pitch angle managed automatically. This approach would provide a 'unified control' concept as each inceptor controls the same orthogonal motion throughout the flight envelope, without discrete mode changes.

(d) An adaptation of the two-inceptor strategy that transfers the left-hand control function onto a discrete thumb-controlled switch on the right-hand inceptor (the one-inceptor option).

(e) An adaptation of the above strategies to allow translational rate control (TRC) during very slow-speed flight. TRC allows the pilot directly to command manoeuvres in the horizontal plane (inertial axes) for the final landing task.

It was determined from early VAAC simulation studies that some lateral/directional control augmentation was required in order not to adversely affect handling-qualities assessments in the longitudinal plane. Therefore, a single set of control laws was developed for use with the production auto-stabilizer servos to give limited-

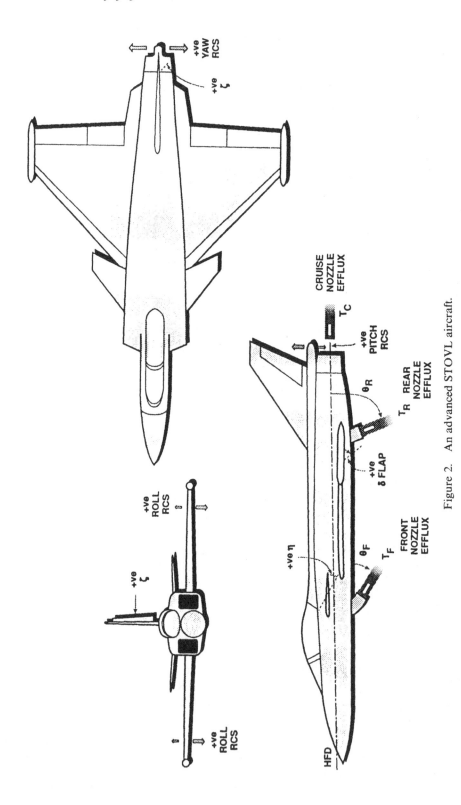

Figure 2. An advanced STOVL aircraft.

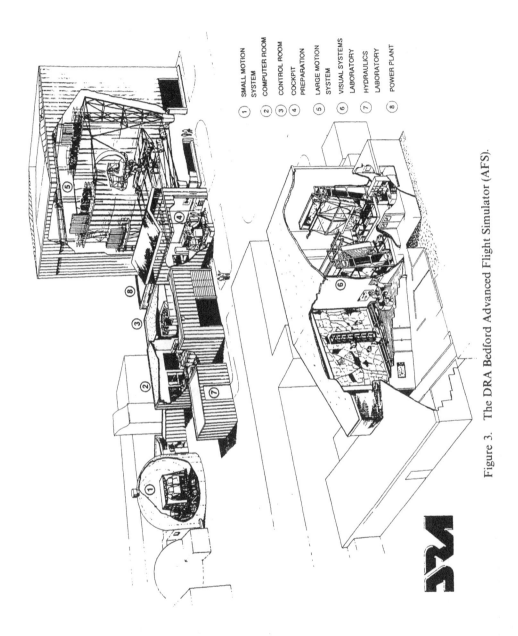

①	SMALL MOTION SYSTEM
②	COMPUTER ROOM
③	CONTROL ROOM
④	COCKPIT PREPARATION
⑤	LARGE MOTION SYSTEM
⑥	VISUAL SYSTEMS LABORATORY
⑦	HYDRAULICS LABORATORY
⑧	POWER PLANT

Figure 3.   The DRA Bedford Advanced Flight Simulator (AFS).

authority lateral/directional augmentation. These lateral/directional control laws were used in conjuction with all experimental pitch control laws. In roll, the lateral control law gave a roll-rate response scheduled with the pilot's lateral stick position (with an explicit bank-angle hold term when the undercarriage was down and the stick was centred). In yaw, the directional control law performed automatic sideslip suppression at higher speeds, blending to yaw-rate demand, proportional to the pilot's pedal input, at low speed. When commanding yaw rate, the pilot could also invoke an explicit heading hold by centralizing the pedals.

Control laws were developed to implement some promising control strategies and initially assessed, with the appropriate HUD display formats, on the DRA Advanced Flight Simulator (AFS) complex and other industry-operated facilities. The AFS, Fig. 3, has a large motion system with five degrees of freedom and pro-

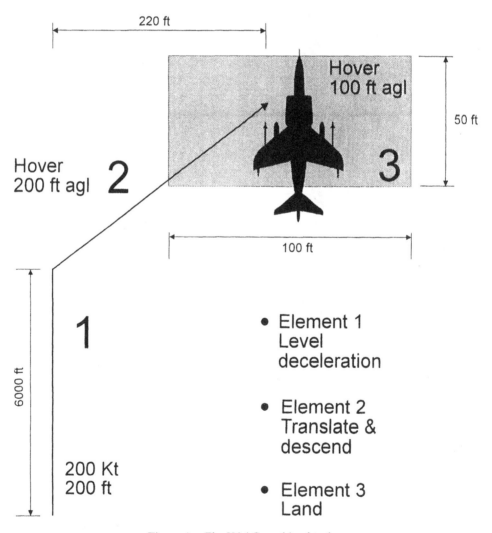

Figure 4.   The VAAC workload task.

vides ± 5 m of vertical travel, with 10 m s^{-2} acceleration. The simulator cockpit has an active 'g-seat' to provide additional motion cues, a programmable inceptor force system and cockpit displays. Outside-world visual cues can be generated from a closed-circuit television camera moving over high-resolution terrain model belts or from computer-generated imagery.

A range of piloting tasks were used to perform the simulator-based assessments. These included: vertical takeoff and landing, short takeoff and landing, transitions (accelerating and decelerating), and height and speed captures. Control laws selected for testing on the VAAC Harrier were subsequently assessed against a similar set of tasks in flight, although tasks involving takeoffs and non-vertical landings were not performed. The final assessment involved a specifically designed workload assessment task, depicted in Fig. 4 (Griffith 1994). This workload task comprised a deceleration from wing-borne flight to the hover, manoeuvring while in the hover and a vertical landing on a pad. Guest pilots were invited to fly the workload task and provide Cooper–Harper ratings (for task elements and overall); they consisted of approximately equal numbers of Harrier and non-Harrier pilots, with all the participating Harrier pilots having at least 1000 flight hours on type.

5. Control strategies and response types

The longitudinal plane control strategies and response types addressed by the VAAC programme are summarized in Fig. 5. The flight trials programme was structured to expand, progressively, the sophistication of both the control strategy and the response type. Figure 5 illustrates this progressive expansion from the 'Digital Harrier' implementation, through an enhanced three-inceptor strategy, to the more complex two- and one-inceptor control strategies. Each control law was given a unique identifier, of the style CL00x, at the start of its design and development.

The following sections describe the control law designs, the associated manned simulation trials, and the results and lessons learned from the flight trials assessment of the Digital Harrier, CL002 and CL003. Two further two-inceptor control laws, CL001 (Rawnsley *et al.* 1994) and CL005 (Shanks *et al.* 1994), were also successfully developed and flight tested. Briefly, the design method used for CL001 included an algorithm which can be regarded as a 'non-linear static inverse' of the aircraft. Feedback loops were designed with classical methods to achieve the required handling qualities. CL005 was designed using an H$_{\infty}$ method (Hyde and Glover 1992) to provide robust stabilization. The controller was written as an exact plant observer with state feedbacks (Hyde and Glover 1993), which provided both a visible structure and a straightforward method to gain-schedule and compensate for motivator authority limitations. During flight testing, both CL001 and CL005 demonstrated good handling qualities from wing-borne flight down to the hover. The flight testing of CL001 and CL005 is not discussed further here, but the results from both sets of tests were encouraging and correlated well with the CL003 results described below.

6. Digital Harrier

This open-loop, pitch control law implemented models of the mechanical control circuits of the basic aircraft (the auto-stabilization system of the standard aircraft was not represented). Thus it provided a fly-by-wire equivalent of the standard, mechanically controlled aircraft – hence the name 'Digital Harrier' – and was the

Figure 5.   Summary of control strategies and response types.

CONTROL LAW	PILOT'S INCEPTORS							CONTROL BLEND REGION (Knots)
	THROTTLE (LEFT-HAND)		NOZZLE	PITCH STICK (RIGHT-HAND)		PITCH STICK TRIMMER		
	LOW SPEED	HIGH SPEED		LOW SPEED	HIGH SPEED			
Digital Harrier	————— As Basic Harrier —————					---		---
CL002	————— As Basic Harrier —————			PRCAH	PRC	---		---
CL003	U_G	As Basic Harrier	Not Used	HRC	PRC	θ^{**}		60 - 130
CL001	U_G	ACVH	Not Used	HRCHH or VACHRH	FRCFH	θ^*		25 - 35
CL005	U_G	VCVH	Not Used	HRCHH or VACHRH	PRCFH	θ^*		80 - 100

FRCFH   Flightpath Rate Command Flightpath Hold
HRCHH   Height Rate Command Height Hold
PRCAH   Pitch Rate Command Attitude Hold
PRCFH   Pitch Rate Command Flightpath Hold
ACVH   Acceleration Command Velocity Hold
VCVH   Velocity Command Velocity Hold
VACHRH Vertical Acceleration Command Height Rate Hold

PRC   Pitch Rate Command
HRC   Height Rate Command
U_G Groundspeed
T_x Horizontally Resolved Thrust
*   Low Speed with U/C down
**   Left Hand Function, Single Inceptor Variant

baseline against which the evaluation pilot could assess the other control options. Although the control law was very limited in its capability, it was extremely useful for:

(a) providing a vehicle to assess the characteristics of the basic VAAC FCS and, in particular, the response time delays;

(b) supporting the VAAC FCS flight clearance, and the evaluation of the instrumentation and telemetry recording system;

(c) safety pilot evaluation of the VAAC FCS disconnect mechanisms;

(d) structural coupling measurements, between all the VAAC actuators and the sensors, to establish signal conditioning requirements.

The evaluation pilots reported no discernible difference between the handling provided by the Digital Harrier and the basic aircraft with mechanical controls. The effective time delay introduced by the experimental FCS (including a contribution from the actuators) was approximately 80 milliseconds. Completion of the activities (a) to (d) provided the confidence to proceed to the evaluation of the advanced control laws.

7. Three-inceptor control, CL002

7.1. *Control philosophy*

This was an extension of the standard Harrier three-inceptor control strategy. With this strategy, the pilot used the centrestick (or right-hand inceptor (RHI)) to demand the aircraft's pitch rate, and the throttle and nozzle levers (or the left-hand inceptors (LHIs)) to demand thrust magnitude and direction (nozzle angle), respectively. The core of CL002 was proportional-plus-integral control of the pitch rate, via the tailplane. It also incorporated an automatic flap schedule and decoupled pitch rate from flap and nozzle angle changes. At low speed, with the undercarriage down, pitch attitude was held explicitly when the RHI was within a central deadband. The LHIs provided open-loop control, just like the Digital Harrier. Because the Harrier's thrust line is always close to the centre of gravity, control authority was not a problem and the pilot could determine the required control strategy with acceptable workload.

7.2. *Design strategy*

CL002 was designed using classical techniques. As for all VAAC control laws, the design studies were based on the DRA Harrier wide-envelope model (WEM), a full, rigid-body, non-linear simulation of the VAAC Harrier. The WEM comprises models of the aircraft's aerodynamics, the Pegasus engine and FCS components (e.g. actuators and sensors). As a result of the classical design process, the control law contained a limited number of frequency-shaping filters and gain schedules.

In addition to providing major handling improvements relative to the basic aircraft, one of the benefits of CL002 was its simplicity. At the design stage, further sophistication, particularly to the thrust magnitude and direction channels, was not considered worthwhile.

7.3. *Manned simulation trials*

The AFS was used extensively for pilot assessment, particularly to evaluate the pitch attitude stability through the transition region to the hover, i.e. the elimination of the usual cross-coupling effects due to changes in throttle and nozzle demands. Hence, refinement to the design was possible in an interactive way with pilots, prior to the flight trials evaluation. The HUD symbology appropriate to this response type was designed and developed using the Sea Harrier format as a baseline.

The other major role of the AFS was for training evaluation pilots before flight trials. This role was common to all the control laws that were ultimately flight tested. This control-law-specific training was of particular importance prior to the flight evaluation of the more advanced control strategies.

7.4. *Flight trials results*

The preliminary flight trials assessment concentrated on refining the handling characteristics and, in particular, on harmonizing the control strategy and response type with the existing cockpit systems and the inceptors. Optimization of the control law was performed in flight, with an on-line 'variable-gain' facility (this facility was available to all experimental control laws, see § 8.5.1). The pitch attitude hold mode was optimized with some further developments to the HUD symbology.

Flight trials of CL002 also revealed deficiencies in the characteristics of the RHI, which was part of the basic aircraft's equipment and, hence, not designed for ACT systems. The handling assessment was dominated by the RHI's lack of accurate self-centring, the main impact of which was that the pitch attitude hold mode could not be reliably activated without significant pilot workload. The workload problem was compounded by a similar lack of accurate self-centring in roll, which resulted in a small roll-rate demand (without pilot input) during the assessment of acceleration and deceleration tasks. With modifications to the software-controlled stick shaping, and additional HUD symbology, the impact of the centring problem was mini- mized, but not entirely eliminated. Recommendations were made to replace the existing inceptor and feel system; a modern RHI assembly is now being installed.

Figure 6 illustrates flight trials results for a straight-in, decelerating recovery task from 200 knots to the hover. The response decoupling between pitch attitude and nozzle angle is good, despite pilot activity with the throttle and nozzle. Figures 7(*a*) and 7(*b*) present crossplots of RHI pitch and roll inputs during a typical decelerating recovery task; results from both CL002 and the Digital Harrier are presented for comparison. The reduced RHI pitch activity for CL002, in the same moderately turbulent conditions, is indicative of the workload reduction achieved. The variation in pitch attitude during the task with CL002 is small compared with the Digital Harrier, as shown in Figures 8(*a*) and 8(*b*). This difference is a result of the attitude- hold mode working through the underlying pitch-rate stabilization loop. The small variations illustrated in Fig. 8(*b*) are due to occasional fine tracking adjustments by the pilot. Turbulence rejection was generally good in pitch but, beyond modest turbulence levels, the overall handling-qualities rating was reduced by the restricted capability of the limited-authority lateral/directional system.

Although CL002 used a three-inceptor control strategy, with a conventional pitch-rate command/pitch attitude hold response type and no closed-loop control of

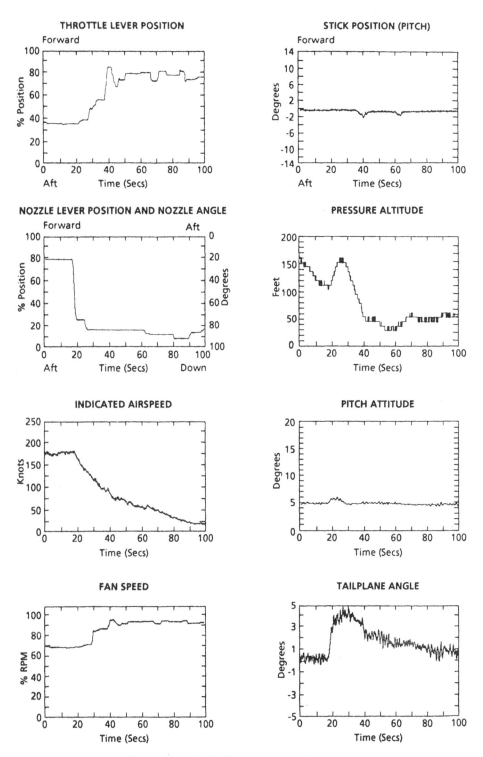

Figure 6.   CL002 flight data: time histories.

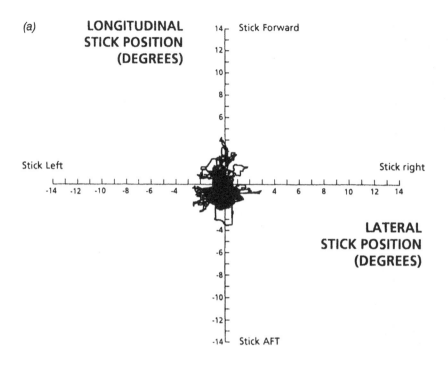

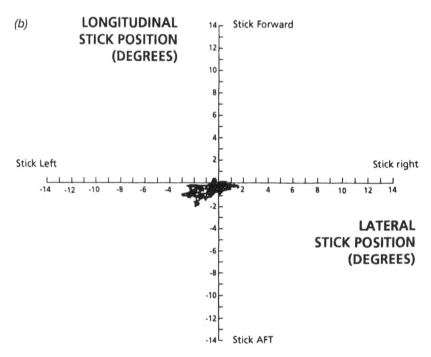

Figure 7. (*a*) Digital Harrier flight data: RHI activity crossplots; (*b*) CL002 flight data: RHI activity crossplots.

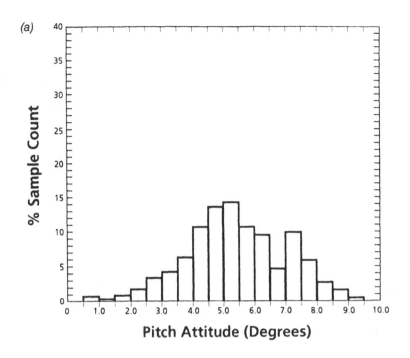

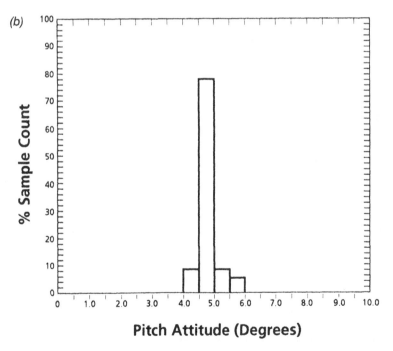

Figure 8. (*a*) Digital Harrier flight data: pitch attitude variation; (*b*) CL002 flight data: pitch attitude variation.

the thrust vector, it was successful in transforming the aircraft's handling qualities as follows:

(a) The transition to and from the hover could be performed easily at constant attitude, even in turbulence, with minimal pilot intervention, thus allowing an RHI 'hands-off' approach (in pitch).

(b) The interaction between the changes in thrust direction (nozzle angle) and pitch attitude was eliminated.

(c) Harrier pilots performing the workload task with CL002 gave mainly upper 'Level 2' handling-qualities ratings, whereas comparable results from the Digital Harrier were only marginal 'Level 2'. It was anticipated that non-Harrier pilots flying the same task would give worse ratings for CL002, owing to their lack of training on the basic, three-inceptor Harrier. However, they also gave predominately 'Level 2' ratings, which was attributed to the enhanced pitch attitude stability afforded by CL002's pitch attitude-hold feature.

(d) The HUD format included response-type-specific symbology to assist the pilot in his or her tasks and was particularly useful to indicate the FCS mode. Appropriate symbology also contributed to the reduction in pilot workload.

Flight and simulation experience showed that it was not worth while developing either the throttle or the nozzle channels further. CL002 was therefore used as a second, more capable, standard to judge the other, more advanced, control laws. Its control strategy and response type were broadly representative of the Harrier II, although the VAAC FCS had a larger control authority and the basic aircraft had a different aerodynamic/powerplant configuration.

7.5. *Projection of the design to future STOVL aircraft*

CL002 was successful in providing a considerable reduction in pilot workload (relative to the Digital Harrier) for Harrier pilots, with very modest sophistication. Pitch-rate command with pitch attitude hold was confirmed as an effective response type for the low-speed handling and control of future STOVL aircraft. However, optimization of the pilot's inceptors would be needed to obtain the maximum performance.

A comparison of the handling qualities between the aircraft and the manned simulator, for various task assessments, was extremely close. This validation, and the experience gained with manned simulation, increased confidence in the ability to predict the handling characteristics of the two- and one-inceptor control strategies prior to flight trials.

8. Two-inceptor control, CL003

8.1. *Control philosophy*

This advanced, pitch-plane control law, designed by British Aerospace (BAe) and subsequently developed through flight tests by the DRA (Fielding *et al.* 1994), was based on a two-inceptor control strategy (which would allow a 'Hands-on-throttle-and-stick' (HOTAS) implementation). A one-inceptor variant was developed during the latter stages of the flight testing (see § 8.5.1). CL003 provided a unified pitch control strategy by blending in a 'seamless' manner, i.e. without any discrete mode changes, from a pitch-rate response type in wing-borne flight to a height-rate response type at low speed.

The control law was designed primarily for performing decelerating transitions from wing-borne to jet-borne flight and vice versa. The control law, shown in simplified form in Fig. 9, provided integrated management of the aircraft's thrust vector and aerodynamic surfaces in response to commands from two cockpit pitch inceptors:

(a) The LHI commanded aircraft axial thrust in wing-borne flight (like a conventional throttle) and the horizontal thrust component, with closed-loop groundspeed control, in jet-borne flight. The mode change was continuously blended as a function of airspeed.

(b) The RHI commanded pitch rate, via the tailplane, in wing-borne flight and height rate, via the vertical thrust component, in jet-borne flight. To maintain unified control throughout the flight envelope, these two modes were automatically blended together in the transition region (as a function of airspeed). Once the control law had fully blended into the height-rate mode, the pilot could only adjust the aircraft's pitch attitude through the RHI trim button, although this had no effect on flight path as the control law decoupled pitch attitude from the thrust vector in this speed regime.

The inceptor functionality was scheduled with airspeed, as shown in Fig. 10.

To maintain precise control of airspeed and flight path with acceptable workload, fixed-wing pilots normally fly approaches using the left hand to control the aircraft's descent rate and the right hand to control airspeed; this is known as the 'back-side' technique, whereas, in up-and-away flight, the pilot uses the opposite, or 'front-side' technique. The reason for employing the back-side technique on the approach is to avoid the inherent airspeed deviations (and the resulting increase in workload) that would be caused by the pilot varying the aircraft's pitch attitude to control its flight path. An important result of the unified control law philosophy, described above, is that all the flying is done using the front-side technique. Whilst the CL003 strategy could thus be regarded as unconventional, it was considered that significant workload benefits could only be obtained by employing the flexibility of ACT to implement such a unified control strategy. This radical change in flying technique is made possible by ACT, because effective speed hold is now achievable and, once the landing speed is selected, the pilot only has to 'point' the aircraft where he or she wants to land, using the RHI.

8.2. *Design strategy*

A visible and logically arranged control law architecture was derived using simple, well-proven, control law functions (e.g. filters and gain schedules) with a view to readily satisfying the practical issues of implementation, verification and flight clearance. Known physical effects were separated out within the control law to simplify the subsequent design of gain schedules; non-linear control functions were introduced into the control law, based on studies of the aircraft's physical attributes (e.g. thrust vector geometry). The importance of this stage of the design was that significant non-linearities in the basic vehicle characteristics could effectively be 'linearized' by the non-linear part of the control law, thus providing a sound basis for

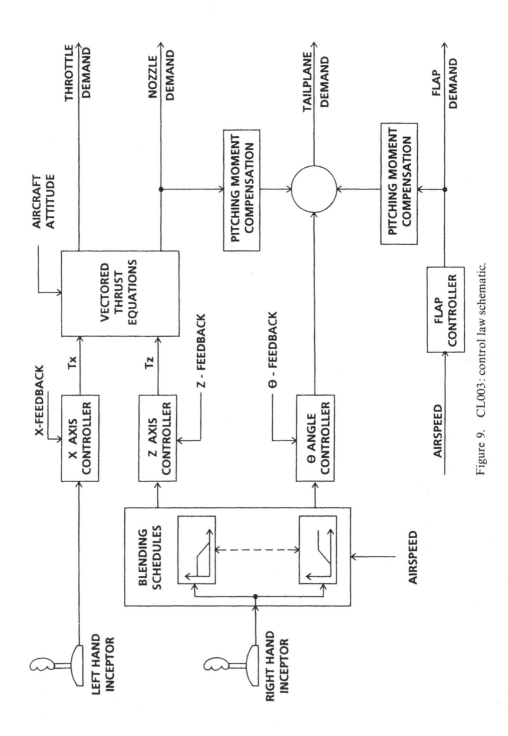

Figure 9. CL003: control law schematic.

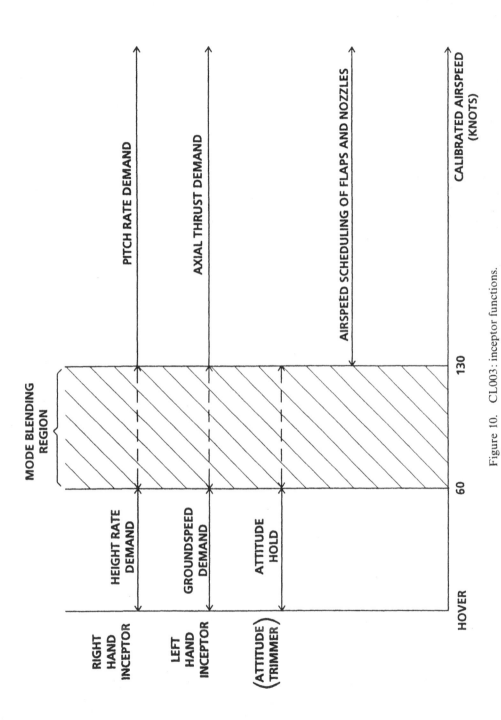

Figure 10. CL003: inceptor functions.

gain schedule design and resulting in simple gain schedules. Once the gain schedules had been designed, further non-linear design was necessary to ensure that the system performed satisfactorily when the thrust vector commands exceeded the available capability. This involved holding integrators to prevent wind-up and assigning a control hierarchy to ensure that the system achieves the optimum control strategy, despite the system's partially saturated state.

CL003 was designed loop by loop, with the most important loop being closed first, followed by the next most important, etc. This approach had significant benefits in that the designer used a systematic approach to the design, with insight being gained at each stage. The effects of opening a control loop could be fully understood and then used to great advantage as part of the overall design; for example, to satisfy control-hierarchy requirements in the event of thrust vector saturation.

In wing-borne flight, non-linear command filtering was designed using the BAe 'Gibson criteria' (Gibson 1991) to provide accurate pitch-tracking characteristics for small pilot inputs on the RHI. The filtering was designed such that a rapid normal acceleration response could be achieved for large pilot inputs, as the filter modulates and becomes 'transparent'. In jet-borne and partially jet-borne flight, high-order handling-qualities design criteria were not available so a significant degree of engineering judgement had to be used, supported by pilot comments from simulation trials. Parallel piloted simulation research studies were undertaken at BAe, in order to derive criteria suitable for STOVL aircraft handling-qualities design. These essentially extended the concepts of the highly successful 'Gibson criteria' for high-order flight control systems to the low-speed regime. It was shown, by retrospective analysis, that the control law conformed to these new criteria.

8.3. *Operational characteristics*

As the aircraft decelerated in wing-borne flight the flaps were lowered by an open-loop airspeed schedule. A flap-to-tailplane crossfeed minimized flap-induced pitch attitude changes. As airspeed further decreased, the reduction in aerodynamic lift was compensated for by introducing a vertical thrust component via an airspeed-dependent schedule. This simultaneously rotated the nozzles and increased the thrust. A crossfeed from the nozzle angle to the tailplane minimized nozzle-induced pitching activity. The aircraft was then flown like a conventional aircraft for the initial part of the transition, using incidence to control the flight path. A crossfeed from the RHI to the vertical thrust component was used to prevent excessive incidence demands during manoeuvres and to reduce the steady-state AoA. The vertical thrust component was corrected for bank-angle changes during lateral manoeuvres.

During the later stages of the decelerating transition, the response type blended from pitch-rate demand to height-rate demand on the RHI, and from open- to closed-loop speed control on the LHI. The pitch rate and attitude were held by a proportional, integral and derivative (PID) controller in the pitch-rate loop down to jet-borne flight. Control of the tailplane and pitch reaction control valves was progressively transferred to the pitch attitude control loop and the aircraft's pitch attitude automatically changed to the datum landing attitude. In jet-borne flight, the LHI demanded the fore/aft, or 'x component', of thrust, in the inertial axes, which blended into closed-loop groundspeed control; allowance has been made for the thrust x component to be negative, to facilitate limited rearwards flight in the hover.

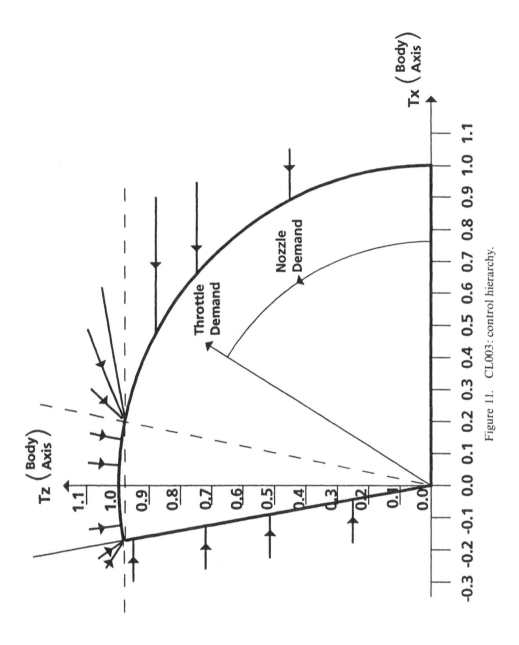

Figure 11. CL003: control hierarchy.

The RHI demanded the orthogonal up/down, or 'z component', of thrust, also in the inertial axes, which translated into height-rate demand. The z component was governed by a PID controller, the gains of which were scheduled with airspeed, and were sufficiently high to provide aircraft height hold in the absence of external disturbances (to which the system has been shown to be insensitive). Figure 11 shows the 'control-hierarchy' function which ensured that the z component of thrust took priority in the event of the thrust vector demand exceeding the achievable limits. The x and z components of the thrust vector were converted to throttle and nozzle angle demands by a conversion from rectangular to polar coordinates, giving 'decoupled control' of the thrust vector components in the inertial axes.

8.4. *Manned simulation results*

The most significant result from the initial pilot-in-the-loop simulation phase at BAe Warton and DRA Bedford was that although the system was found to be easy to fly by non-V/STOL combat aircraft pilots, difficulty was experienced by Harrier pilots who had to 'unlearn' their training for low-speed flying (similar feedback was obtained from assessments of other advanced VAAC control laws like CL001 and CL005). For straight-in approaches with minimal manoeuvring, the system performed well. For more aggressive approaches, such as are sometimes flown in the Harrier, the original system received some adverse pilot comments because the control law had reduced the performance capability of the aircraft in terms of maximum deceleration and had not made efficient use of the available aerodynamic lift. These issues were addressed and resolved during subsequent development of the control law. In terms of the functionality of the control law, the early simulation trials identified two design aspects which needed addressing:

(a) In jet-borne flight, CL003 needed to provide automatic protection against descent rates which could not be easily reversed by thrust modulation.

(b) CL003 used an automatic, discrete mode change as the aircraft decelerated through 130 knots, with a significant change in handling qualities.

Both of these aspects were resolved during development, by introducing automatic sink-rate limits and continuous airspeed scheduling to achieve the blended mode change described in § 8.1. The final phase of manned simulation trials was rated as providing 'Level 1' handling qualities.

The piloted simulation and off-line system analyses showed that the two-inceptor pitch control strategy had been successfully developed to a standard which was considered to be suitable for in-flight assessment.

8.5. *Flight trials results*

CL003 was installed in the VAAC Harrier's FCS without encountering any significant problems. It was first engaged, in flight, on 23 February 1993. In the following 13 months, 97 test flights were flown, with the control law engaged for approximately 39 flight hours of development, assessment and demonstration, including 30 vertical landings.

CL003 was flown in a wide range of ambient conditions (all-day VMC): in temperatures ranging from $-2°C$ to $+26°C$ and in winds from flat calm to 25 knots, gusting 35 knots. The control law performed well in all conditions, and has resulted in a very low pilot workload during the transition and in the hover. Moreover, during their first flight in the VAAC Harrier, pilots with no previous Harrier experience have been able to bring the aircraft to a stable hover, after a well-controlled deceleration from wing-borne flight, and then land vertically, with high accuracy, on a pad.

8.5.1. *In-flight control law development.* During the flight test development, a number of changes were made to the control law. These were not only to improve its basic performance, but also to assess changes in its functional design:

(a) The FCS had a 'variable gain', or perhaps more accurately, a 'variable scaling factor', facility whereby the test pilot in the rear cockpit could adjust key control law parameters in flight, by operating a simple keypad and a switch on what was the throttle lever. This facility was used extensively throughout the development process, primarily to optimize changes in the control law's functionality, but also to compensate for modelling deficiencies in the Harrier WEM (most notably the lack of sensor noise models).

(b) The sink-rate demand limiter introduced during the earlier piloted simulation (see § 8.4) was found to be too restrictive; in particular there was a significant reduction in the pilot's ability to achieve acceptable descent rates during the blend from pitch-rate to height-rate demand. The explicit sink-rate limit designed during the piloted simulation was also insensitive to ambient conditions and the aircraft's instantaneous weight, which, of course, have a major effect on the aircraft's ability to recover from large rates of descent at low altitudes. Instead, a dynamic throttle and nozzle limiter strategy, using information from the FCS's safety monitoring system, was adopted. This feature worked well; it allowed the pilot to command the descent rate required, with the knowledge that the control law would prevent irrecoverable situations (from the safety pilot's perspective) developing rapidly, without undue restriction of the aircraft's capabilities. Both pilots were informed of limiter action by warning lights in each cockpit.

(c) Pilot workload was found to be relatively high in the early stages of the transition, when compared with the lower end of the transition and hover. This characteristic was particularly apparent in turbulent conditions, where, at higher speeds, the pilot had actively to monitor and control the aircraft's vertical velocity through pitch rate, whereas at lower speeds the pilot was commanding vertical velocity directly. Accordingly, an explicit flight path hold feature and an automatic pitch-rate compensator for turning flight were added to the basic pitch-rate demand system to solve the problem.

(d) As a further workload-reducing measure, with potential applicability to all future fixed-wing combat aircraft, a one-inceptor variant was implemented and flight tested. This option, which is pilot selectable, transferred the function of the LHI to the trim switch on the RHI. The trim switch commanded a fixed-rate input, scheduled with speed, that was integrated over time to give a signal analogous to that from the LHI. Flight testing has shown that the one-inceptor control strategy results in a further workload reduction, particularly for precise positioning around the hover, and leaves the pilot with one hand completely free to perform other system management tasks;

indeed, it became the preferred method of flying the aircraft for the majority of pilots.

(e) In order to address concerns from Harrier pilots that a two- or one-inceptor control law reduces the operational flexibility of V/STOL aircraft at low speed, two outer-loop, low-speed modes were added: a pitch attitude trim facility with the two-inceptor option (in the original design, but not implemented initially, see § 8.1) and a braking, i.e. fully deflected, nozzle pitch-up for maximum deceleration, with the one-inceptor variant. Both these modes were successfully tested. These developments were implemented extremely rapidly, owing to the flexibility afforded by the experimental FCS.

8.5.2. *Flight test observations*. In addition to the control law changes described above, a number of observations have arisen from the flight experience with CL003:

(a) CL003 was rated as having predominantly 'Level 1' handling qualities, using both two- and one-inceptor options, for all pilots performing the workload task. This result was achieved in spite of some initial scepticism from Harrier pilots; indeed more than 80% of the ratings given by Harrier pilots for the one-inceptor option were Cooper–Harper 2 or better.

(b) At low speed, CL003's strategy of controlling fore/aft motion by changing the thrust vector angle, without changing pitch attitude, was well received (the usual Harrier practice is to adjust pitch attitude with the thrust vector angle fixed relative to the fuselage). This result is consistent with the success of the Sea Harrier's open-loop, speed trim system and may have considerable relevance to the design of low-speed flight/powerplant control systems of future STOVL aircraft.

(c) For future STOVL aircraft with ACT, traditional carefree handling concepts, such as automatic flight envelope limiting, need to be extended down to the hover, including the landing. One possible method of achieving this limiting, by applying an appropriate dynamic limiting strategy to the thrust vector demands, was developed and employed by CL003.

(d) Relative to the Harrier's existing arrangement, where the pilot has to control thrust magnitude and direction independently, significant workload reductions and acceleration/deceleration performance improvements could be achieved by integrated control of the thrust vector. Figure 12 shows pilot control activity during two similar decelerating approaches to the hover: the three time histories on the left illustrate the level of pilot control activity required to control the Digital Harrier in a conventional approach, whereas the two traces on the right show activity during an approach flown with the one-inceptor option of CL003.

(e) The control strategy employed by CL003 not only provides an important baseline for consideration during the design of the flight control system of a future STOVL aircraft, but could also be applied to a possible future derivative of the Harrier aircraft, and would result in further reductions in pilot workload during low-speed flight. This would be of particular relevance to the Harrier's night attack role.

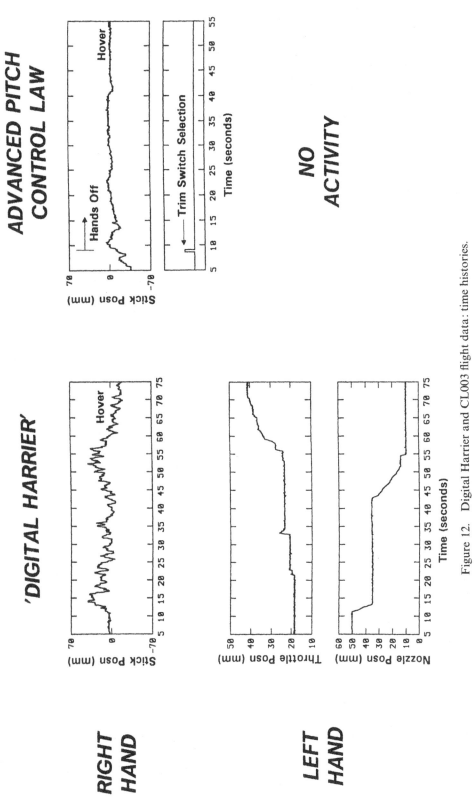

Figure 12. Digital Harrier and CL003 flight data: time histories.

(*f*) The consistent, 'front-side' response strategy, where the LHI always commands faster/slower and, similarly, the RHI always commands up/down, appears very natural and is easy to fly for non-V/STOL pilots. Moreover, the workload assessment showed that even very experienced Harrier pilots (with more than 1000 hours on type) found the control law easy to fly. This potentially raises major savings in the requirements for training the pilots of a future STOVL aircraft, although the experience of pilots familiar with the Harrier's inherent operational flexibilities should not be ignored unwittingly just, for example, to reduce training costs.

8.5.3. *Lessons learned.* Criteria need to be developed and validated further, to provide a guide for the design aims for the integrated systems on future aircraft configurations. These will, of course, also need to reflect the future operational requirements. Specifically, work is required to extend and investigate the handling-qualities criteria used for fixed-wing aircraft, to cover STOVL operation and to define the requirements for carefree handling at low speed and in the hover.

Early standards of CL003 used the aircraft's airspeed to trigger a discrete change between the pitch-rate and height-rate demand modes, with associated signal equalization. This proved to be unnecessarily complicated and introduced a discrete, and undesirable, change in handling qualities. Instead, the control law was developed to use airspeed to blend smoothly between the modes, leading to a significantly easier implementation and providing continuity of handling characteristics. The resulting unified control strategy was, generally, very well received by test pilots, pilots and non-pilots flying the VAAC Harrier.

The demonstrated reduction in pilot workload was mainly due to the automatic axis transformations inherent in the control law. The pilot's thrust vector commands were in inertial-axis rectangular coordinates (where appropriate) rather than body-axis polar coordinates. The former allows the pilot directly to command the flight variables of primary interest and provides decoupled aircraft control. This was the major benefit and could be used in conjunction with either a one-, two- or three-inceptor pilot interface. However, the two-inceptor pilot interface would enable HOTAS throughout transitions between wing-borne and jet-borne flight regimes, thereby further reducing pilot workload.

The flexibility of the VAAC Harrier FCS enabled a one-inceptor pilot interface to be developed and flight tested. The one-inceptor option demonstrated further reductions in pilot workload, particularly at low speed. It has also raised fundamental questions as to how thrust should be controlled in future combat aircraft; for example, it potentially alleviates the need for HOTAS by leaving one hand completely free to operate other aircraft systems. This was the first real opportunity to redesign radically part of the pilot/aircraft interface; it is too early to say if the one-inceptor cockpit is the optimum arrangement as further refinement is required, but, as a concept, it deserves serious thought (Griffith 1994).

CL003's control strategy and response type are considered to be appropriate to future STOVL configurations. Although research under VAAC is ongoing, similar strategies have already been applied to other BAe STOVL projects (simulation phase only) and are being applied to an RULS (Remote Unaugmented Lift System) aircraft as part of the MOD-sponsored Integrated Flight and Powerplant Control Systems (IFPCS) programme (Fielding 1994), which is addressing the cost-effective rig demonstration of the key control system technologies required for the next generation of combat aircraft.

9. Conclusions

A range of control strategies and response types have been described for potential application to future short takeoff/vertical landing (STOVL) configurations. While current operational Harrier experience is undoubtedly important in the assessment of these advanced control strategies, it must be recognized that few pilots of future STOVL aircraft will have flown a Harrier, and that the airframe/powerplant geometry may impose control limitations not experienced in the Harrier. Moreover, the capabilities of active control technology flight and powerplant control systems can be exploited to implement advanced control strategies that significantly reduce pilot workload throughout the flight envelope. Accordingly, one-, two- and three-inceptor control strategies, with various response types, have been flight tested on the VAAC Harrier aircraft. The results obtained from the flight testing of all control laws closely match those predicted from manned simulation trials.

Flight trials of a three-inceptor control strategy, which had a pitch-rate command/attitude-hold response type, have shown a significant decrease in pilot workload relative to the basic VAAC Harrier's three-inceptor arrangement during the transition from wing-borne to jet-borne flight, hovering and vertical landing. A sample of both Harrier and non-Harrier pilots gave predominantly 'Level 2' handling-qualities ratings for this system. Flight trials of a two-inceptor control strategy have shown a further, significant reduction in pilot workload. The same sample of Harrier and non-Harrier pilots rated both the two-inceptor system and its one-inceptor variant as predominantly 'Level 1'; notably, Harrier pilots gave entirely 'Level 1' ratings for the one-inceptor variant.

During the extensive, simulator-based, STOVL flight control research at the Defence Research Agency and British Aerospace, and, more importantly, during 3 years of flight testing in the VAAC Harrier, it has been conclusively demonstrated that the biggest reduction in pilot workload is achieved by eliminating all conscious mode changing (unified control). This entails making it transparent to the pilot that he or she is flying a STOVL aircraft; as far as the pilot is concerned, he or she is flying a conventional aircraft that can hover. If this can be achieved using a single inceptor, even larger benefits may be available. This reduction in pilot workload should be a design goal for the flight control system of the next generation of combat aircraft.

The VAAC Harrier's unique safety pilot arrangement has enabled a wide range of 'pilots', comprising fixed-wing pilots (Harrier and non-Harrier), rotary-wing pilots and non-pilots, to assess safely the full range of experimental control laws. An important result was that the one- and two-inceptor control strategies were easy to fly, irrespective of piloting background; this must fundamentally influence the way in which future STOVL combat aircraft are controlled. Such aircraft will have significantly more complex airframe/engine configurations and more demanding operational requirements, and must benefit from the conclusions of research under the VAAC programme.

It is intended that the results of this programme will feed the handling and control requirements specifications for future STOVL aircraft to ensure a design approach which provides minimum pilot workload.

ACKNOWLEDGEMENTS

The VAAC programme is sponsored by the UK Ministry of Defence.

The authors would also like to acknowledge the invaluable work of all the engineers and pilots who, over a number of years, have contributed to the success of the VAAC programme.

REFERENCES

ADS-33C, 1989, Handling qualities specification for military rotocraft.

COOPER, G. E. and HARPER, R. P., 1969, The use of pilot rating in the evaluation of aircraft handling qualities. NASA-TN-D-5153.

FIELDING, C., 1994, Design of integrated flight and powerplant control systems. *AGARD Conference Proceedings 548*, Paper 9.

FIELDING, C., GALE, S. L. and GRIFFITH, D. V., 1994, Flight demonstration of an advanced pitch control law in the VAAC Harrier. Paper presented at the AGARD Flight Mechanics Symposium, Turin, Italy, 9–12 May 1994.

GIBSON, J. C., 1991, The development of alternative criteria for FBW handling qualities. *AGARD Conference Proceedings 508*, Paper 9.

GRIFFITH, D. V., 1994, The VAAC Harrier – a time for change. Paper presented at the SETP Symposium, Los Angeles, September 1994.

HYDE, R. A. and GLOVER, K., 1992, Development of a robust flight control law for a VSTOL aircraft. *IFAC Aerospace 92 Conference Proceedings*, Ottobrunn, Germany.

HYDE, R. A. and GLOVER, K., 1993, The application of scheduled H_∞ controllers to VSTOL aircraft. *Automatica/IEEE Transactions on Automatic Control–Special Issue on Control Applications*, **1438**(7).

Mil-F-83300, 1986, Flying qualities of piloted V/STOL aircraft.

Mil-F-9490D, 1975, Flight control systems, design installation and test of piloted aircraft, general specification.

Mil-Std-1797(USAF), 1987, Flying qualities of piloted vehicles.

MOD (UK) Defence Standard 00-970, 1992. Design and airworthiness requirements for service aircraft.

RAWNSLEY, B. W., ANDREWS, S. J. and D'MELLO, G. W., 1994, Future STOVL flight control: development of a two-inceptor trimmap-based pitch plane control law for the VAAC Harrier research aircraft. *IEE Conference Publication* No. 389.

SHANKS, G. T., 1991, The VAAC Harrier In-Flight Simulation Facility. *International Symposium, In-Flight Simulation for the 90's, Conference Proceedings*. DGLR-91-05.

SHANKS, G. T., FIELDING, C., ANDREWS, S. J. and HYDE, R. A., 1994, Flight control and handling research with the VAAC Harrier. *International Journal of Control*, **59**(1), 291–319.

PART THREE

Transport Aircraft

7

The design and development of flying qualities for the C-17 military transport airplane

ERIC R. KENDALL

1. Airplane description

The C-17 is a long-range, air-refuelable, turbo-fan-powered, high-wing, heavy military cargo aircraft built around a large, unobstructed cargo compartment. It has a swept wing that uses supercritical airfoil technology and winglets to achieve good long-range cruise performance. A photograph of the airplane flying in its cruise configuration is shown in Fig. 1. Technologies that combine to achieve the C-17's exceptional short-field landing/performance are the large externally blown flaps (EBFs), full-span leading-edge slats, spoilers, high-sink-rate landing gear, anti-skid braking, thrust reversers, head-up displays and sophisticated fly-by-wire flight control system. Figure 2 is a photograph of the airplane just prior to touchdown and Fig. 3 shows the pilot using throttles and head-up-display (HUD) guidance to perform the precision landing.

2. The cockpit

Important to pilot acceptance of an airplane's flying qualities are the displays and control manipulators provided for the task to be performed. A perspective view of the C-17 cockpit is presented in Fig. 4. This shows the centerstick controller which replaces the wheel normally used on more conventional transport aircraft. The numerous switches, buttons and trigger on the stick-grip are shown in Fig. 5.

3. Head-up display

In the cockpit, the HUDs allow pilots to fly their planned approach with head up and eyes out of the cockpit by superimposing the necessary cues on their outside field of view. Using the HUD, the approach path can be aligned to intersect a selected aim point on the runway, which reduces touchdown dispersion. The EBF design enables pilots to approach the runway at airspeeds as slow as 115 knots and at a relatively steep 5° approach angle, with landing sink rates of up to 15 feet (4·57 m) per second. This allows the aircraft to touch down consistently on the runway aim point.

Figure 1.   C-17 in cruise configuration.

4. Flight controls

The C-17 uses a quad-redundant flight control system employing digital electronic fly-by-wire technology. Redundancy is incorporated throughout the flight
control system.

Control of aircraft flight path and attitude is accomplished through the move-
ment of conventional control surfaces. Four elevators generate pitching moments
for longitudinal control, and the movable horizontal stabilizer provides pitch trim.
Two ailerons and eight spoilers generate rolling moments for lateral control. Upper
and lower double-hinged rudders generate yawing moments for directional control.
The four large flaps and the spoilers are used together as in-flight speed brakes.
With flaps retracted, the C-17 exhibits conventional swept-wing handling qualities.
However, when its four large EBFs are extended into the engine thrust flow field,
the aircraft's powered lift gives it unique handling characteristics. With powered lift,
thrust changes result in flight path changes. An increase in power causes the airplane to ascend, while a decrease causes a descent. Elevator inputs result in speed
changes. This powered-lift mode of flight allows steep and precise approaches to
landing and is a major contributor to the C-17's excellent short-field capabilities.

5. Approach characteristics

A diagram showing this capability and defining the 'back-side' control technique
is presented in Fig. 6.

Figure 2.   C-17 landing.

The aeropropulsion characteristics that make this technique possible are shown in Fig. 7 for a full-flap, all-engines, heavy-payload steep approach. At a steep flight path angle of $-5°$ and at a speed of $1·2V_s$ increase or decrease of engine thrust with pitch attitude held constant at $3°$ significantly affects the flight path angle with little effect on airspeed. As the thrust change arrows indicate, the flight path angle can be increased from $-8°$ to about $-2°$ at a constant pitch attitude with airspeed changing only from about 109 knots to 113 knots.

Flying-qualities analysis development to produce this important 'back-side' characteristic is described in more detail in § 9.4.1 of this chapter.

6. Flight control surfaces (see Fig. 8)

A full-authority electronic flight control system (EFCS) interfaces with the 29 flight control surfaces on the C-17. Normally, these surfaces are electronically controlled and hydraulically actuated. A mechanical backup with hydraulic actuation is provided for critical flight control surfaces.

- The horizontal stabilizer is movable to provide pitch trim.

- Four elevators are mounted on the horizontal stabilizer to provide pitch control.

- Two articulated rudders provide directional (yaw) control. The forward half of each rudder is powered, while the rear half is moved by a mechanical linkage.

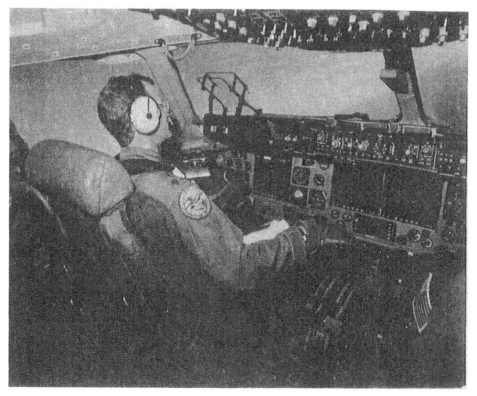

Figure 3.   Pilot using HUD and throttle for precision landing.

- One aileron and four spoilers on each wing provide roll control. Spoiler travel is from zero down, when stowed, to 60° up. The spoilers are set to 9° for steep approaches and are automatically modulated by throttle inputs to provide crisp flight path control by compensating for the thrust-to-throttle lag of the cruise-efficient high-bypass fan engine. The spoilers are set to 6° for conventional approaches. At high speeds, the outboard spoilers are held down to limit airframe stress during rapid maneuvering.

- Four leading-edge slats and two flap sections on each wing provide high-lift augmentation. Each slat in the full-span slat system is operated by two hydraulic cylinders, and each is supplied by a separate hydraulic system. They have two positions: extend and retract.

7. Electronic flight control system

The EFCS is a full-time fly-by-wire control system. It provides pilot commands to four separate flight control computers (FCCs), using stick position and force sensors in the pitch and roll axes and pedal-force sensors in the yaw axis. The FCCs condition the pilot's input signals, combine them with other sensor data indicating aircraft configuration and flight conditions, and apply servoposition commands to the actuation systems of the aerodynamic control surfaces. Elevators, rudders, ailerons, spoilers, flaps, slats, horizontal stabilizer, and the engines are all controlled by the digital fly-by-wire system.

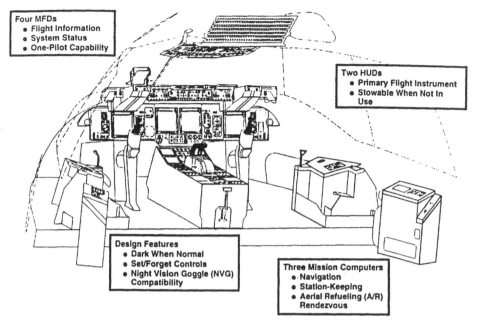

Figure 4.   Perspective view of C-17 cockpit.

The system is fail-operational in all critical functions as well as in most other functions. Fail-operational means that each axis is fully functional after any two EFCS component failures and fails in a passive manner (i.e. no adverse inputs) upon a third critical failure.

Two spoiler control electronic flap computers (SCEFCs) control the position of the spoilers, flaps and slats, and the operation of the flap/spoiler speed brake. Loss

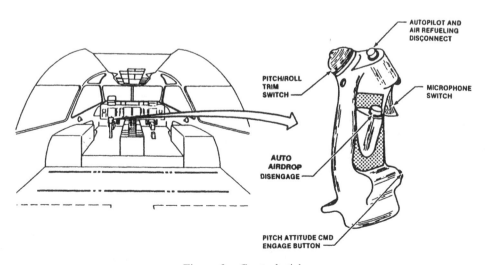

Figure 5.   Control stick.

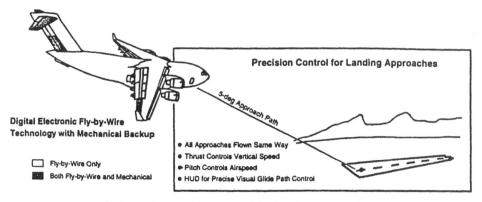

Figure 6.   Precision control for landing approaches.

of one computer results in a fail-operational transfer of control to the remaining SCEFC. Loss of the second computer results in fail-passive shutdown of capability.

A diagram of the interfaces between the EFCS and the airplane control surfaces is presented in Fig. 9.

8. Flight control augmentation system

The flight control augmentation systems comprise:

- the three-axis stability and control augmentation system (SCAS);
- the angle-of-attack limiting system (ALS); and
- the spoiler control/electronic flaps control computer (SCEFC).

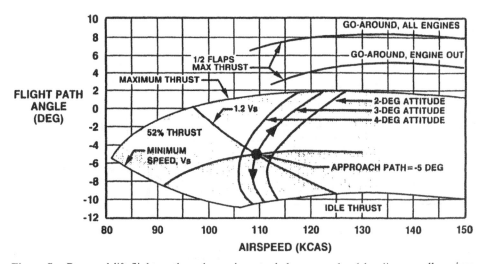

Figure 7.   Powered lift flight path and speed control: heavy payload landing on all engines with full flaps.

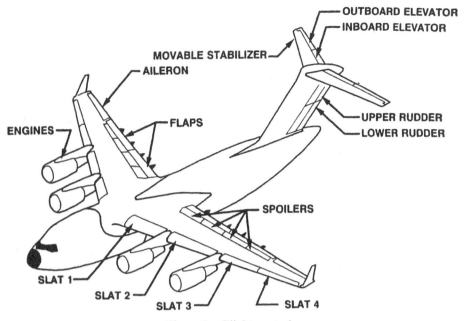

Figure 8. Flight controls.

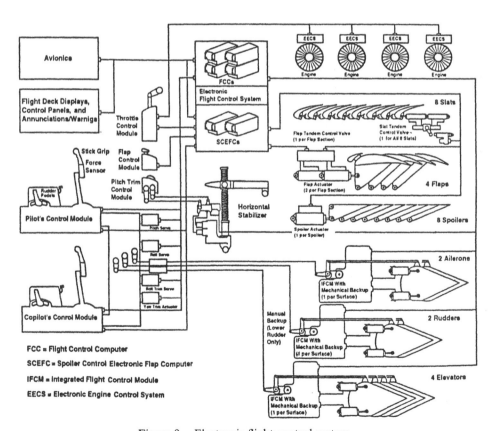

FCC = Flight Control Computer

SCEFC = Spoiler Control Electronic Flap Computer

IFCM = Integrated Flight Control Module

EECS = Electronic Engine Control System

Figure 9. Electronic flight control system.

8.1. *Pitch-axis SCAS*

A simplified linear block diagram of the pitch SCAS, which includes the ALS, is shown in Fig. 10. Counting the ALS, the four modes of pitch SCAS are as follows.

8.1.1. Takeoff mode. This provides stick position proportional control command inputs to the elevator until logic detects that lift-off has occurred and that the pilot has relaxed his or her force on the control column.

8.1.2. Pitch-rate command/attitude hold. Pitch stick commands normal acceleration and pitch rate through lag and lead/lag filters such that short-term pitch rate and long-term normal acceleration result from the pilot's application of pitch-axis control. Pitch attitude reference synchronizes with pitch attitude until pilot force is relaxed at the desired pitch attitude.

8.1.3. Pitch attitude command/attitude hold. Selection of pitch-attitude-command mode via the Pitch Attitude Command Engage button on the pilot's control grip (see Fig. 5) freezes the attitude reference and provides an 'attitude error from reference' feedback to pilot control stick position input. In this mode, pilot input changes attitude from the reference value which is restored when stick force is relaxed. The pitch attitude reference may be adjusted through use of the electronic pitch trim control available to the pilot (see Fig. 5).

8.1.4. Angle-of-attack limiting. This ALS mode activates automatically if airplane angle of attack exceeds prescribed levels. A lower-level 'soft limit' is encountered when the angle of attack is increased to a typical stall warning level. If the angle of attack is increased beyond the 'soft limit', an 'angle-of-attack error' feedback is introduced which requires significant pilot stick displacement (i.e. force against feel spring) to reach the upper level 'hard limit'. At the hard limit, ALS ignores pilot input and produces trailing-edge down elevator to prevent a deep stall maneuver.

8.2. *Roll-axis SCAS*

A simplified linear block diagram of the roll SCAS is shown in Fig. 11.
The two modes of roll SCAS are as follows.

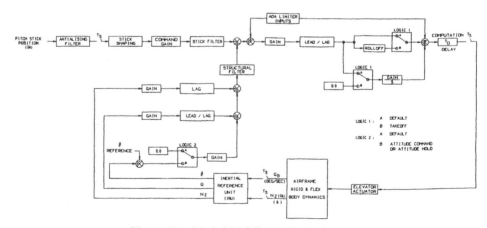

Figure 10. Pitch SCAS linear block diagram.

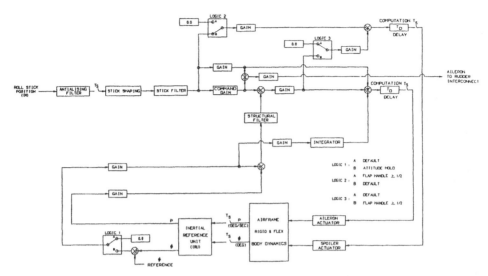

Figure 11. Roll SCAS linear block diagram.

8.2.1. Takeoff mode. This provides stick position proportional control commands to the ailerons and spoilers until logic detects that lift-off has occurred and that pilot roll force is relaxed.

8.2.2. Roll-rate command/attitude hold. Roll stick position commands roll rate. Roll attitude reference is synchronized with roll attitude until pilot force is relaxed at the desired roll attitude.

A roll reference attitude may be set through use of the reference roll trim controls available to the pilot. (See the trim switch on the pilot's grip – Fig. 5.)

8.3. Yaw-axis SCAS

The yaw-axis SCAS provides yaw damping and turn coordination. Rudder pedal-force sensors provide pilots' commands. Feedbacks are lateral acceleration, roll rate, yaw rate, and bank angle. Interconnects from aileron and spoiler command are implemented.

8.4. Spoiler control/electronic flap control (SCEFC)

The SCEFC provides the following functions:

- Roll augmentation and outboard spoiler lockout at high speed.
- Automatic ground spoilers for landing and rejected takeoff.
- Direct lift control for approach configurations.
- Speed brake control for cruise configurations.
- Throttle inputs to spoilers for approach path control augmentation.
- Anti-ballooning spoiler inputs when flaps are lowered into the engine exhaust.
- Flap and slat position commands.

9. Design methods

Design of the flight control augmentation systems followed a process in which available data and appropriate methods were used to accomplish each major step in the development program.

The principal steps were as follows.

9.1. *Proposal analysis*

Control augmentation systems similar to those developed during earlier flight tests on the Douglas YC-15 proof-of-concept airplane were analyzed to produce classical stability and control characteristics of the proposed Model D-9000 design described in the McDonnell Douglas C-X program proposal.

9.2. *Pre-CDR design*

Nonlinear off-line simulations and piloted simulator studies formed the basis for much of the design and analysis which led to a quadruplex fly-by-wire highly augmented control system design in 1987, one year before the June 1988 air vehicle critical design review (CDR). MIL-F-8785B was the flying-qualities basis referenced in the air vehicle specification (AVS). Equivalent second-order system matching was specified as the basis for defining the dynamic characteristics of the augmented system but no maximum value was specified for the flight control system time delay. Quantitative analysis requirements for pilot-induced oscillation (PIO) susceptibility were limited to showing that a 250 millisecond pilot delay would not cause instability if the pilot reverted to controlling sensed normal acceleration (phase lag between pilot's station normal acceleration and pilot's control force to be less than $180° - 14\cdot3\omega_R$, for lightly damped modes in the frequency range between 1 and 10 rad s^{-1}). LAPES, aerial refueling and low-altitude cruise flight control laws were to be analyzed in this way to assess pitch PIO margins. A typical result is shown in Fig. 12.

Since the specification allowed the contractor to determine technically how best to minimize risk of PIOs, available research on the effect of time delays on transport aircraft control tasks was reviewed in detail. This work and its connection with some early test results is described in the next section.

9.3. *Time delay research*

In preparation for the design of control laws for the precision flying required for Category 'A' maneuvers, a careful study was made of available experimental data on the effects of control system time delays on airplane flying qualities. In-flight research conducted during the 1970s provided support for the allowable delays specified in MIL-F-8785C. Other experiments conducted during the 1980s explored the effects on larger Class III airplanes flying lower-bandwith tasks. These data are summarized in Fig. 13. The somewhat inconsistent nature of the results virtually disappears when they are plotted against a 'best-fit' task bandwith on lines of constant flying-qualities level. The curves proposed by Chalk (1980) shown in Fig. 14 are extended to lower bandwidth as shown in Fig. 15. Tests performed at a 'controlled' bandwidth of $0\cdot4$ rad s^{-1} to confirm the assumed correlation of allowable time delay with task bandwidth were reported by Hodgkinson *et al.* (1991) in

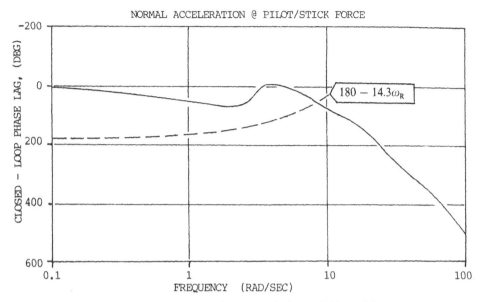

Figure 12. Closed-loop phase lag relative to $180° - 14·3\omega_R$.

AGARD-CP-513, and points for Cooper–Harper ratings of 3·5 and 6·5 are shown in the figure.

These data served to guide control law development for the C-17 Category 'A' maneuvers as shown in Fig. 16. Air refueling flight tasks conducted to support heavyweight flutter envelope expansion used an early version of software for which the equivalent low-order system time delay was about 200 ms. Flight test data analysis indicated that pilots flew this somewhat difficult task at a bandwidth of

SOURCE	MIL-F 8785C	CALSPAN SCR RESEARCH		CALSPAN TIFS	D.A.C. IRAD STUDY		LOCKHEED GA. FDL CONTRACT		NASA LANGLEY	
TYPE OF TEST	FLIGHT				M B S					
DATE	1980/ 1982	1980		1981	1986	1987	1983		1985	
AXIS	All	Long	Lat/Dir	All	Long	Lat/Dir	Long	Lat	Pitch	Roll
F.Q. LEVEL	TIME DELAY AT. F.Q. LEVEL BOUNDARY (SECONDS)									
1	0.10	0.12	0.17	0.20	0.36	--	1.10	0.40	0.5	0.3
2	0.20	0.17	0.24	0.27	0.55	--	1.60	0.60	1.0	0.7
3	0.25	0.21	0.28	0.43	--	--	2.20	0.70	1.3	0.9
CLASS	IV				III					

Figure 13. Experimental data on effect of control system time delays on airplane flying qualities.

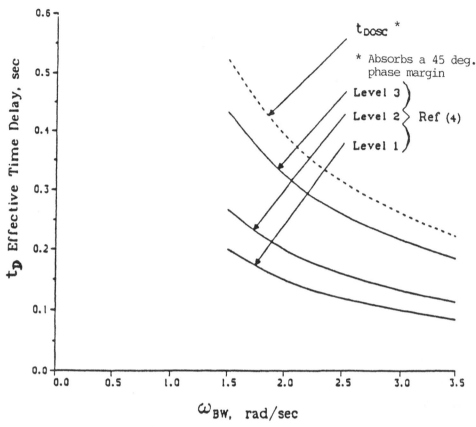

Figure 14.   Time delay vs. bandwidth.

about 2 rad s^{-1}, giving it ratings of borderline 2-to-3 FQ Level. This flight test experience agreed well with the Class III airplane research data and provided confidence that the reduced time delay in the Category 'A' maneuver controls shown in Fig. 16 was well within the Level 1 region. Earlier similar experience on the F-18 program is shown for reference. Here, the time delay was reduced to 70 ms from 140 ms to obtain Level 1 flying qualities at the higher task bandwidth. Also shown in the figure is the ability of handling-qualities during tracking (HQDT) testing to indicate the proximity of sustained oscillations at increased pilot's bandwidth that might occur due to reaction to any emergency situation. During HQDT tests on the C-17 in which pilots aggressively tracked close behind the tanker's boom which was being suddenly moved by the test boom controller, oscillations could be excited, but were easily terminated at will by the pilot. C-17 characteristics were rated similar to those for a C-141 which was flown in identical HQDT tests by the same pilots.

HQDT tests in the power/approach configuration were specified by AFFTC as a prerequisite to performing specification compliance crosswind landing tests. The intention was to minimize the risk of a pilot-induced oscillation (PIO) close to the ground where abandoning the task might not be possible.

These tests were performed by the C-17 tracking the A-37 airplane at low speed as it made sudden lateral and vertical flight path displacements. The crosswind

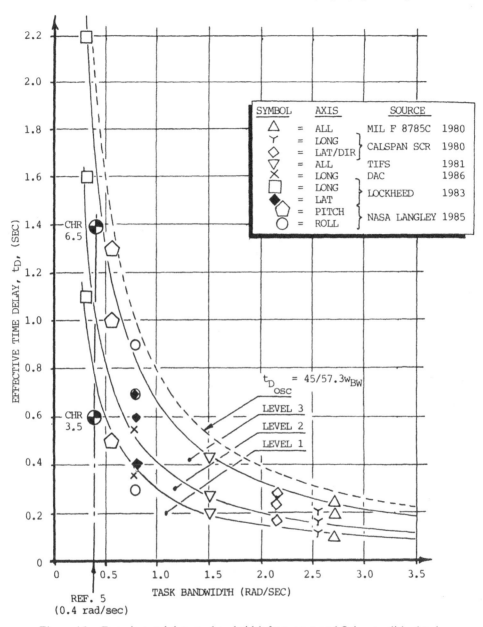

Figure 15. Experimental data vs. bandwidth frequency and flying-qualities level.

landing envelope was eventually expanded to 30 knots without incident and with no related control law change.

9.4. *Piloted simulation*

The specification criteria supplemented with the time delay research for transport-sized aircraft guided the formulation of the stability and control augmentation system control laws to be further developed by piloted simulation. The

201

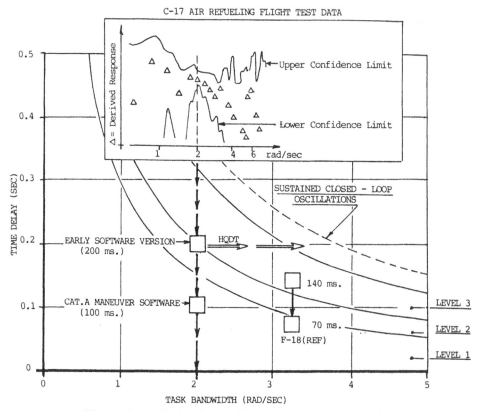

Figure 16. C-17 air refueling flight test time delay correlation.

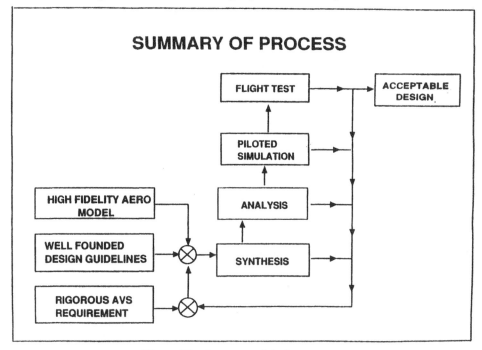

Figure 17. Summary of design process.

process then followed was as shown in Fig. 17. Comments from pilot evaluations and control problems identified by the simulation studies were addressed using some well-founded synthesis methods and design guidelines to supplement the basic AVS design requirements. An example cited below describes the use of Stapleford's recommendations for good 'back-side' approach characteristics.

9.4.1. *Stapleford/pitch attitude boundaries*. During piloted simulation of the steep approach to an SAAF landing pronounced airspeed coupling to throttle movement in some flight conditions gave rise to adverse pilot comments. In resolving this issue, it was found that the work by Stapleford *et al.* published in 1974 gave valuable guidance. Figure 18 shows the type of criteria boundaries used where $dV/d\gamma$ limits suggested by the STOL research are combined with orthogonally intercepting pitch attitude limits specific to the C-17's geometry and flight characteristics. At the upper extreme of approach pitch attitudes a 'pilot's pad' is provided between the highest acceptable value of 5·5° and the angle at which essential symbology begins to exceed a head-up-display (HUD) limitation. At the lower extreme, the intent is to avoid any potential requirement for the pilot suddenly to pull back on the stick to prevent a nose-wheel-first touchdown. The smaller rectangle limited by the Stapleford Level 1 $dV/d\gamma$ limit = 1·25 knots/deg and having approach pitch attitudes between 2·5° and 4·5° was considered as one where pilot acceptance of the required back-side approach characteristics could be virtually guaranteed.

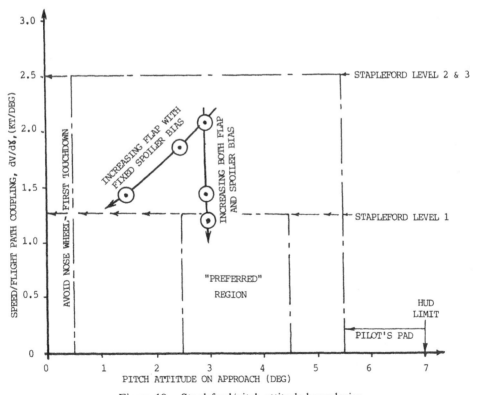

Figure 18. Stapleford/pitch attitude boundaries.

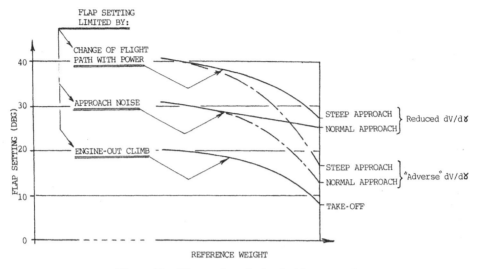

Figure 19.   Flap settings for back-side approach.

Then combinations of increased flap settings and up-spoiler bias settings were explored which moved ($dV/d\gamma$, θ) pairs away from the region, where adverse comments had been received, towards the 'preferred' region. Rejected approach performance requirements and approach noise requirements were maintained by the configuration choices selected.

Typical effects of flap setting, spoiler bias and approach speed on the migration of ($dV/d\gamma$, θ) pairs across the Stapleford/pitch attitude chart are shown in the figure. The resulting increase in approach flap settings selected for the airplane are shown in Fig. 19. These provided acceptable back-side characteristics and resulted in a

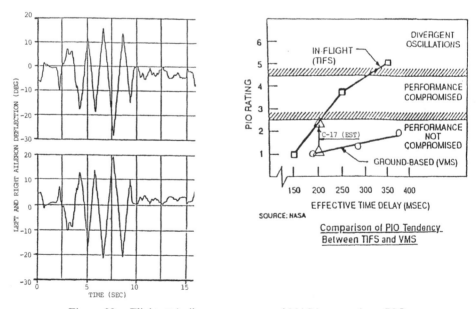

Figure 20.   Flight #4 aileron response and NASA research on PIO.

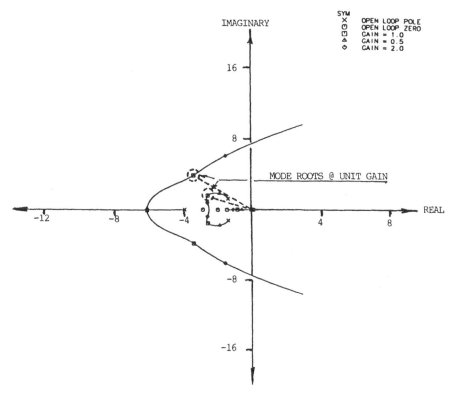

Figure 21. Open-loop elevator root locus for early pitch SCAS (cruise configuration, pitch-rate command).

significant beneficial field performance side-effect of the reduced approach speeds at the higher airplane gross weights. Increased spoiler bias selections were limited by consideration of stability margins of existing autothrottle and autopilot designs.

9.5. *Flight test*

As is usual flight test development revealed a number of flight control system issues. The extent and type of analysis performed to resolve such issues is probably best exemplified by the considerable effort made to reduce susceptibility to PIOs, mild forms of which had occurred in roll on approach and in pitch during aerial refueling with the earlier software versions.

9.5.1. *Control gains.*
Pilots unanimously complained of an oversensitive roll control axis throughout the early flights. One pilot started calling the airplane the 'F-17!' On Flight #4 of T-1, a mild PIO in the roll axis was experienced during approach. Aileron displacements are shown in Fig. 20. No such experience had been experienced on the simulator and, as indicated in the right-hand part of Fig. 20, the experience seems to confirm the results of NASA research. In identical tests on the Vertical Motion Simulator (VMS) and on the Total In-Flight Simulator (TIFS) pilots' PIO ratings for a given time delay were found to be significantly worse in

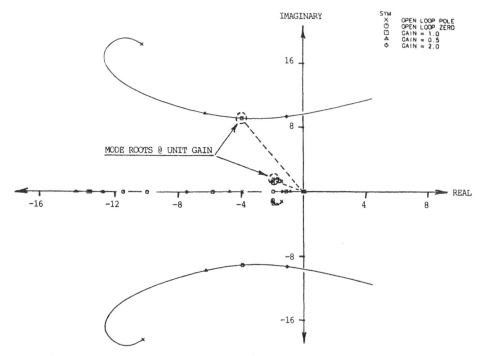

Figure 22. Open-loop elevator root locus for improved pitch SCAS (cruise configuration, pitch-rate command).

flight. It was noted from flight test responses that the roll angles did not exceed about 2° and that the frequency was about 3·5 rad s^{-1}. This is remarkably close to the frequency for sustained closed-loop oscillations that would be indicated by Fig. 16 for time delays of about 200 ms. Pilots continued to complain of roll sensitivity until a revised software version was implemented with roll gains reduced by more than a factor of two. Here we see confirmation also of the 'rule of thumb' in MIL-STD-1797, Appendix 'A', which is to use no more than half the optimum sensitivity determined in a ground-based simulator (AGARD-CP-333). This rule had not been followed. We had not learned our history and so were destined to relive it.

9.5.2. *Early air refueling.* An interesting example is one where it was found from a root-locus analysis (see Fig. 21) of an early software version in flight conditions appropriate to those for air refueling that open-loop elevator-mode roots were in close proximity at nominal system gain settings (roots corresponding to a gain of 1·0 have been highlighted on the figure). This was considered to be a principal cause of poor low-order system fits to the high-order system, which had resulted in high residuals (payoffs between 30 and 70) and high equivalent system time delays (208 to 229 ms). Separation of these roots by root-locus synthesis (see Fig. 22) led to much improved HOS/LOS fits (payoff between 4 and 8) and much reduced equivalent system time delays (80 to 124 ms). The control anticipation parameter (CAP) also showed significant improvement with ($\omega_n^2/n_z/\alpha$) values increasing by about 30% from below the Category 'A' Level 1 boundary to a location closer to it. As noted from Figs 23 and 24, Gibson plots indicated significant improvement.

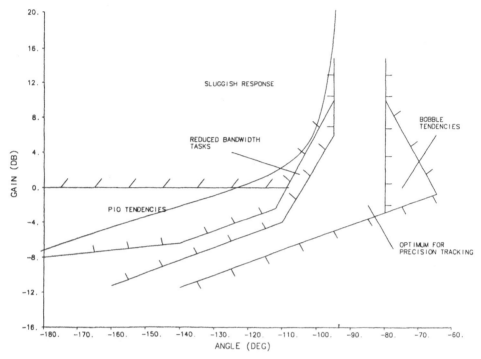

CRUISE CONFIGURATION, PITCH RATE COMMAND
GIBSON ATTITUDE BOUNDARIES FOR 0.3 HZ CROSSOVER FREQUENCY
CATEGORY A

K• THETA /PSTICK
K TO GIVE 0.3 HZ CROSSOVER FREQUENCY

Figure 23.   Early pitch SCAS Gibson plot.

The plot for a 0·3 Hz crossover frequency is much further from the '− 1,0*j*' point. Also, it occupies more of the optimum 'corridor' for precision tracking at low frequency and much less of the 'PIO tendencies' region at higher frequencies.

9.5.3. *LAPES.* The low-altitude parachute extraction (LAPES) maneuver in which heavy payloads slide along the cargo deck before exiting the rear cargo door presented no significant flying-qualities issues. Occurring late in the flight test program, it had the benefit of the Stapleford analysis for back-side steep-approach characteristics and of all the lessons learned from PIO criteria trends during earlier software development. Also, since LAPES was flown manually without autopilot or autothrottle, AFCS stability considerations did not restrict the flying-qualities engineers from selecting the flap/spoiler bias combination indicated by the Stapleford criterion. To provide a more positive round-out at a $4\frac{1}{2}$ foot (1·4 m) wheel height AGL the steep-approach spoiler bias was increased from 9° (for SAAF landings) to 12° (for LAPES). Given the good back-side (low dV/dγ) characteristics the rate of descent was arrested at constant pitch attitude by a throttle push at about 30 feet (9 m) AGL to bring the flight path vector on the HUD up to the level-flight indication. Spoiler bias was reduced automatically as throttles were pushed up giving good positive control of rate of sink reduction.

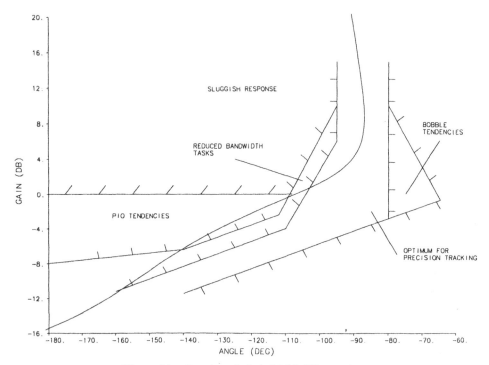

Figure 24.   Developed pitch SCAS Gibson plot.

Typical load extractions and cg shifts are compared with those of the C-130 in Fig. 25. When simulating the drop of some 60 000 lbs in the Calspan Total In-Flight Simulator (TIFS), pilots were able to release the pitch control stick and allow the pitch-axis SCAS to control pitch attitude as the load was extracted. This usually demanding task previously performed on other aircraft only by expert pilots is easy to learn on the C-17, and is released for AMC operations after training by any of the line pilots as they convert to their new military transport aircraft.

10. Conclusions

The design and development of flying qualities on the C-17A military transport airplane has been described with some examples of work performed in the analysis of issues found during piloted simulation and flight test.

Focused efforts and stringent reviews by all program participants have resulted in a high level of flying qualities being developed for the C-17 airplane. It seems fitting to conclude with a picture (Fig. 26) of the airplane performing a LAPES drop, a maneuver which challenges the flying-qualities engineer to deliver an easy-to-perform task to the average line pilot.

Principal lessons learned were:

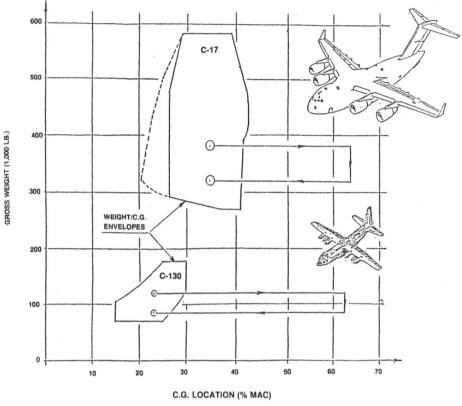

Figure 25. LAPES maneuver cg travel.

- More research is needed to develop reliable design criteria for the avoidance of PIO.

- Equivalent system time delay of approximately 100 ms should be adhered to in the design of control systems of large Class III airplanes that are to be flown precisely with relative pilot ease.

- Back-side characteristic requirements as recommended by Stapleford should be considered in the early stages of high-lift configuration choices for STOL airplanes.

ACKNOWLEDGEMENTS

Acknowledgement is given to all at McDonnell Douglas in Long Beach and St Louis, the Air Force Systems Program Office in Dayton, Ohio, the Air Force Flight Test Center at Edwards Air Force Base and the Air Force Air Mobility Command Initial Squadron at Charleston Air Force Base who have during particular program phases participated in and supported the almost 15 year task of designing for and developing excellent flying qualities on McDonnell Douglas's first fly-by-wire tactical transport aircraft, the Globemaster III. Particular thanks are due to Mr 'Obi' Iloputaife who provided the root-locus and Gibson plot data and to Ms Oke Sweet for her patience in typing and editing the several versions of the original text.

Figure 26.   C-17 performing LAPES.

REFERENCES

CHALK, C. R., 1980, Calspan recommendations for SCR flying qualities design criteria. Calspan Report No. 6241-F-5, January (for NASA).

HODGKINSON, J., ROSSITTO, K. F. and KENDALL, E. R., 1991, The use and effectiveness of piloted simulation in transport aircraft research and development. Douglas Paper MDC 91K0049, presented to the 79th AGARD Flight Mechanics Panel Symposium on Simulation, Brussels, Belgium, October 7, 1991 (contained in AGARD-CP-513).

McDONNELL DOUGLAS CORPORATION, 1981, *C-X Program – Proposal 80D-290-Volume 2 – Technical*, Chapter 7: Flying qualities, 15 January.

McDONNELL DOUGLAS CORPORATION, 1993, C-17 Globemaster III – technical description and planning guide. Report No. MDC 93K0048, July.

STAPLEFORD, ROBERT L., *et al.*, 1974, A STOL airworthiness investigation using a simulation of deflected slipstream transport. Vol. 1: Summary of results and airworthiness implications. NASA TM X-62392, FAA-RD-74-143-I, October.

8

Fly-by-wire for commercial aircraft: the Airbus experience

C. FAVRE

1. Introduction

European experience in fly-by-wire application is now some 20 years old. With the entry into service of the A320 (see Fig. 1), a new standard of fly-by-wire was defined in the flight control law and system integration areas.

This experience was directly used throughout the A321 and A330/A340 programmes (see Fig. 2), such that effort could be focused on system integration and methodology improvements.

Figure 1. A320 aircraft.

Figure 2.   A340 aircraft.

The objective of this paper is to provide an overview of the flight control law characteristics and constraints, when integrated in a fly-by-wire system for commercial aircraft, emphasizing the lessons learned during these two major programmes.

2.   Control law design

2.1.   *Objectives*

The general objective of the flight control laws integrated in a fly-by-wire system is to improve the natural flying qualities of the aircraft, in particular in the fields of stability, control and flight domain protections.

In a fly-by-wire system (see Fig. 3), the computers can easily process the

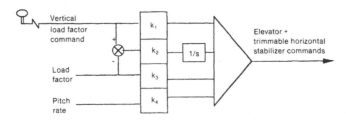

Figure 3.   Control law general structure.

anemometric and inertial information as well as any information describing the aircraft state. Consequently, control laws corresponding to simple control objectives could be designed.

The stick inputs are transformed by the computers into pilot control objectives which are compared with the aircraft actual state measured by the inertial and anemometric sensors.

Thus, as far as longitudinal control is concerned, the sidestick position is translated into vertical load factor demands, while lateral control is achieved through roll rate (p_1), sideslip (β), and bank angle (ϕ_1) objectives.

This structure allowed us to use the principles of active control where the control surface commands are obviously functions of the pilot's commands.

As one control surface is theoretically sufficient to control as many modes as independent measurements are available, increased aircraft stability characteristics can be obtained, while degrees of freedom remain available to optimize other chosen criteria.

For instance, in the case of lateral control, four independent variables describing the aircraft state are available:

> sideslip (β) or lateral load factor (n_y)
>
> roll rate (p_1)
>
> yaw rate (r_1)
>
> bank angle (ϕ_1)

As two independent commands are available (roll$-\Delta_p$ and yaw$-\Delta_r$) the four rigid-body modes (the generally low-damped dutch roll mode, comprising two conjugated complex modes, the roll mode, rapidly convergent real mode, and the spiral mode, slowly convergent or divergent real mode) can be controlled.

This stability augmentation improves the aircraft flying qualities and contributes to the aircraft safety: as a matter of fact, the aircraft remains stable in the case of pertubations such as gusts or engine failure due to a very strong spiral stability, unlike conventional aircraft. Aircraft control through these objectives significantly reduces the crew workload: the fly-by-wire system acts as the inner loop of an auto-pilot system, while the pilot represents the outer loop in charge of objective management.

Finally, protections forbidding potentially dangerous excursions out of the normal flight domain can be integrated in the system. The main advantage of such protections is to allow the pilot to react rapidly without hesitation, since s/he knows that her/his action will not result in a critical situation.

2.2. *Constraints*

Although it is theoretically possible to modify strongly the natural flying qualities of an aircraft, different types of constraints limit the effective degrees of freedom to design flight control laws.

The first limitation concerns the inputs of the control laws, i.e. the sensors. In fact, the movement of the aircraft can be modelled by a differential equation of the form:

$$(\mathrm{d}x/\mathrm{d}t) = Ax + Bu$$

where x is the state vector: $(\beta, p_1, r_1, \phi_1)^t$ for lateral control; u is the control vector: $(\Delta_p, \Delta_r)^t$ for lateral control; A is the matrix representing the aircraft dynamics; and B is the matrix representing the control surface contribution.

A and B are functions of the aircraft state (speed, high lift configuration, altitude, weight, inertia, flexibility...) and can be estimated through sensors and computers (as well as the state vector) to design a control law ($u = f(x$, pilot commands)).

The sensor failures must be considered. In particular, sufficient redundancy must be available to eliminate possible faulty sources and degrade the control law status if necessary. The difference between the aircraft behaviour with the so-called normal laws and the aircraft behaviour with the degraded laws, closer to the 'natural' aircraft behaviour, must be limited to avoid drastic pilot control changes in the case of failure.

The physical limitations of the control surfaces and associated servo-controls (maximum amplitudes and rates, fatigue) must also be considered: it is intuitive that the more the closed loop behaviour is different from the natural aircraft behaviour, the more the servo-controls are solicited.

Other constraints are a result of the control law interaction with the aircraft structure, such as passenger comfort, flutter and loadings. For example, a dutch roll pulsation which is strongly increased compared with the natural aircraft, may augment the yaw effect of a lateral gust and, in turn, modify passenger comfort.

Finally, human factors must be considered in the control law design. The general aircraft behaviour must be in accordance with the usual visualizations and sensations expected by the pilots whether in 'normal' law or degraded laws. The role of ground-based simulators and flight tests with close feedback from the pilots is fundamental and is described in § 4.

2.3. *Control law structure*

2.3.1. *Longitudinal law.* The normal longitudinal law controls the vertical load factor (N_z) and is derived from the so-called C^* law (NASA study for the Space Shuttle). This law ensures the short term platform stability, incorporates the load factor limitation, includes the auto-trim function and allows steady turns with sidesticks at neutral up to 33° bank angle.

The longitudinal control law structure is described in Fig. 4. The pitch attitude rate and load factor feedbacks participate in the short period mode control, while precision is achieved through the integrated term. The controlled modes are assigned to be close to the natural aircraft modes to minimize the control surface activity and to minimize the difference in aircraft behaviour

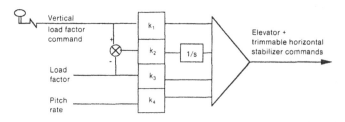

Figure 4.   Longitudinal control law structure.

between the normal law and the degraded (direct) law. As a consequence of this control law, the static stability of the aircraft is almost neutral, the phugoïd mode being transformed into a highly damped mode and an almost zero mode. Static stability is restored at the limits of the normal flight domain when the following protections are activated:

(i) angle of attack protection at low speeds;

(ii) high-speed/high-Mach protections;

(iii) pitch attitude protection complementing the two above protections to minimize aircraft accelerations and decelerations.

A homogeneous law, ensuring aircraft behaviour independent of the flight conditions and, in particular, independent of the centre of gravity location, is achieved by tabulating the gains as a function of the computed air speed, high-lift configuration and centre of gravity location.

2.3.2. *Lateral law* (Farineau and Letron 1990). The stability objectives were to increase the dutch-roll damping coefficient beyond 0·6 (the natural damping coefficient being lower than this value in some flight conditions) without significantly affecting the pulsation, to keep the roll mode unchanged and to increase the spiral stability. Again, the controlled modes are assigned to be close to the natural aircraft modes to minimize the control surface activity and to minimize the difference in aircraft behaviour between the normal law and the degraded (direct) law.

The remaining degrees of freedom are used to decouple the roll angle and the sideslip so that lateral gusts induce low roll responses and, conversely, so that dutch roll has an imperceptible influence on the roll angle.

Once the stability is specified, pilot control objectives must be defined. It was decided that side-stick inputs would be translated into a roll rate demand (like a mechanically controlled aircraft) but at constant zero sideslip to provide automatic turn co-ordination and, therefore, reduce pilot workload. The spiral stability is perfectly zero as far as pilot control is concerned (the bank angle is kept constant with side-sticks at neutral) and very high in the case of perturbation. Finally, it was decided that the rudder pedals would command a combination of sideslip and roll angle to restore some of the conventional aircraft behaviour.

The main constraints relative to the lateral law design were the engine failure case at low speed and decrab with strong lateral winds. As the side-stick position is not representative of the control surface position it was necessary to help the pilot in case of engine failure. In fact, the performance objective in case of engine failure at take-off is to keep the roll control surfaces at zero to minimize drag. Consequently, a sideslip objective (the so-called 'β target') is visualized on the primary flight display (PFD) and allows the pilot to achieve his/her performance objective through an instinctive rudder pedal action.

Static spiral stability is restored when the bank angle reaches 33° bank angle: if bank angles above this limit are reached, the aircraft automatically comes back to a 33° bank angle with the side-stick at neutral. The maximum achievable bank angle is limited to 66°, to be consistent with the 2·5 g vertical load factor limit of the longitudinal law in steady stabilized turns.

The lateral control law structure is described in Fig. 5.

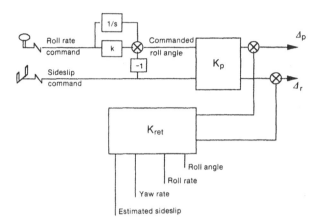

Figure 5.   Lateral control law general structure.

The stability objectives of the control law are achieved through the estimated sideslip, roll rate, yaw rate and roll angle feedbacks. The sideslip estimation is founded on the lateral acceleration measurement by solving the lateral force equation. The gain matrix K_{ret} (dimension 2×4) is entirely and uniquely deduced from the stability and roll angle/sideslip decoupling objectives. This matrix depends on flight conditions (computed airspeed and high lift configuration).

As far as control objectives are concerned, the roll rate command is integrated and converted into a roll angle command. The direct gain k is designed to accelerate the aircraft response. The rudder pedals provide the sideslip and roll angle commands. The pre-command matrix K_p is designed effectively to achieve the commanded sideslip and roll angles in steady-state without perturbation.

3.   System integration

3.1. *General architecture*

The term fly-by-wire has been adopted to describe the use of electrical rather than mechanical signalling of the pilot's commands to the flying control actuators. One can imagine a basic form of fly-by-wire in which an aeroplane retained the conventional pilot's control columns and wheels, hydraulic actuators (electrically controlled) and artificial feel, as experienced in the 1970s with the Concorde programme; and the fly-by-wire system would simply provide electrical signals to the control actuators that were directly proportional to the angular displacement of the pilot's controls, without any form of enhancement.

In fact, the design of the A320, A321, A330 and A340 flight control systems takes advantage of the potential of fly-by-wire to incorporate control laws that provide extensive stability augmentation and flight envelope limiting. The positioning of the control surfaces is no longer a simple reflection of the pilot's control inputs and, conversely, the natural aerodynamic characteristics of the aircraft are not fed back directly to the pilot.

The side-sticks, now part of a modern cockpit design that includes a large visual access to instrument panels, can be considered as the natural result of fly-by-wire, since mechanical transmissions with their pulleys, cables and linkages can be suppressed, along with their associated backlash and friction.

The induced roll characteristics of the rudder provide sufficient roll manoeuvrability to design a mechanical back-up on the rudder alone for lateral control. This permitted us to retain the advantages of the side-stick design and to be rid of the efforts required to drive mechanical linkages to the roll surfaces.

Looking for minimum drag leads us to minimize the negative lift of the horizontal tail plane and, consequently, diminishes the aircraft longitudinal stability. For the Airbus family, it was estimated that no significant gain could be expected with rear c.g. positions beyond a certain limit; this allowed us to design a system with a mechanical back-up requiring no additional artificial stabilization.

These choices were obviously fundamental to establish the now classical architecture of the Airbus fly-by-wire systems (see Fig. 6): namely, a set of five full-authority digital computers controlling the three pitch, yaw and roll axes, which are completed by a mechanical back-up on the trimmable horizontal stabilizer and on the rudder (two additional computers, part of the auto-pilot system, are in charge of rudder control in the case of the A320 and A321 aircraft).

3.2. *Computer arrangement*

The five computers are simultaneously active. They are in charge of control law computation as a function of the pilot inputs as well as individual actuator control, thus avoiding specific actuator control electronics. The system incorporates sufficient redundancies to provide nominal performance and safety levels with one failed computer, and it is still possible to fly the aircraft safely with only one active computer.

As a control surface runaway may affect the aircraft safety (elevators in particular) each computer is divided into two physically separated channels; the first one, the control channel, is permanently monitored by the second one, the monitor channel. In the case of disagreement between control and monitor, the

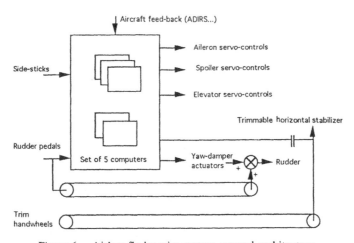

Figure 6.   Airbus fly-by-wire system general architecture.

computer affected by the failure is passivated, while the computer with the next highest priority takes control. The computer repartition and priorities are dictated by the safety analysis.

Despite the non-recurring costs induced by dissimilarity, it is fundamental that the five computers be of a different nature to avoid common mode failures, which could lead to the total loss of the electrical flight control system. Consequently, two types of computers may be distinguished:

two elevator and aileron computers (ELACs) and three spoiler and elevator computers (SECs) on the A320/A321, and

three flight control primary computers (FCPCs) and two flight control secondary computers (FCSCs) on the A330/A340.

3.3. *Control law reconfiguration*

The control laws implemented in the flight control system computers have full authority and must be elaborated as a function of consolidated information provided by at least two independent sources in agreement.

Consequently, the availability of control laws using aircraft feedback (the so-called normal laws) is closely related to the availability of the sensors. The Airbus aircraft fly-by-wire systems use the informations of three 'air data and inertial reference units' (ADIRUs) as well as specific accelerometers and rate gyros. Moreover, in the case of the longitudinal normal law, analytical redundancy is used to validate the pitch rate information when provided by a single IRS: the load factor is estimated through the pitch rate information and compared with the available accelerometric measurements to validate the IRS data.

After double or triple failures, when it becomes impossible to compare the data of independent sources, the normal control laws are reconfigured into laws of the direct type, where the control surface deflection is proportional to the stick input.

To enhance the dissimilarity, the more sophisticated control laws with aircraft feedback (the normal laws) are integrated in one type of computer, while the other type of computer incorporates the direct laws only.

4. The A320 experience

The entry into service of the A320 (see Fig. 1) has provided a major milestone in fly-by-wire application. The experience gained during this program mainly concerns:

(*a*) the general structure of control laws and system reconfigurations adapted to transport aircraft, as described in the previous sections;

(*b*) the design, development and validation procedure;

(*c*) new certification methods adapted to fly-by-wire.

4.1. *Experience accumulated before the A320*

Previous experience accumulated by Aerospatiale in electrical flight control systems and the use of simulation, both ground-based and in flight, has largely contributed to the success of the development process.

(*a*) Since 1969, Concorde has been flying with a three-axis full-authority analogue flight control system with a mechanical back-up on each surface.

(*b*) In 1978, the first production Concorde airplane, belonging to Aerospatiale, was fitted with a side-stick by the left-hand seat and several flight hours with different crews were achieved. The (already) electrical flight control system of Concorde proved to be easy to adapt to this type of control and an elaborated pitch control law (*C**) was even introduced. This flight test campaign confirmed the very good acceptability of 'flying through a side-stick' in all flight phases as well as the advantages of the *C** type control law for precise flight path control, autotrim and behaviour in turbulence.

(*c*) In 1981, the concept of a 'forward facing crew cockpit' was introduced in the Airbus A300 B4 programme: at the same time a fully digital dual/dual autopilot system (two computers, each with a command and a monitor channel) was introduced.

(*d*) Digital computers were introduced on Airbus A310 and A300–600 to control spoilers, flaps and slats.

(*e*) The A300B S/N 3 was used during two flight test campaigns, in 1983 and 1985, to validate some of the concepts selected for the A320: confirm the side-stick characteristics, evaluate the priority logic software in lieu of mechanical coupling between the side-sticks, test the angle of attack protection control law, test a specific pitch control law adapted for flare in the landing phase and test a lateral control law. For that purpose, a specifically modified digital autopilot computer was fitted on the airplane, together with left and right side-sticks (at least for the 1985 test campaign). The flight control laws were adapted to the A300 characteristics with the same rules and methods that would be used to tune the A320 flight control laws. This experiment gave an opportunity to more than 50 pilots from various origins to test and approve the main principles of future A320 flight control laws and side-stick control. The main causes of potential pilot induced oscillations (PIO) close to the ground were also analysed during this experiment. At the same time, parallel work on a fixed-base simulator was performed in order to derive a global PIO methodology. Once the basis of such a methodology has been validated by actual flight testing on this A300 airplane, it can be applied to any other type of airplane through a fixed-base simulator.

(*f*) In the case of pre-go-ahead A320 activities, some Aeroformation (the Airbus Training centre) and CEV (Centre d'Essais en Vol, the French flight test centre) full-flight (moving cabin) simulator hours were used to assess the possibility to fly and land an aircraft with a completely failed electrical flight control system relying only on the selected mechanical back-up architecture. These tests showed encouraging results leading to the selection of an architecture including mechanical back-up for rudder and horizontal tail plane controls.

(*g*) In addition, the increasingly complex systems of Concorde, Airbus A300 and later A310 and A300–600 were an opportunity for Aerospatiale to design, build and operate efficiently ground-based simulators connected to very comprehensive test benches, including most of actual system equipment (the so-called 'Iron bird').

4.2. *Design, development and validation procedure* (Chatrenet 1991)

Simulation codes, full-scale simulators and flight tests were extensively used in a complementary way to design, develop and validate the A320 flight control laws in particular, and the A320 fly-by-wire system in general.

4.2.1. *Development tools.*

A 'batch'-type simulation code called OSMA (Outil de Simulation des Mouvements Avion) was used initially to design the flight control laws and protections, including the nonlinear domains, and also for general handling quality studies. This type of simulation tool is in general use in handling quality departments. OSMA was developed with the objective of standardizing the models (aerodynamics, engines, ground roll, etc) and the associated simulation software used in the various simulation facilities. As a guarantee of high fidelity, the models used and supplied to the training simulators within the data packages were selected.

A development simulator was then used to test the control laws with a 'pilot in the loop' as soon as possible in the development process. This simulator is fitted with a fixed-base faithful replica of the A320 cockpit and controls, and a visual system; it was put into service as early as 1984, as soon as a set of provisional A320 aero data, based on wind tunnel tests, was made available. The development simulator was used to develop and initially to tune all flight control laws in a closed-loop co-operation process with flight test pilots.

Three 'Integration' simulators were put into service in 1986. They include the fixed replica of the A320 cockpit, a visual system for two of them, and actual aircraft equipment including computers, displays, control panels, warning and maintenance equipment. One simulator can be coupled to the 'Iron bird' which is a full-scale replica of the hydraulic and electrical supplies and generation, and is fitted with all the actual flight control system components including servo-jacks. The main purpose of these simulators is to test the operation, integration and compatibility of all the elements of the system in an environment closely akin to that of an actual aircraft.

Finally, flight testing remains the ultimate and indispensable way of validating a flight control system. With the current state of the art in simulation, simulators cannot yet fully take the place of flight testing for handling quality assessment. On this occasion a specific system called Système Pour Acquisition et Traitement d'Informations Analogiques ARINC et Logiques (SPATIAAL) was developed to facilitate the flight test campaign. This system allows the flight engineer to:

(i) record any computer internal parameter;
(ii) select several pre-programmed configurations to be tested (gains, limits, thresholds, etc);
(iii) inject calibrated solicitations on the controls, control surfaces or any intermediate point.

The integration phase complemented by flight testing can be considered as the final step of the validation side of the now classical V-shaped development/validation process of the system.

4.2.2. *Interest of ground-based simulators.* The quality and associated safety and reliability of operation of a critical system like a digital flight control system mainly relies on a two-step process:

(1) quality of the specification;
(2) quality of equipment and complete consistency between specification and equipment (software in particular).

The last step is guaranteed through the use of the very stringent rules associated with level 1 software which ensure that the software embodied in the flight control system computers is strictly consistent with its specification.

The first step is still somewhat less formalized and more difficult to assess fully: how can it be guaranteed that the specification on which software is based fulfils all the performance objectives and offers adequate functioning in every forseeable configuration of the environment of the system? In this area, simulation constitutes an invaluable tool for analysis or checking of huge numbers of potential cases, or combinations of cases, which are obviously out of the scope of flight testing: for example, parameters like weight, centre of gravity location, altitude, speed (inside and outside the normal flight envelope) aircraft configuration, wind, turbulence (including windshear) have been systematically covered by simulation at every major step of the flight control system design. We may even say that, owing to the considerable number of inputs to the system (several hundreds), checking all the combinations of these inputs, if they were considered to be independent, would be practically impossible. In this respect, a good simulator providing faithful simulation of all these inputs to the system as well as overall aircraft behaviour, allows for a significant reduction of the number of potential cases to be analysed: all inputs are no longer fully independent parameters, and combinations which are not possible are automatically eliminated.

Even if the nominal functioning and operation of a commercial airliner already provides for a wide scope of various environmental conditions, abnormal operation is still more complex. The A320 simulators have been extensively used to develop and check all the logic embodied in the flight control system specification. Areas of particular interest in this respect include:

(i) runaway of inputs from other systems (ADCs—air data computers, IRSs—inertial reference systems, etc.),
(ii) oscillatory failures,
(iii) runaway of servo-controls,
(iv) hydraulic failures,
(v) mechanical failures (jamming, disconnection),
(vi) electrical supply transients,
(vii) effect of lightning induced disturbances,
(viii) EMC compatibility.

A thorough assessment of system behaviour in the case of abnormal conditions is clearly impossible by pure analysis or by flight testing.

On the other hand, A320 simulators have certainly made flight testing both safer and more productive. Here are some examples which illustrate how simulators have proved valuable aids to flight testing.

(a) Flight crew training before first flight.

(b) Reduction of the scale of test programmes: prior selection on a simulator allows the most significant scenarios to be tested.

(c) Aircrew familiarization in the case of tricky test events.

(d) Systematic test on the integration simulators of any new version of software; this test was mandatory to allow any new version of the flight control computer to be fitted on a development aircraft for flight testing.

(e) Debugging in the case of unexpected failure during flight testing: by playing back the conditions of the incident, varying the suspected parameters or locally increasing the scale of sensitivity of the instrumentation, a detailed set of facts can rapidly be built up to trace and correct the anomaly with a minimum delay in the flight test programme.

(f) Use of the simulator in place of flight testing for very severe or critical failures; some very severe failures have been thoroughly tested on the simulators, in addition to limited flight testing. This is the case when the probability of such a failure or combination of failures produces hazardous consequences according to the certification requirements; this was also the case whenever the aircraft, voluntarily forced into this failure state, became particularly vulnerable to an additional unexpected failure.

4.3. *Certification procedure*

This section discusses some of the major (Denning and Whittle 1988) certification aspects of fly-by-wire civil aircraft in the field of handling qualities. In fact, it is the control laws with extensive stability augmentation and flight envelope limiting, rather than the use of fly-by-wire *per se*, that has demanded a review of airworthiness requirements in this field.

4.3.1. *Normal load factor limiting.* The structural requirements define a manoeuvre envelope of $2.5\,g$ to $-1.0\,g$ that must be designed for in the clean configuration. There were no handling requirements in JAR/FAR 25 (European/US regulations) that demanded the ability to manoeuvre in the range of such values. Moreover, as a protection against overstressing the aircraft, JAR 25 specifies a stick force of not less than 50 lb to reach the structural limit strength in the *en route* configuration. These requirements had to be adapted to cope with the presence of automatic limiting associated with side-stick control.

On the basis of the argument that the rate of application of 'g' was as significant as the minimum 'g' obtained, that $2.5\,g$ applied rapidly was as good as $3\,g$ or $3.5\,g$ applied slowly, and that the presence of automatic limiting would give the pilot the confidence to manoeuvre rapidly, it was concluded that automatic load factor limiting did not conflict with existing airworthiness requirements. A special condition was raised that required that any such limiting should not be more restrictive than the manoeuvre envelopes of the structural requirements.

4.3.2. *Abnormal flight conditions*. Thanks to the flight envelope limiting functions (load factor, attitude, angle of attack, etc.) it is not possible for the aircraft to perform outside the defined envelopes by pilot inputs or as a result of any reasonable atmospheric disturbance. However, the authorities remained concerned that should the aircraft get beyond these envelopes, despite these protections, the recovery should not be affected by the automatic functions.

A special condition was raised on this topic and the outcome was the introduction of abnormal attitude laws, which adapt the automatic control functions, should the aircraft get significantly beyond the limit values of the normally protected parameters.

4.3.3. *Overspeed protection*. The minimum V_{MO}-to-V_D margin (typically 50 knots) was dictated by the upset manoeuvre specified in JAR/25.335, which called for the aircraft to be dived from V_C at $-7.5°$ flight path angle for 20 s without exceeding V_D. The A320 incorporates overspeed protection control functions that progressively reduce the pilot's nose-down authority as V_{MO} or M_{MO} is exceeded. (M_{MO} is the maximum operating Mach number, V_{MO} is the maximum operating limit airspeed, V_D is the design dive speed, and V_C is the design cruising speed.)

The concern was that the overspeed law might be unduly tailored to the arbitrary upset manoeuvre defined in JAR 25.235.

The outcome of this was a special condition that added an extra case to JAR 25.235. This required the aircraft to be dived through V_C at $-15°$ flight path angle, but permitted recovery action to be taken 2 s after the onset of a high-speed warning.

It was checked on an A320 simulator that this extra case resulted in a similar speed rise to the $-7.5°/20$ s upset manoeuvre on an unprotected aircraft and that the overspeed protection reduced this rise in speed down to approximately 30 knots.

4.3.4. *Longitudinal static stability*. The A320's C^* pitch control law is a manoeuvre demand law: the pilot's control inputs are interpreted as a demand for a given level of manoeuvre rate and the control system provides the surface deflection needed to generate this rate. Releasing the side-stick commands flight path stability. This causes a potential non-compliance with the airworthiness requirements for static stability, which are based on angle-of-attack stability.

It appears that the piloting task increases with greater instability, to correct the airspeed divergence that would otherwise occur in response to any minor disturbance. Thus, the transition from static stability to instability does not represent a boundary of controllability, but results in increased pilot workload in maintaining the desired flight condition.

It was concluded that the A320 pitch control law eliminating the need for constant manual retrimming reduced the workload involved in flight path control to a very low level, which compared favourably with stable aircraft.

4.3.5. *High angle of attack protection*. The angle of attack protection function part of the A320 fly-by-wire system prevents a pilot-induced stall and reduces the risk of a stall due to atmospheric disturbances. In handling terms it is all good news.

The difficulty for certification arises from the fact that the stall speed is the traditional foundation block from which take-off and landing speeds are determined ($V_2 > 1 \cdot 2V_S$; $V_{REF} > 1 \cdot 3V_S$, $V_S > 0 \cdot 94V_{S1}$ g) and that the aircraft cannot be intentionally flown at that speed with the system active.

If the minimum speed achievable with the system operating was treated as being the stall speed and the traditional factors were applied to this speed to set the minimum values of V_2 and V_{REF}, the 'reward' for providing this safety beneficial system would be a commercially intolerable performance penalty. It was therefore necessary to carry out a fundamental review of the requirements for determining minimum values of stall-speed related operating speeds. (V_2 is the take-off speed, V_S is the calibrated stalling speed, V_{REF} is the lowest selectable landing speed, V_{sig} is the stabilized speed at maximum lift.)

5. The A340 experience

The general design objective, relative to the A340 (see Fig. 2) fly-by-wire system, was to reproduce the architecture and principles chosen for the A320 as much as possible for the sake of commonality and efficiency, taking account of the A340 peculiarities (long-range four-engine aricraft).

Consequently, the Aerospatiale innovation efforts concentrated on system integration and methodology improvements in the system, control law and software fields taking account of the 320 in-service experience.

5.1. *Fly-by-wire integration*

5.1.1. *System.* As is now common for each new program, the computer functional density was increased between the A320 and A330/A340 programs (see Fig. 7): The number of computers was reduced to perform more functions and control an increased number of control surfaces.

5.1.2. *Control laws.* The general concept of the flight control laws described in the previous sections was maintained, adapted to the aircraft characteristics and used to optimize the aircraft performance.

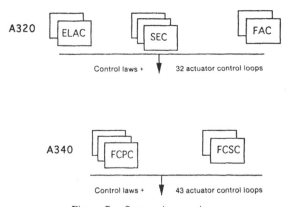

Figure 7.   System integration.

(*a*) The angle of attack protection was reinforced to cope better with the aerodynamic characteristics of the aircraft.

(*b*) The dutch roll damping system was designed to survive against rudder command blocking thanks to an additional damping term through the ailerons and to survive against an extremely improbable complete electrical failure thanks to an additional autonomous damper. The outcome of this was that the existing A300 fin could be used on the A330 and A340 aircraft with the associated industrial benefits.

(*c*) The take-off performance (V_{MU} limited) could be optimized by designing a specific law that controls the aircraft pitch attitude during the rotation.

(*d*) The flexibility of fly-by-wire was used to optimize the V_{MCG} speeds. In fact, the rudder efficiency was increased on the ground by fully and asymetrically deploying the inner and outer ailerons on the side of the pedal action as a function of the rudder travel: the inner aileron is commanded downwards and the outer aileron complemented by one spoiler is commanded upwards.

5.1.3. *Turbulence damping function.* A first step in the direction of structural mode control through fly-by-wire was made for the A340 programme.

In fact, it appeared that some structural modes could be excited in turbulence conditions and that uncomfortable vibration levels could be generated in the cockpit and in the rear part of the fuselage.

As the aircraft response in turbulence is the superposition of rigid body modes and structural modes, both of these may participate in the 'comfort or discomfort' of the aircraft. The manual flight control laws and the auto-pilot were firstly designed to remove the interaction with structural modes. The next step was to include an additional function, the so-called 'turbulence damping function' in the fly-by-wire system to attenuate the fuselage response to turbulence in the frequency range of the relevant modes.

Assuming the direct correlation between the fuselage acceleration levels in turbulence and the damping of the relevant fuselage modes, the objective to attenuate the fuselage response to turbulence was considered equivalent to increasing the damping of the fuselage modes.

The general principle of the function is to add a specific turbulence damping command to the flight control law orders. There are three main modes to damp. The system is therefore composed of three lanes. One lane controls a longitudinal mode through the elevators as a function of the information of an accelerometer situated in the aircraft nose. Two lanes control one lateral mode each through the rudder.

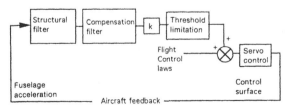

Figure 8.   Turbulence damping: principle of one lane.

The three lanes have a similar structure (see Fig. 8). The fuselage movement is measured through the relevant accelerometer. The frequency range of the mode to be damped is isolated through a band pass structural filter, while gain and phase are adjusted through a compensation filter to create a control surface movement approximately in phase with the speed of fuselage displacement.

As the decision to implement this function was taken late in the programme, the first challenge was to integrate this new function into the already existing flight control system with a minimum of modifications, taking account of the available computing power and of the actual servo-control performance. In particular, the equipment specific to the function was reduced to the two lateral accelerometers and one activation switch. The vertical accelerometer already used for the normal longitudinal law was also used for the longitudinal lane of the turbulence damping function.

The second challenge was to define a complete methodology to define, tune and certify a robust and performing control law showing the additional following characteristics:

(i) independency of flight condition and aircraft weight distribution,

(ii) large stability margins in the whole flight domain (± 6 dB, $\pm 60°$),

(iii) no impact on safety in case of misfunctioning,

(iv) no interference with flying qualities.

These objectives were achieved using the following methodology.

The general structure of the law was first determined by using conventional Nyquist linear methods on a dynamic model reduced to a limited number of modes. The complete law was then validated on the complete flutter model. Several law tunings were then flight tested in different flight and weight conditions in open and closed loop to adjust the models, improve performance and evaluate the stability margins. To do this, the aircraft was successively excited through wing tip-vanes, the elevators and the rudder.

The stability margins of the final control law were demonstrated by using the classical flutter analysis on the basis of the aeroelastic model. The non-interference of the function with the flying qualities could be demonstrated by comparing the aircraft response to calibrated inputs with and without the turbulence damping function.

The performance of the function was then evaluated by harmonic analysis of the closed loop flight test, by simulation using the aeroelastic model with turbulence models, and finally by flight tests in tubulence (Figs 9–11). The order of magnitude of the gain produced by the function is 50% of the natural vibration level and depends on flight conditions, weight distribution, gust intensity, etc.

5.2. *Methodology*

The A320 experience showed the necessity to be capable of detecting possible errors as early as possible in the design process to minimize the debugging effort along the development phase.

Consequently, it was decided to develop tools as shown in Fig. 12 that would enable the engineers actually to fly the aircraft in its environment to check that

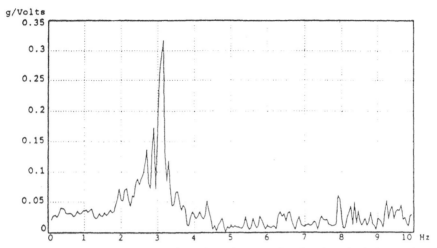

Figure 9. Nose vertical acceleration/gust intensity transfer function (turbulence damping function OFF).

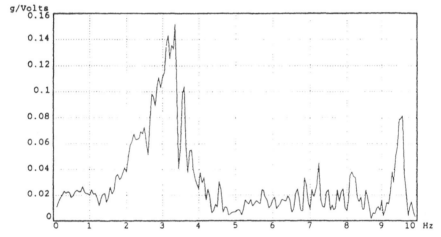

Figure 10. Nose vertical acceleration/gust intensity transfer function (turbulence damping function ON).

the specification fulfils the performance and safety objectives before the code even exists.

The basic element of this project is the so-called SAO specification (Spécification Assistée par Ordinateur), the Aerospatiale graphic language defined to specify clearly control laws and system logics, and developed for A320 programme requirements. The specification is then automatically coded for engineering simulation purposes in both control law and system areas.

In the control law area, OCAS (Outil de Conception Assistée par Simulation) is a real-time simulation tool that links the SAO definition of the control laws to the already mentioned aircraft movement simulation (OSMA). Pilot orders are entered through simplified controls, including side-stick and engine thrust levels. A simplified primary flight display (PFD) visualizes the outputs of the control law. The engineer is then in a position physically to judge for

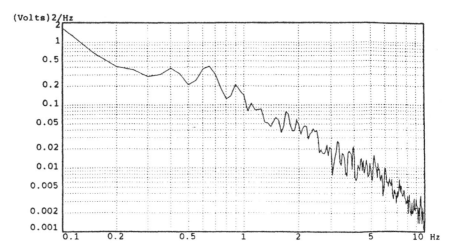

Figure 11.   Typical turbulence power spectrum.

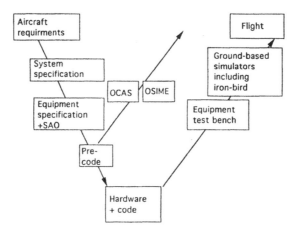

Figure 12.   Validation methodology.

him/herself the quality of the control law that has just been produced, in particular with respect to law transition and nonlinear effects. In the early development phase, this very same simulation was used in the full-scale A340 development simulator with a pilot in the loop.

In the system area, OSIME (Outil de Simulation Multi Equipement) is an expanded time simulation that links the SAO definition of the whole system (control law and system logic) to the complete servo-control modes and to the simulation of aircraft movement (OSMA). The objective was to simulate the whole fly-by-wire system including:

(*a*) the three primary computers (FCPC),

(*b*) the two secondary computers (FCSC),

(*c*) the servo-controls, in an aircraft environment.

This tool contributed to the functional definition of the fly-by-wire system, to the system validation and to the failure analysis. In addition, the behaviour of the system at the limit of validity of each parameter, including time delays, could be checked to define robust monitoring algorithms. Non-regression tests have been integrated very early into the design process to check the validity of each new specification standard.

Once validated, both in the control law and system areas using the OCAS and OSIME tools, a new specification standard is considered to be ready to be implemented in the real computers (automatic coding) to be further validated on the test bench, simulator and on the aircraft.

6. Conclusions

With the A320 experience, a general control law philosophy and an associated system architecture applicable to transport aircraft were defined. The extensive use of ground-based simulators, complemented by flight tests, was integrated into the development process and the certification procedure was adapted to take account of the benefits of fly-by-wire.

On the A330/A340 programme, all of the above principles were reproduced. New functions were integrated and, in particular, a first step in the direction of structural mode control through fly-by-wire was made. Upstream validation tools were developed both in the control law and system areas. This method certainly contributed to defining a mature system very early in the programme's development.

REFERENCES

CHATRENET, D., 1991, Use of ground-based simulators and in-flight simulation for development of the A320 flight control system. *DLR In-flight Simulation Conference*.

DENNING, J. L. F., and WHITTLE, D. A., (C. A. A.), 1988, Aerotech Fly-by-wire certification.

FARINEAU, J., and LETRON, X., 1990, Qualités de vol latéral d'un avion de transport civil équipé de commandes de vol électriques. Expérience de l'Airbus A320. *AGARD Symposium on flying qualities, AGARD Conference Proceedings*, Paper No. 508.

9

Practical control law design for aircraft using multivariable techniques

JAMES D. BLIGHT, R. LANE DAILEY and DAGFINN GANGSAAS

Nomenclature

e	elevator
F	stick, pedal force (lb)
lat	lateral
le	leading edge flap
lon	longitudinal
n_z	normal (vertical) acceleration (g)
p	roll rate ($\deg\,s^{-1}$)
ped	pedal
pt	pitch thrust vector
Q	dynamic pressure ($lb\,ft^{-2}$)
q	pitch rate ($\deg\,s^{-1}$)
r	yaw rate ($\deg\,s^{-1}$)
rud	rudder
str	strake
T	thrust (lb)
te	trailing edge flap
tv	thrust vector
u, V	airspeed ($ft\,s^{-1}$ or knots)
yt	yaw thrust
α	angle of attack (deg)
β	sideslip angle (deg)
δ	effector deflection (deg)
θ	pitch attitude (deg)
ϕ	bank angle (deg)
ψ	heading, track angle (deg)

1. Introduction

Stability and control is one of the major technical challenges in the design of an aircraft. The failure of many aircraft projects in the past was the result of inadequate solutions to the stability and control problem. New aircraft designs, which typically are driven by requirements for reduced operating cost, in the case of civil transport aircraft, and reduced radar signature, in the case of military aircraft, present an increasing challenge to the flight control engineer.

The importance of feedback control in furnishing the required stability and control characteristics for aircraft has been firmly established. All high-performance aircraft produced today employ some form of feedback control. This may be in the form of an autopilot, a limited authority command and stability augmentation system or a full authority fly-by-wire system.

Feedback control is playing an increasingly important role in meeting cost and performance objectives for new civil and military aircraft. Typical control functions are: command and stability augmentation for superior flying qualities and manoeuvrability; automatic flight trajectory and speed control for operations in all weather conditions; manoeuvre and gust load reduction for extended service life; ride quality enhancements for crew and passenger comfort; aerodynamic and propulsion performance optimization for reduced operating cost; flutter suppression and failure or battle damage reconfiguration for enhanced safety. The control laws must function properly at all flight conditions and aircraft states and ensure safety of flight.

New aircraft employ multiple coupled control loops to achieve the required performance and stability. Measurements of rigid and flexible-body motion states, the dynamic states of subsystems, such as the engines and the external environment are combined with pilot inputs or guidance and navigation information and fed back to multiple control effectors. The latter may comprise many aerodynamic control surfaces, and propulsion controls including inlet and nozzle controls. Typically, each control function requires multiple measurements and control effectors, resulting in strong interactions between the various control loops.

For most aircraft flying today the control laws have been developed using predominantly classical single-loop frequency response and root locus design techniques. This approach to aircraft control design is essentially the same as that outlined by Bollay (1951). These methods have been used successfully for both single loop and multiloop control problems and are generally accepted within industry.

Over the last 20 years new multivariable control law synthesis and analysis techniques have been proposed. These techniques have their roots in the theories of optimal contol developed by Pontryagin (Pontryagin *et al.* 1962) and Bellman (Bellman 1957). There has been a proliferation of extensions and variations which have kept the academic community well occupied. The proponents of multivariable control theory claim that the new techniques can handle multiloop control problems in a formal and systematic manner. However, in spite of the availability of good computational algorithms and software, the practising control engineers in industry have been reluctant to adopt and use the new techniques. They still rely predominantly on the classical one-loop-at-time frequency response and root locus techniques.

The first significant applications of multivariable control techniques started at Boeing in 1978. The work was sponsored by NASA and performed under a project entitled 'Integrated Application of Active Controls (IAAC) Technology to an Advanced Subsonic Transport.' The results, presented in NASA CR-159249 (Boeing Commercial Airplane Company 1980) clearly demonstrated that multivariable control law design techniques offer significant advantages over classical techniques in the solution of multiloop control problems. Command and stability augmentation, gust load reduction, and flutter suppression control functions were successfully implemented using linear quadratic optimal control techniques. A companion activity using classical single loop techniques produced control laws offering significantly less performance and robustness.

Motivated by the initial success, practical multivariable design techniques have been developed further and successfully applied to a wide range of control problems at Boeing over the last 15 years. This success is primarily due to: (1) the combination of multivariable techniques with classical frequency-domain interpretation of control loop performance and sensitivity properties; (2) establishment of practical guidelines for transforming design requirements into the mathematical formulation of solutions; (3) active training of engineers in the use of multivariable techniques; and (4) the availability of highly user-friendly software packages. In support of military airplane design, multivariable control law synthesis and analysis have become a necessary part of the iterative airplane design cycle, and thus influence the development of the airframe, propulsion, and other subsystem configurations. This is in contrast to past practices where the control design would mainly accommodate an *a priori* airplane design.

This paper highlights some of the multivariable control design experience at Boeing. Section 2 describes the successful redesign of the lateral autopilot control laws for the Boeing 767 commercial transport. Repeated attempts at using classical single-loop techniques failed to produce a satisfactory solution. The redesigned control law is currently in commercial service. Section 3 describes the development of a control law for the longitudinal axis of a transport airplane with reduced inherent longitudinal stability. Section 4 describes the development of the control laws for a highly agile tactical fighter with post-stall manoeuvring capability. This represents a highly coupled multiloop control problem with strong nonlinearities in the airplane dynamics. A large number of successful piloted air-to-air simulations were flown to evaluate the utility of post-stall manoeuvring. Use of the Gibson flying qualities template is also discussed. Section 5 describes the highly user-friendly software environment that is necessary to produce quality control laws in a timely fashion.

2. Flight test experience with practical LQR: improvement of 767 lateral autopilot

This example addresses the elimination of a small amplitude limit cycle instability experienced on the Boeing 767 commercial transport airplane (Fig. 1). The problem was associated with the heading and track hold autopilot, called the lateral autopilot, and was not solved after repeated attempts using classical synthesis techniques. The solution involved a good understanding of the control problem combined with a straightforward application of linear quadratic regulator (LQR) theory. The latter furnished the necessary insight, in terms of required feedback signals and corresponding gains, to eliminate the limit cycle

Figure 1.   Boeing 767.

instability without compromising the performance of the autopilot heading and track hold functions. The success can be attributed to the excellent robustness properties of the regulator (Safonov and Athans 1977). The data presented here are summarized from earlier papers (Bruce and Gangsaas 1984 and Gangsaas *et al.* 1986).

2.1. *Problem statement*

Occasional ride discomfort was reported during early passenger service of the Boeing 767 commercial jet transport. It was due to a small-amplitude, sustained yawing oscillation that occurred only during high altitude cruise flight when both the yaw damper and lateral autopilot were engaged. The yaw damper increases the damping of the dutch roll mode (involving yaw and roll angle oscillations) of the aircraft using the rudder as a single control. The yaw damper is normally engaged both in manual and automatic flight. During automatic flight, the lateral autopilot is engaged and it controls heading or track angle using the combination of left and right ailerons as a single control.

Flight testing showed that if the yaw damper was disengaged, that is, the rudder control loop was opened, but the lateral autopilot was engaged, the aircraft did not exhibit the limit cycle instability. However, with this non-standard configuration of the yaw damper and lateral autopilot, the aircraft dutch roll mode was lightly damped. Analysis of flight test data showed that engaging the lateral autopilot tended to reduce the damping of the dutch roll mode particularly with the yaw damper disengaged. Figure 2 shows the

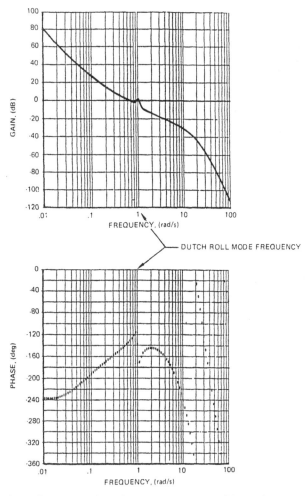

Figure 2.   Aileron open-loop frequency response with rudder loop open.

open-loop gain and phase characteristics in the aileron control loop with the rudder loop open. It is clear that the stability margins are very small and that very small gain and phase variations would lead to instabilities at the dutch roll mode frequency. Based on this, it was hypothesized that dead band and hysteresis in the rudder control loop combined with relatively small variations in aerodynamic control effectiveness in the aileron loop could cause the observed limit cycle oscillations.

Tests had shown that all of the aerodynamic parameters and nonlinearities were well within normal and predicted values for those aircraft exhibiting the limit cycle behaviour. Thus, it appeared that due to adverse coupling between the aileron and rudder control loops, there was high sensitivity to small nonlinearities and variations in aerodynamic parameters. It was further hypothesized that this sensitivity could be reduced if the destabilizing effect on the dutch roll mode from the lateral autopilot was eliminated, or even better, turned into a stabilizing effect. Thus, the design problem was to improve the dutch roll damping and improve stability margins with the lateral autopilot engaged and the yaw damper disengaged.

The autopilot control law had been synthesized using standard root locus and frequency response techniques with sequential loop closures on the various feedback sensors. The latter comprised almost the full state vector except for sideslip angle β and yaw rate r. Yaw rate was sensed, but not used for feedback. Root locus analysis showed that dutch roll damping could be improved, but only at the expense of reduced heading or track mode stability that led to significant degradation in lateral autopilot performance. Extensive root locus analysis failed to produce a set of gains that offered significant improvements in dutch roll mode stability while maintaining the required lateral autopilot performance. It was then decided to use full-state feedback synthesis in an attempt to establish whether or not a better solution existed.

2.2. *Objectives and constraints*

The objectives were to: eliminate perceptible residual yaw oscillations without affecting lateral autopilot performance; reduce RMS lateral accelerations and aileron deflections due to gust inputs; make changes only to the lateral autopilot control laws; control either heading angle or track angle without gain changes; have a system insensitive to nonlinearities and variations in aerodynamics; have a control loop bandwidth and high-frequency gain no greater than the existing design.

2.3. *Design method*

Figure 3 shows a block diagram of the plant model used for analysis and synthesis. It represents the aileron actuation system, flight control computer time delay, airplane dynamics, and antialiasing filters on the sensor signals. In addition, the yaw damper control laws, computational delay, sensors, and rudder servo and actuator dynamics were modelled. This total model was expressed in state-space form at various flight conditions. The state vector comprised over 50 elements. The rudder control loop was open for control law synthesis (see Fig. 3); however, it was closed for performance analysis with the yaw damper engaged.

The control law design was based on the airplane model for the nominal cruise flight condition, using LQR synthesis. Full-state feedback gains were calculated based on the following cost function

$$J = (\tfrac{1}{2})E[Q_{r}(\dot{\psi}_{c} - r) + Q_{\psi}(\psi_{c} - \bar{\psi})^2 + Q_{I\psi}(\textstyle\int(\psi_{c} - \bar{\psi}))^2$$
$$+ Q_{dr}(z_{dr})^2 + (\delta_{ac})^2] \tag{1}$$

where Q_{r}, Q_{ψ}, $Q_{I\psi}$ and Q_{dr} are the penalty weightings on yaw rate error, complemented heading or track error, integral of complemented heading or track error, and dutch roll mode displacement, respectively. For this problem, measured yaw rate r is approximately equal to heading rate $\dot{\psi}$ (Babister 1961). The dutch roll mode displacement z_{dr} is related to the states through the eigenvectors in a standard modal decomposition. δ_{ac} is the input to the aileron actuators. The subscript c in (1) refers to command values.

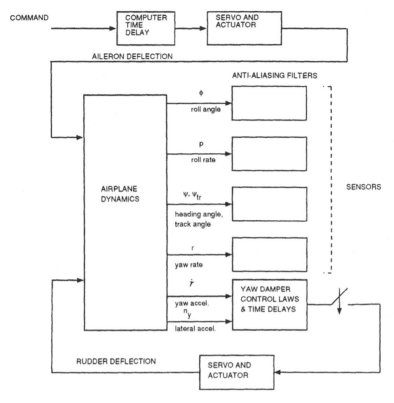

Figure 3. Plant model for lateral autopilot improvement.

To meet the requirement that heading angle ψ or track angle ψ_{tr} should be controlled interchangeably without control law gain changes, it was necessary to close the proportional and integral loops on complemented heading or track angle, $\bar{\psi}$, as defined in Fig. 4. Angles ψ and ψ_{tr} are related by

$$\psi_{tr} = \psi + \beta \qquad (2)$$

The dutch roll mode dominates the sideslip β response. Thus, if ψ_{tr} were substituted directly for ψ there would be a significant impact on the dutch roll mode stability requiring control law gain changes from a redesign. By setting the break frequency ($a = 0 \cdot 2 \, \mathrm{rad\,s}^{-1}$) of the complementary filter in Fig. 4 well below the dutch roll mode frequency of $1 \, \mathrm{rad\,s}^{-1}$, there is sufficient attenuation of the β response at this frequency to ensure minimal impact on dutch roll mode

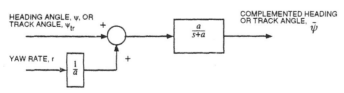

Figure 4. Complemented heading or track angle.

stability when ψ_{tr} is substituted for ψ. The yaw rate r input to the complementary filter ensures that good heading and track mode stability is maintained. This is a good example of how frequency domain loop shaping can be used to help satisfy apparently conflicting design requirements.

Reflecting a standard rate, proportional and integral control structure, the penalties Q_r Q_ψ and $Q_{1\psi}$ were adjusted to obtain the same heading or track mode damping, bandwidth, and integral time constant as the classical design. Next, the damping of the dutch roll was increased by increasing the dutch roll mode weighting Q_{dr}. This did not affect the damping of the other modes. Q_{dr} was adjusted to give a set of gains that included a gain of zero on β since β was not available as a feedback sensor. Even with this zero gain, the dutch roll mode damping was increased from 0·01 to 0·08 as compared with the classical design.

Figure 5 compares the gains obtained from the LQR synthesis with those of the classical design. Only the significant gains from the full-state LQR design were retained. The remainder were set to zero without any impact on stability and control performance. The proportional heading and integral heading gains were approximately the same for both designs. However, there are significant differences in the two designs for the yaw rate r, roll angle ϕ, and roll rate p, gains. The classical design had a zero gain on yaw rate while the LQR design has a relatively high gain. This gain maintained a fixed ratio to the heading gain for all designs having a well-damped heading mode. There were no combinations of weights in the cost function (1) that would produce a roll angle gain as large in magnitude as that of the classical design. Originally, the rationale for this large gain was to ensure good tracking performance for heading angle. In co-ordinated flight—that is, with the sideslip angle close to zero—roll angle and yaw rate are related kinematically (Babister 1961) and therefore are equivalent feedback signals. However, that is not true when there are significant sideslip oscillations as in the case of a lightly damped dutch roll mode. It is interesting to note that the LQR synthesis provided the insight that yaw rate feedback rather than roll angle feedback would give a much better trade-off between robustness and control performance, as will be seen later. The fact that yaw rate feedback had been excluded in the earlier work using the root locus technique accounts for the failure to find an acceptable solution.

2.4. *Performance*

Prior to flight test the control law performance was evaluated by analysis at ten different flight conditions. These reflected the full range of gross weight, centre of gravity location, and speed expected in high altitude cruise flight. Data from the four worst flight conditions are presented here. Analysis was per-

Gains	K_ψ	K_r	K_ϕ	K_p	$K_{1\psi}$
Control laws:					
Original	7.14	0	3.03	1.52	.05
New	7.84	7.13	0.89	1.21	.05

Figure 5. Gains used in flight test.

formed with the yaw damper both engaged and disengaged and with control of heading angle or track angle. All combinations produced satisfactory results (Bruce and Gangsaas 1984). However, only data for heading control with the yaw damper disengaged will be presented here since it represented the most difficult design problem.

The redesigned control law had a significant reduction in the sensitivity to variations in flight condition as well as improvements in damping of the dutch roll and heading modes. In fact, for all cruise flight conditions the system exhibited excellent stability without any gain scheduling (Bruce and Gangsaas 1984). The new design shows considerably less sensitivity to the yawing moment derivative due to aileron deflection, $C_{n\delta a}$, and also shows considerable reduction in RMS lateral acceleration and aileron deflection due to turbulence.

2.5. *Flight test results*

The new lateral autopilot control law was implemented in the flight control computers. This entailed modifying the autopilot gain schedules to provide the redesigned gain values of Fig. 5 at the cruise flight conditions and adding a yaw rate feedback to the aileron command input. The performance of the original and new control laws were evaluated during flight test. Figure 6 shows the light dutch roll damping with the original control law. Figure 7 demonstrates the significant improvement in dutch roll damping offered by the new control law. The improvement was demonstrated over a wide range of flight conditions with and without the yaw damper engaged. The particular test aircraft had never exhibited the limit cycle behaviour with the original control law. However, it was conjectured that this demonstrated improved dutch roll mode damping would eliminate the problem from those airplanes exhibiting limit cycle oscillations in service.

The new flight control law was incorporated on airplanes that earlier had exhibited the limit cycle oscillations. Pilots who flew with the modified autopilot control law gave favourable comments and said that they now did not detect any

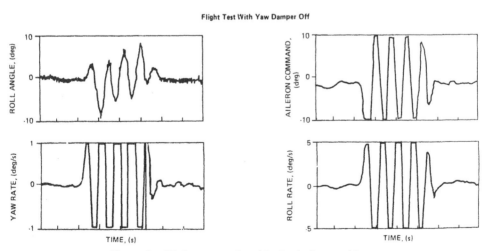

Figure 6.   Flight test results with classical control law.

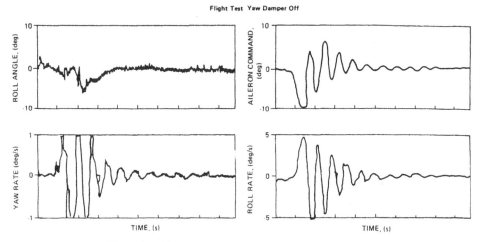

Figure 7.   Flight test results with modern control law.

limit cycle oscillations. They considered performance of the autopilot with the new gains to be a significant improvement. The modified control law is now incorporated as a permanent change to the autopilot.

3.   Air transport longitudinal flight control design and piloted simulation

We present a practical application of integral regulator theory to an aircraft control system design. This example has been reported previously by Blight *et al.* (1986) and Gangsaas *et al.* (1986).

The application involves the synthesis of a command and stability augmentation control law for a transport airplane with relaxed requirements for inherent longitudinal stability. This airplane is typical of the next generation of transports. The design involved one application of frequency-shaped linear quadratic Gaussian (LQG) synthesis, and the control law performed well during nonlinear piloted simulations. It was derived from a single-point design and achieved good control performance and robustness properties over the full flight envelope and centre of gravity range using minimal gain scheduling.

3.1. *Problem statement*

Traditionally, transport airplanes have been designed to have a certain level of inherent longitudinal stability. This and other control requirements dictate the size of the horizontal tail and restrict the permissible aftmost location of the centre of gravity (c.g.). The efficiency of these airplanes can be improved by decreasing the horizontal tail size and moving the c.g. aft. The corresponding reductions in weight and trim drag from the decreased tail size and trim load on the tail can yield a significant reduction in fuel consumption (Boeing 1980). However, these airplanes will have unsatisfactory longitudinal stability and control characteristics within part of their c.g. and flight envelopes.

The stability and response characteristics for such an airplane were evaluated at the four flight conditions listed in Fig. 8. Figure 9 shows, for a range of c.g. locations, typical normal acceleration and pitch rate time responses to a step

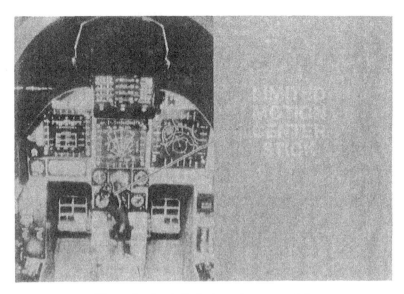

Figure 7. Lavi: limited-motion centrestick.

opinion: 3·3 in (84 mm) for aft travel and 2 in (51 mm) for forward and lateral stick travel were chosen. A single spring in each plane gives the desired level of artificial feeling.

The main components of this system include the following.

(a) Two flight control computers (FLCCs), each having two DFCS channels and one (plus one monitor) EFCS channel.

(b) Three angle-of-attack (AoA) sensors, two of them on the left side of the aircraft.

(c) Four rate gyros for each of the three axes – pitch, roll and yaw (total 12 RGs).

(d) Four accelerometers for longitudinal and four for lateral axes (total eight ACCs).

(e) Three anemometric (static and impact pressure) sensors.

(f) Seven quad-redundant fly-by-wire (FBW) servo-actuators (SAs) for primary control: four elevons (elevator + aileron), two canards and one rudder.

(g) Two leading-edge flap drive systems for secondary control.

(h) Mechanical stick/artificial feel system with quad-redundant LVDTs (travel measurement), both in pitch and roll.

(i) Mechanical pedal system with quad-redundant LVDTs

Figure 8 shows the FCS main elements.

The digital system is based on a Zilog Z8000 CPU. The programming languages were 'C' and 'Assembler'.

A cross-channel data-link (CCDL) enabled voting and monitoring of the system's inputs. Special treatments were given to the AoA and anemometric systems,

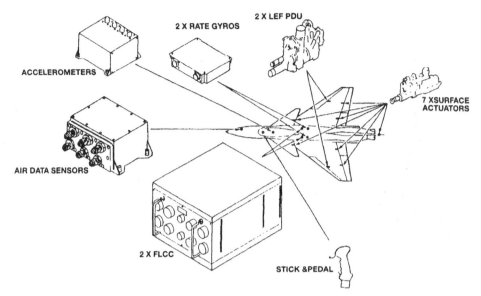

Figure 8.   Lavi: FCS main elements.

as these were triple mechanically redundant. The system was incorporated with output command monitoring logics, channel fail logics, preflight and in-flight built-in tests (BITs) and with channel synchronization algorithms.

The EFCS was basically simple, with only rate feedbacks (P, Q, R), and a few gains being scheduled.

The system enabled flying with one EFCS channel (EFCS-1 or EFCS-2), as available. In this chapter we shall limit our discussion to the main DFCS control laws.

2.2. *Control law design philosophy*

The control laws were designed by classical design methods. These methods provided better insight into the system dynamics and were found to be better than other design methods in handling the different, and sometimes contradicting, design requirements. Optimal design methods were used during the preliminary design phase. Elements like backup gains for anemonetric system failure and elevon separation gains were designed by these methods.

MIL-F-8785C and other published papers on flying qualities served as guidelines and criteria for CL design. The 'equivalent-system' method was used to represent the actual high-order system (HOS) dynamics.

The main tool for flight control law (FCL) evaluation was the flying-qualities simulator. Major changes had been made in the CL during these simulator sessions. Many gains were reduced and some changes were made in the CL architecture.

The flying-qualities simulator sessions have proved that CL should be designed to achieve good pilot plus vehicle dynamics. Minor changes to CL (in roll sensitivity) were made during the NT-33 in-flight simulations for the approach and landing tests.

Our main conclusions regarding the CL design are that:

(*a*) the classical method is intuitive and time consuming but it is preferable for manned air vehicles' control system design;

(*b*) the flying-qualities simulator serves a vital role in CL design – simulation should start from the preliminary design phase;

(*c*) interaction between structures and FCS (mainly aeroservoelasticity) should be handled very carefully, when designing high-gain, fly-by-wire systems.

2.3. *Flight control law development*

The first FCL set was very simple and had a basic structure. This was done intentionally in order to be able to concentrate on *understanding* the flying-qualities requirements using 'pilot-in-the-loop' simulations. One of the major problems in developing FCL is that it involves many engineering skills, far beyond the classical control theory. In order to achieve good results, the FCL engineer should have knowledge of the following areas:

(*a*) control theory (classical and modern);

(*b*) control system architecture (sensors, stick and pedal, FLCCs, etc.);

(*c*) aerodynamics;

(*d*) aircraft dynamics (including at high angles of attack);

(*e*) aero- (and aeroservo-) elasticity;

(*f*) aircraft loads;

(*g*) weight and balance;

(*h*) simulation and modelling methods (SDF, landing gear, wind, gusts).

In addition, the FCL engineer should understand how the aircraft flies, from the pilot's point of view. In practice, such an engineer who *really* knows all of these areas does not exist, which is why, even today when more modern control design methods are on hand, it is advised to avoid complexity and concentrate on the main issues of concern.

3. Flying-qualities and other requirements

Every FCL design starts with an in-depth review of the requirements. MIL-F-8785 served as a guide together with several other requirements from IAF and the open literature. It was hard to establish a very precise set of requirements, which, in our opinion, do not exist, even today.

An example, often discussed in many publications, is the famous requirement for the short-period frequency (versus n/α). It had already been determined that these requirements should be met with an equivalent system.

However, several questions arose.

(*a*) As is known, different transfer functions (q/F_s, α/F_s) yield different values of equivalent ω_{nsp}. Which one should be dominant? (Some say both, others α/F_s and some q/F_s.)

(b) Should the HOS include all dynamic elements, including structural modes? If the answer is yes, matching such a system to second order is quite impractical.

(c) The most important question was: why should we try to build a modern fighter according to old conventional aircraft behaviour? As is known, this requirement reflects the natural behaviour of a conventional aircraft.

The answers to these questions, which were given during the Lavi programme development process, are as follows:

(a) The q/F_s transfer function was considered as the dominant one.

(b) All dynamic elements were included (like sensors dynamics, H/W filters, etc.). The structural modes were excluded. Matching the HOS to low-order systems (LOSs) was done up to 10–12 rad s^{-1}, only.

(c) A compromise was taken: the ω_{nsp} should increase with n/α, but with a moderate slope. An example of a matching of the HOS to the LOS, of pitch rate to stick input, is shown in Fig. 9.

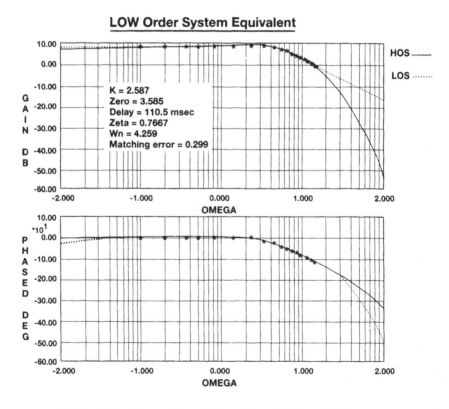

LAVI - TD : Example of Matching HOS to LOS for Q/Fs Transfer Function.
Mach = 0.6, Alt = 20 Kft.

Figure 9. Lavi T/D LOS to HOS matching.

In any case, meeting the MIL-F-8785 requirements was quite easy, not only for ω_{nsp}, but for other requirements as well. The IAF added some new requirements, but still we were looking for more modern or 'tougher' requirements.

Examples of other requirements that were considered are the 'Neal–Smith' criteria, the 'Bandwidth' criteria and the 'YF-17' frequency-response criteria. The only one supported with a lot of data and reasoning was the 'Neal–Smith' criterion. It was used extensively for evaluating the Category 'C' CL.

The 'YF-17' criterion was one which the IAF insisted on. Meeting this requirement was one of the reasons why the pilots rejected the Category 'A' CL, flying the simulator. The requirement was obviously misleading (see Fig. 10). Until these days, there had been no real 'cook-book' for flying qualities. The FCL engineer had therefore to participate in extensive 'pilot-in-the-loop' simulations in order to complete the CL design before the first flight.

During the early days of the Lavi CL development, it was decided to have relatively large stability margin requirements: at least 60° PM and 10 dB GM. The reason was the large amount of uncertainty in all areas, but especially in the need to attenuate the structural modes. The wings and some parts of the fuselage and tail were made from composite materials, causing doubts about their structural modes. Later in the programme, when additional models were inserted, the margins became smaller.

The first flight was conducted with an expected PM of 45° and 8 dB GM in some flight conditions. At Mach 0·95 the PM measured in flight was smaller. However, the GM was kept as predicted for all flight conditions and configurations.

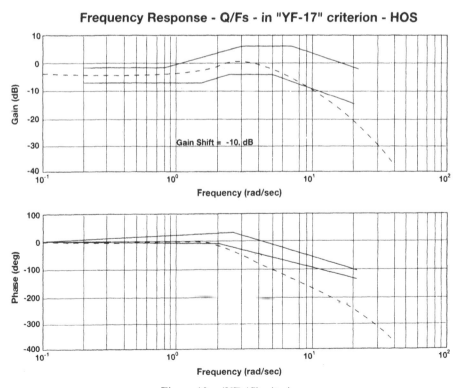

Figure 10. 'YF-17' criterion.

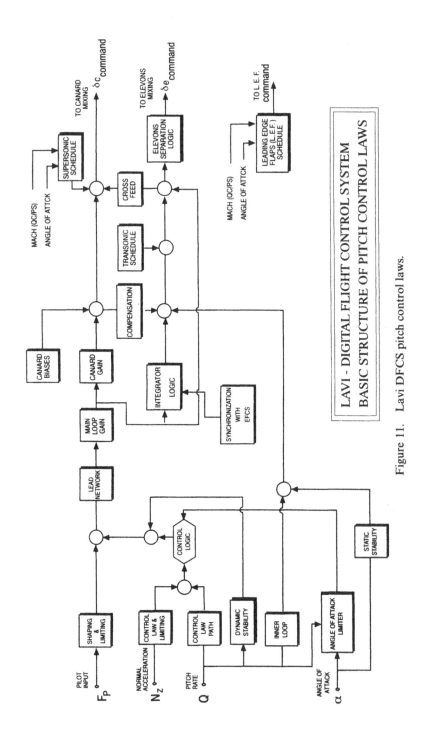

Figure 11. Lavi DFCS pitch control laws.

4. Final structure of the DFCS

Figure 11 presents the pitch loop CL structure. As can be seen, it is quite simple. In fact, the CL of the Lavi had more than 100 scheduled gains and functions. The complexity was in those functions that were built step-by-step through extensive simulations and analyses.

The pitch control laws consisted of a basic blend of pitch rate and normal acceleration (N_z) in the low-speed, and pure N_z for the high-speed regime. The angle-of-attack limiter (AoA-Lim) control law feedback was compared with the low-frequency control law feedbacks. When AoA-Lim became larger, it overrode the q–N_z blend and became dominant.

The inner loop blend of AoA and low-pass pitch rate to the elevons provided stability augmentation at low airspeed, and a destabilizing effect at high airspeed, to augment the aircraft response where it was too sluggish owing to overstability.

Special care was given to the elevon separation logic (ESL), shown in Fig. 12. This logic was designed to maximize control power with minimum control surface deflection. As a result, the outboard elevons were mainly deflected for roll and the inboard elevons mainly for pitch. Inboard elevons served as lift augmentation devices during takeoff, approach and landing, as they were biased towards trailing-edge down. This arrangement enabled slower (15 knots calibrated airspeed) lift-off speeds for configurations that were heavy and had a very forward c.g.

In addition to the elevons, the canards were also biased during Category 'C' and thus reduced even further the approach airspeed and enhanced the pilot's visibility.

All transitions from one set of CL to another (e.g. upon landing gear extraction or upon weight on wheels) were properly smoothed via fading logics.

The design of the lateral directional CL mostly concentrated on two major issues: reducing sideslip during roll and, on the other hand, avoiding large yaw rates in response to lateral control commands. These two requirements were contradictory in most of the flight envelope regimes, as the Lavi is of a 'proverse' yaw nature.

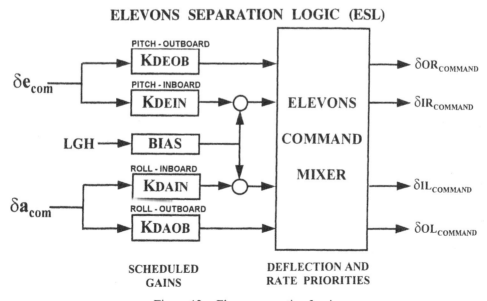

Figure 12. Elevon separation Logic.

The need to suppress the sideslip as much as possible (highly augmented dutch roll), even if it produced large yaw rates, was proved to be wrong mainly for small or moderate roll commands. As a result, the control laws were built to yield a different response for different amplitudes of commands.

5. Simulation

The FCS development phase started in 1981. In mid-1983, this analytical work resulted in a set of classic FCLs. These FCLs were first introduced to the evaluation pilots group in a moving-base flying-quality simulator. This simulator is located in the Netherlands National Aerospace Laboratory – NLR. The objectives of this simulation were to find whether the theoretical FCL design could be accepted and flown by the pilots. The first evaluation determined that an increased level of effort would be required to rework the initial FCLs.

At that time we also built a fixed-base flying-quality simulator. A definite difference was found between the fixed- and the moving-base simulators. It is strongly believed that, in order to develop FCLs for a highly augmented fighter, a moving-base simulator is a necessity. It was found that motion gives the pilot much more information than that which can be gained from a fixed simulator; the final results are far better.

A great deal of basic design can be carried out on a fixed-base simulator, but good flying qualities can best be achieved by using the motion simulator. Increased assets derived from moving-base simulation encompass all high-gain tasks, especially air-to-air tracking, air-to-ground bombing, close-formation flying and landing. In addition to these two ground-based simulators, the Calspan's NT-33 in-flight simulator was used. Owing to the somewhat limited capabilities of the NT-33 to simulate Lavi's dynamics in the entire flight envelope the objectives of this simulation were to confirm the closed-loop high-gain landing approach phase. This in-flight simulation confirmed the initial ground-based simulation. Most of all, it gave the assurance that the FCLs were good.

Pilot involvement in development of the flight control system began $3\frac{1}{2}$ years before the first flight. A total of 835 simulation hours was flown. A major lesson learned during that effort was that a complete aerodynamic, aeroelasticity and aero-servoelasticity data package should be available to the cognizant FCL engineers as soon as possible. Because this data package was not available on time, the set of FCLs had to be updated several times during the simulation phases. Further, final design and weight and balance changes forced us to enlarge the elevons' area in order to maintain a sufficient pitch-down moment at a given angle of attack (AoA) and aircraft configuration. This change resulted in a new wing version with enlarged elevons and caused a significant FCL redesign.

The build-up of applicable simulation methodology was a task in itself. A primary result of our simulation effort was a growing recognition that, to gain full benefit from the programme, each flight had to be executed as a real test flight. Therefore, a team consisting of two flight control system engineers, a flight test engineer, a pilot and a senior FCS engineer supervised each flight. These flights began with a preflight briefing covering specific test objectives and fully detailed testing methods. Flights lasted about an hour and a half and were run on an item-by-item basis duplicating airborne testing. Flights were video recorded because this contributed significantly to the success of the FCL development. It helped the test pilots to explain and show the FCS engineers what they were talking about. A typical post-flight debriefing lasted 3 hours. By having the FCS engineers actually share the pilot's sensations through the video, the pilots were able to explain fully what they wanted.

It is worth mentioning that the time and effort put into the different ground-based and in-flight simulations, including engineering the six degrees of freedom (SDF) computer program, produced very good FCLs and, later on, excellent aircraft flying qualities. It is the authors' opinion that future aircraft and FCL development should use extensive ground and in-flight simulation, minimizing unnecessary flight test risks.

Four IAF and six IAI pilots participated in the $3\frac{1}{2}$ year simulation phase. As all of the pilots involved have combat experience, they brought to the programme a high degree of applicable know-how. Some of the pilots have a degree in aeronautical engineering, are graduates of test pilot school or hold both qualifications. All of the IAI pilots are on active reserve duty. Their background experience covers most of the past and current IAF inventory, namely the A-4, F-4, Mirage, Kfir, F-15 and F-16. In addition, they have experience of transport aircraft, which was found to be very helpful, especially for the development of the autopilot. The availability and utilization of different flying techniques associated with the different background experience produced an important advantage. The 'bandwidth' of the Lavi's flying qualities is wide enough to span the various flying techniques of the Lavi aircraft's future pilots. The operational aspects of flying qualities such as A/A tracking, A/G bombing and air combat manoeuvring (ACM) were an important part of Lavi flight control development. At one point of that development it was decided to stop the 'fine tuning' for A/A tracking.

This decision proved to be very cost effective, as actual flights showed that some aerodynamic coefficients were different from those which had been predicted with wind tunnel data. It was felt that further development would best be accomplished at a later stage, when flight test data would update simulation.

6. Special control law features

6.1. *Pedal-to-aileron interconnect (PAI)*

While trying to decrab the aircraft before touchdown, pilots experienced difficulties in the roll axis control, when a significant crosswind factor occurred during the landing approach. This phenomenon was more aggravated in the A/G load configuration when the moment of inertia around the X axis (I_{xx}) was significantly larger. Holding a constant runway heading with rudder and aileron inputs caused a definite lateral pilot-induced oscillation (PIO).

There was no apparent reason for the PIO that appeared in the simulation. In order to minimize the pilot's lateral inputs a new interconnect was implemented. A rudder pedal input automatically introduced aileron deflection, a 'mirror' image of the ARI. The results were exceptionally good: Level 3 handling-quality ratings rose to Level 1. Figure 13 presents strip chart data on how lateral stick activities and aircraft responses were reduced.

6.2. *DFCS to EFCS auto transfer*

The IAF specified a 'Fail-Op Fail-Op Fail-Safe' concept for the FCS. This specification required a backup. That backup must, at all times, follow the DFCS.

FLIGHT CONTROL LAWS DEVELOPMENT
KPAI
10 KNOTS CROSSWIND LANDING - DECRAB

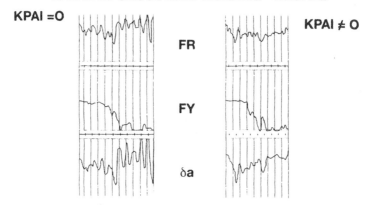

KPAI =O   KPAI ≠ O

FR

FY

δa

FR - ROLL STICK FORCE, FY - YAW PEDAL FORCE, δa - AILERON DEFLECTION

Figure 13.   Lavi KPAI.

During manoeuvring, the stick can be at maximum travel. In this case, N_z or AoA limiters override the pilot command and actually prevent the aircraft from overstress or departure. When an automatic transfer to the backup system occurs, the aircraft could depart or overstress because no limiters exist in this system. In order to eliminate this danger, a special feature was introduced. At the instant of transfer, for 1 s, the stick input to the FLCC is limited. The pilot feels the transfer due to the reduced manoeuvring. The pilot then releases the stick and, after 1 s, can resume command in the backup mode knowing the systems's limitation. Figure 14 presents an example of a manual $3g$ DFCS-to-EFCS transfer.

FLIGHT CONTROL LAWS DEVELOPMENT
DFCS ⇒ EFCS AUTO TRANSFER

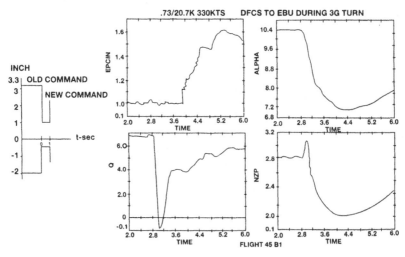

Figure 14.   Lavi DFCS to EFCS auto transfer.

6.3. *Canard schedule*

The 'delta–canard' combination requires different canard positions and responses to various flight conditions. Overall it is activated dynamically to improve the initial response of the aircraft and statically to maximize L/D.

In the supersonic regime, the canard is biased as a function of AoA, releasing hinge moments from the elevons, allowing better manoeuvrability.

In the subsonic regime, the canard is activated together with the elevons in high AoA and aft c.g. to provide maximum pitch-down moment.

In the power approach (CAT-C) the canard is biased in order to decrease AoA, at a given approach speed, so improving forward visibility. On touchdown, the canard moves to pitch down the nose.

6.4. *Roll-command gradient as a function of load factor*

The basic roll-command gradient had a parabolic shape. During the piloted simulations it was found that the overall roll command is too large at a high load factor. In addition, the roll-axis response was too sensitive for precise tracking at high g. For this reason, the roll gradient coefficients were scheduled as a function of load factor. This relatively simple feature (Fig. 15) solved the above-mentioned problems, and resulted in very good pilot comments. Another important benefit was

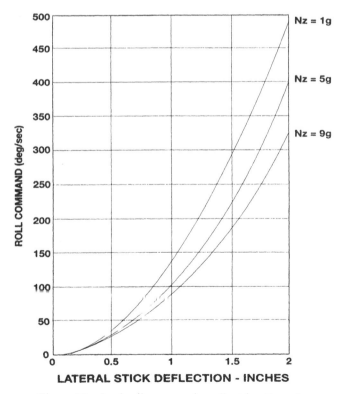

Figure 15.   Lavi roll-command gradient function of g.

the significant reduction in aircraft loads at high *g*. All together, this feature proved to be highly beneficial.

6.5. *Anti-spin*

The so-called 'Anti-spin' mode of the FCL was designed for two purposes. The first was to avoid, with all possible power, aircraft departure from controlled flight. The second was to activate the control surfaces during spin in an anti-spin direction.

Anti-spin mode was designed to be engaged and disengaged automatically, as a function of AoA. Special logics were inserted into this mode to prevent the aircraft from entering into inverted flight (or inverted spin) and, if it happened to recover from this 'unpleasant' situation, to an erect flight.

This mode was intended to be tested during the flight test of the B-4 prototype, which was supposed to be equipped with an anti-spin chute, but this prototype was never built. In the other prototypes that flew (B-1, B-2, T/D) the mode was never activated, as the AoA limiter function very successfully prevented the aircraft from departing.

7. How does it fly?

7.1. *First flight*

The results of the extensive FCL development and test programme promised that the Lavi would offer good handling qualities. However, many pertinent questions remained unanswered. The most critical questions concerned lift-off and touchdown transients. These transients were extremely difficult to simulate owing to the uncertainties in wind tunnel data. Utilizing the number 1 prototype, several high-speed ground runs were performed. These tests included an idle thrust full rotation.

The first flight takeoff was accomplished with aircraft rotation and lift-off achieved at some 20 knots faster than the speeds recommended in the flight manual. Takeoff was both smooth and easy, with aircraft performance similar to simulation results.

A known problem was a pitch-down 1 s after lift-off. Before the flights, the stability of the voted AoA value was in question. Therefore, it was decided to engage the AoA feedback only after lift-off, which caused a slight pitch-down of about 1° θ. After collecting data from several high-speed ground runs, takeoffs and landings, a solution was found. It was incorporated in the technology demonstrator (FCL) which shows a very smooth transition from ground to air and vice versa, in lift-off and touchdown respectively.

Thanks to the enlarged elevons, elevator power is larger than the same coefficient in the B-1/B-2 wings. This elevator power enables the pilot to rotate the aircraft's nose earlier on takeoff and also to hold it up, for aerobraking, until 60 knots during landings.

Both the FCL and aerodynamic configurations, namely the B-1/B-2 and the T/D, achieved, as predicted, very good handling qualities. In fact, the T/D is flown by many operational and test pilots from several countries. The aircraft is going through operational scenarios, both in A/A and A/G. It has very good responses to the pilot's inputs and demands, from *slower* than 80 knots to the open flight envelope of 600 knots at low altitude. Loading the aircraft with two MK-82 bombs

(500 lb (227 kg) each) on the outboard wing stations did not deteriorate the handling qualities, although the aircraft is flown with the A/A FCL configuration only, which shows that the robustness of the task-tailored FCL is good enough to successfully handle the aircraft's moment of inertia changes.

The dual control stick installation in the T/D, with the single-seater artificial feeling spring in the pitch axis, causes slightly high pitch forces. This phenomenon was mentioned by several of the pilots, after performing slow-speed air combat manoeuvring. These manoeuvres were performed at such slow speeds that the AoA feedback and limiter was very pronounced, requiring the pilot to hold the stick in the full aft position.

7.2. *Problems uncovered and solutions*

During the first 15 test flights of the number 1 prototype, two problems were uncovered. Both of the identified problems were related to lateral–directional characteristics. In A/A tracking, small aileron inputs caused too much yaw, which resulted in a snaking motion. The other problem was larger than predicted sideslips during roll reversals.

These problems became more pronounced on the T/D. This aircraft is flown with the instrumentation pod installed on the centreline fuselage station. This

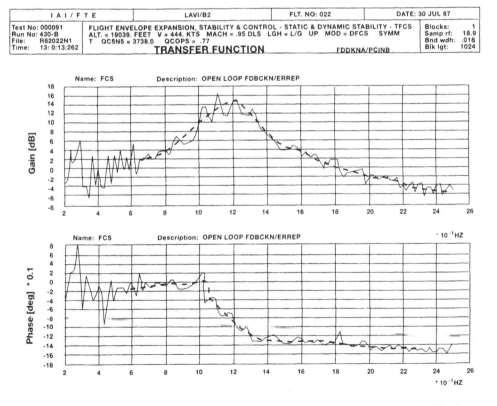

Figure 16. Lavi B-2 open-loop frequency response. Flight Test Results – LAVI B2. Open Loop Frequency Response. Mach = 0.95, Alt = 20 Kft.

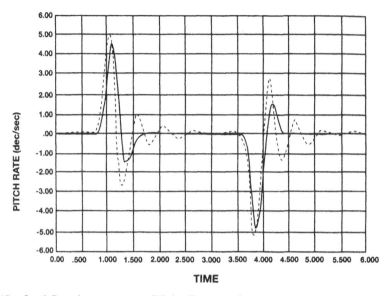

Figure 17.   Lavi B-1 time response. Flight Test Results – LAVI B1. Pitch Oscillations at Mach = 0.95, Alt = 36 Kft. SDF ——, F.T. ------.

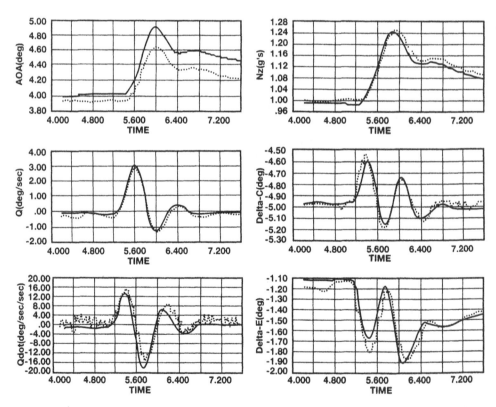

Figure 18.   Lavi T/D time response. Flight Test Results – LAVI-TD. Pitch Oscillations improvement. Mach = 0.95, Alt = 36,000 ft. SDF ——, F.T. ------.

reduces the directional stability ($C_{n\beta}$) and causes even larger sideslip angles (β) than the B-1/B-2 appeared to have, but the pilots did not really have any problem with this. However, it was suspected that the aircraft might depart from this performance if this tendency of low $C_{n\beta}$ was aggrevated at high α. It did not happen. Even with the flight test pod, there is still enough directional stability to manoeuvre freely at low airspeed/high α.

Another phenomenon was discovered in the transonic regime, where lower than predicted pitch damping caused pitch oscillations at Mach 0·95.

This was a real problem that was due to a combination of unpredicted phenomena: that is, the destabilizing effect of damping in pitch (C_{mq}), larger elevator efficiency and much higher static margin – up to 3% more stable than predicted at this Mach number. All these resulted in a near 25° phase margin at low altitude (9500/10 000 ft (3000 m)). This problem was solved on the T/D version, by inserting a lead–lag network in the forward loop. Gains were updated in the pitch damping path, in the elevator power gain and in the static stability path.

Figure 16 is a Bode plot of the open-loop transfer function of pilot oscillations, flying the B-2 at 19 000 ft (5790 m). The PM is near 30°.

A typical time response at 0·95 Mach/36 000 ft (10 975 m) flown on the B-1 is shown in Fig. 17. The dashed curve is the aircraft flight response, and the solid curve denotes the design intention results from the six degrees of freedom (SDF) simulation. Figure 18 shows the improved Lavi T/D time response.

8. Flying-qualities data

Figures 19–24 present the Lavi's flying qualities and characteristics as exhibited during the flight tests. The presented results reflect the aircraft's performance with respect to the MIL-F-8785C and IAF specifications.

Stick force per g versus N_z/α in CAT-A (Fig. 19), ω_{nsp} versus N_z/α from Q/F_s (Fig. 20) and from α/F_s (Fig. 21), are well inside Level 1.

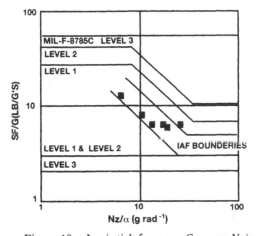

Figure 19.   Lavi stick force per G versus N_z/α.

FLYING QUALITIES DATA CATEGORY A

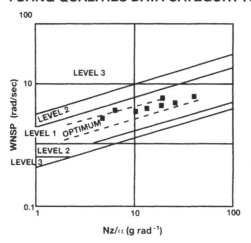

Figure 20.   Lavi ω_{nsp} versus N_z/α from Q/F_s.

FLYING QUALITIES DATA CATEGORY A

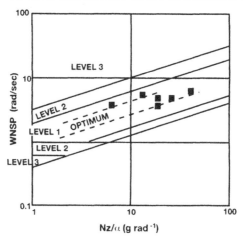

Figure 21.   Lavi ω_{nsp} versus N_z/α from α/F_s.

The time delay, from Q/F_s time history at the given flight condition (Fig. 22), shows a deviation of 20 ms from Level 1. This result takes into account the stick dynamics, which contributes nearly 35 milliseconds to the overall equivalent time delay. In the beginning, the worst case was considered. However, today we believe that only a portion of the stick dynamics' delay should be taken into account. The 20 ms deviation did not deteriorate handling qualities as far as could be judged, which only strengthens the possibility that there was no deviation at all. In the T/D FCL version, the equivalent time delay is smaller by 10 ms.

The results in the lateral–directional axes are similar to the longitudinal axis. Figure 23 presents the dutch roll frequency and damping ratio that are in Level 1,

FLYING QUALITIES DATA

LON BA-A   .50/31K AL   ALFA=11.94   .99G   53.56%

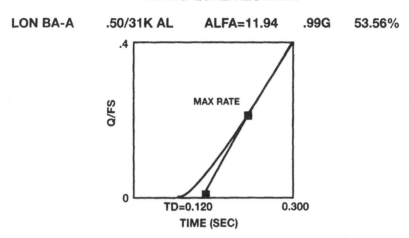

Figure 22.   Lavi time delay from Q/F_s time history.

FLYING QUALITIES DATA

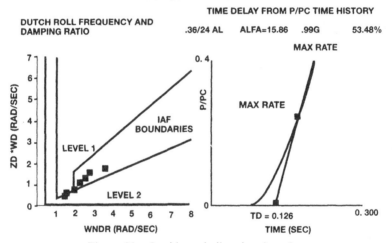

Figure 23.   Lavi lateral–directional results.

TASK / PILOT #	HIGH SPEED TAXI (T.O. & LDG)	ROTATION	LIFTOFF	LANDING	FLARE & TOUCHDOWN
	HDG ±3°	INITIAL θ±2°		ON FINAL θ,α = ±1°	SPOT LDG ± 150'
	1/8 OF RWY WIDTH	MAINTAINING θ ± 1°	θ ± 1°	1/4 RWY WIDTH	α ± 1°
				Vc ± 3 KTS	Vc = 3 KTS
1	2	2, 2	4	2, 1, 3	2, 2-3, 2-3
2	2-3	2, -	6	2, 1, 3	2, 2, -
3	2	4, 3	4	3, 2, 3	3, 3, 4
4	2-3	2-3, 2	6	4, 4, 4	2, 3, -
5		2, -	2	2, 2, 2	2, 2, 2
SUMMARY	**2.3**	2.5, 2.3 **2.4**	**4.4**	2.6, 1.5,2,75 **2.3**	2.2, 2.5, 2.8 **2.5**

Figure 24.   Lavi flying-qualities pilot ratings.

and generally within IAF specification boundaries. The equivalent time delay of the roll axis (Fig. 23) is close to the longitudinal delay.

The stick dynamics effect on the roll-axis delay is similar to the effect on the pitch delay.

A summary of flying qualities pilot ratings is presented in Figure 24.

9. Status

Since the first flight on 31 December 1986, prototypes 1 and 2 (B-1, B-2) flew a total of 81 flights (78 flight hours). Performance envelopes attained during these flights include:

(*a*) a maximum of Mach 1·45 from above 36 000 feet (10 975 m) to 43 000 feet (13 110 m);

(*b*) a maximum of 540 knots at 10 000 feet (3050 m);

(*c*) a minimum speed of 110 knots and 23° true angle of attack, as well as 7·5*g*, were demonstrated.

On the technology demonstrator 113 flights were flown until 31 December 1992, amounting to 129 flight hours. Performance envelopes include the following:

(*a*) maximum of Mach 1·2;

(*b*) maximum 600 knots at low altitude;

(*c*) minimum speed of less than 80 knots was flown during combat evaluation manoeuvring, as well as 25° AoA and 8·2*g*.

A maximum of Mach 1·8, 800 knots, 50 000 ft (15 245 m), 9*g* and 25° true angle of attack was the designed envelope.

The T/D is continuing to fly routinely, testing, evaluating and demonstrating the Lavi's avionics and exceptional airframe capabilities.

10. Summary

Comparing the Lavi's FCL and FCS to the Kfir's Autocommand (which was developed and produced during the early 1970s) shows significant differences. The Autocommand, an analogue pitch control augmentation system, gives the pilot some improvements in areas where the unaugmented aircraft exhibits some handling-qualities problems. Low-altitude high-speed flight characteristics as well as air-to-air and air-to-ground fine tracking have been improved. Very heavy takeoffs, where smooth and precise pitch control is needed, have also been improved.

The basic control laws of pitch rate (Q), load factor (G) and some AoA feedback are being used in both systems.

The technology level, and the precise tailoring of the advanced digital fly-by-wire FCS in the Lavi, is a 'full-time' three-axis control system, providing the pilot with an excellent handling-qualities combat flying machine. The pilot is free to operate the aircraft's combat systems, with minimum compensation and workload in flying the Lavi.

11

Digital autopilot design for combat aircraft in Alenia

ALDO TONON and PIER LUIGI BELLUATI

Nomenclature

AP   Autopilot
FD   Flight Director
PAH   Pitch Attitude Hold
RAH   Roll Attitude Hold
HH   Heading Hold
ALT   Baro Altitude Hold
TRK   Track Acquire and Hold
HDG   Heading Acquire and Hold
TH   Attitude Hold
AH   Altitude Hold
BH   Bank Hold
AA   Altitude Acquire
HA   Heading Acquire
AC   Autoclimb
RMS   Root Mean Square
FCC   Flight Control Computer
FCS   Flight Control System
A/C   Aircraft
C/L   Control Law

1. Introduction

Alenia, recently formed by the merger of Aeritalia and Selenia, has inherited the extensive experience of the Aeritalia-Defense Aircraft Group in the design and development of combat aircraft. In the last 20 years Aeritalia (now Alenia) has been involved with the development of autopilots for modern combat aircraft, mainly associated with three major programmes currently in three different stages of development.

    (a)   For the Tornado programme, whose development was completed in the mid-1980s, Aeritalia has performed extensive assessment work in support of MBB for the development of the Automatic Terrain-Following System.

(b)   The AMX programme is in the full production phase, but development of some advanced system functions is ongoing. The aircraft was developed together with Aermacchi of Italy and Embraer of Brazil, Alenia being the prime contractor. The AMX is a single-seat, single-engine subsonic aircraft. It was designed to give close air support to land or naval forces. Therefore the AMX has to operate mainly at low altitude and high subsonic speed, and has to carry a large amount of air-to-ground and air-to-air weapons. AMX has a high battle damage tolerance due to its configuration. The FCS is based on a fly-by-wire technology, which incorporates a hybrid analogue and digital FCC, and performs stabilization and control augmentation on pitch, roll and yaw axes. The digital AP integrated in the FCS is provided to reduce the pilot's workload throughout the flight envelope.

(c)   EF 2000 is an air superiority fighter. The high instability of the airframe is stabilized by a quad-redundant, fly-by-wire, full-authority FCS. The EF 2000 programme recently reached a major milestone having started flight testing. Three companies are involved with Alenia in the EFA development: BAe (UK), CASA (Spain) and DASA (Germany). Concerning the FCS design, Alenia has responsibility for the design of the basic AP control laws. The EF 2000 AP is integrated in the FCS and is designed to alleviate pilot workload and to perform fully automatic procedures during flight, combat and approach phases.

2. AMX flight control system (FCS)

The AMX FCS provides stability augmentation and control in pitch, roll and yaw axes. A general layout is provided in Fig. 1 and Fig. 2.

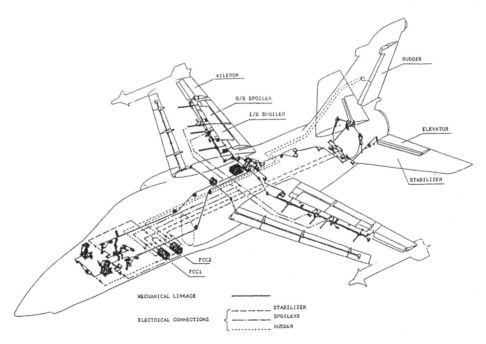

Figure 1.   AMX FCS layout.

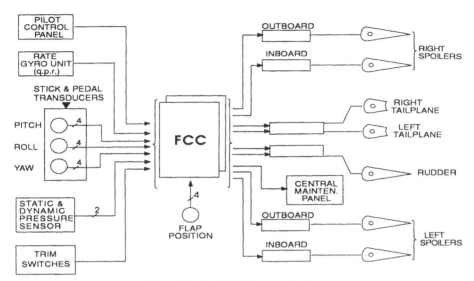

Figure 2.   AMX FCS general scheme.

Pitch and roll control is provided by a conventional mechanical system with a fly-by-wire electronic augmentation (EFCS) which guarantees full performance capability (Level 1 handling qualities). This allows reversion from powered to manual mode in the case of total loss of hydraulic power, guaranteeing aircraft safe re-entry (Level 3 handling qualities) after failure of both the hydraulic and both electrical circuits. Yaw-axis control is provided only by the EFCS with no mechanical backup. The primary flight control system is managed by the pilot by means of a conventional control stick and rudder pedal. All controls are powered by two independent hydraulic systems.

The EFCS controls the movements of stabilizers, spoilers and rudder providing pitch, roll and yaw damping and trim capability. It assures adequate responses throughout the flight envelope. Secondary flight controls are flaps and slats while the spoilers are also used as airbrakes and lift dumpers.

Control stick displacements are mechanically transmitted through conventional rods and cables to four hydraulic actuators that move, respectively, the two aileron surfaces and the elevator surfaces.

With this configuration the FCS is still able to operate safely in the event of a second electrical or hydraulic failure.

2.1. *Flight control computer (FCC)*

The FCC was designed by Alenia Avionic Equipment Division and GEC Avionics. It is a dual duplex self-monitoring system which provides full operation following a first failure. Each FCC contains one analogue lane and two digital lanes:

Analogue lane. This is used for the actuator control loops and to compute the primary command functions.

Digital lane. The digital lanes are arranged to perform different roles. Digital lane A has a dual role: it generates the digitally implemented control functions and monitors the analogue command computing, while lane B computes a complete model of the analogue and digitally implemented functions, and provides an independent monitoring function to both the analogue command lane and digital command lane A.

The monitoring is arranged such that each digital lane acts as an independent monitor of the analogue command computing. The processors are not synchronized in any way and therefore each processor requires one frame to sample the command computing and make a valid comparison.

The architecture and monitoring have been configured to enable the EFCS to operate safely and minimize failure disconnection transients. This has been achieved with a minimum use of analogue components by making full use of digital computing techniques.

2.2. *Autopilot function*

The digital AP function is integrated within the FCC and provides command signals, via the FCC analogue channels, to the stabilizer and spoiler actuators.

2.2.1. *Autopilot modes.* These are divided into the two categories indicated below:

(*a*) *Basic modes:*

- Pitch attitude hold (PAH).
- Roll attitude hold (RAH) or heading hold (HH), depending on the value of the bank angle at the time of engagement.

(*b*) *High-level modes:*

- Baro altitude hold (ALT).
- Track acquire and hold (TRK).
- Heading acquire and hold (HDG).

In addition the following functions are available:

Autopilot override capability. This facility provides automatic, temporary AP disengagement in both pitch and roll axes by pressing an appropriate switch located on the hand grip. When this switch is released the AP is re-engaged in the basic modes; in addition if the ALT mode was previously engaged it is re-engaged if its re-engagement conditions are satisfied.

Autotrim capability on the longitudinal axis. This function is provided in order to trim the aircraft properly according to the flight condition.

Datum adjust facility on the longitudinal axis. This facility is provided in order to update the longitudinal reference parameters. By operating the normal pitch trim switch, this allows, with a limited authority, the pitch angle datum reference to be varied during the PAH mode operation, or it permits the altitude datum reference to be varied during the ALT mode operation.

Instinctive cut-out (ICO). This facility provides immediate deselection capability for all the AP and FD functions.

2.2.2. *Autopilot design criteria.* Principal consideration, which entered into the design of the AP function were:

- Pilot workload reduction.
- Integration with the FD function.
- The AP function is not mission critical.

These considerations imply: AP disengagement after the first failure of each FCC or avionic equipment such as the main computer, air data computer, inertial navigation system, secondary attitude and heading system; limited AP demand authority.

2.2.3. *System architecture.* A general description of the AP function architecture is shown in Fig. 3, and block diagrams for the AP pitch and the roll axis are shown in Fig. 4 and Fig. 5, respectively.

Inside the FCCs, both the modes selection and the control laws are implemented in the digital lanes.

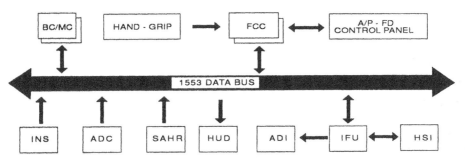

Figure 3.   AMX autopilot functional architecture.

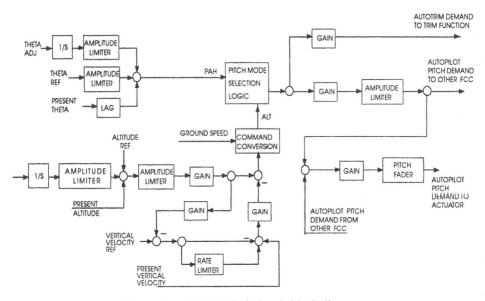

Figure 4.   AMX FCS pitch-axis block diagram.

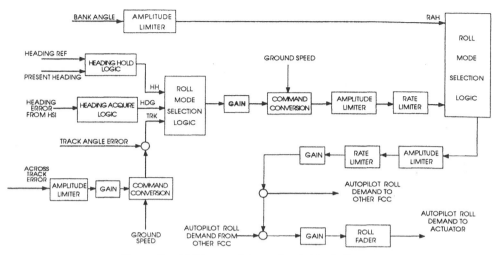

Figure 5. AMX autopilot – roll-axis block diagram.

The AP is integrated with and receives signals from the following sensors: the inertial navigation system, secondary attitude and heading reference system, air data computer; from the heading situation indicator; and from the attitude direction indicator. These signals are sent to the FCC by the main computer/bus controller via the 1553 data bus.

In addition the AP receives signals to select/deselect either AP or FD functions from the AP/FD control panel and the hand grip, through a hardwired connection.

Each digital lane of both FCCs is connected to the AP control panel and to the hand grip switches. Each signal is compared between the two lanes, and the consolidated signal is used in order to undertake the relevant actions.

The AP demands, both for the pitch and the roll axis, are crossfed between the two FCCs in order to minimize the mismatch between the actuator demands of the two FCCs and monitor the AP function performance within each FCC.

All duplicated data, which are available from different sources, are compared prior to being used. In addition the data sources used in the AP function are monitored for validity.

2.2.4. Autopilot preflight BITE capability. The preflight built-in-test (BITE) software function, implemented into each FCC, provides on-condition maintenance relative to the whole EFCS, with failure status being displayed on the central maintenance panel. Relative to the AP function BITE performs some tests involving data bus 1553 input/output data receive/transmit and input/output hardwired signal.

2.3. Autopilot software development

2.3.1. Software requirements definition. The software requirements are defined by means of:

- block diagrams relevant to the control laws;
- status diagrams relevant to the modes selection logic.

These requirements were implemented first on the flight simulator, in order to perform an assessment phase and later an optimization of the control laws. After this assessment phase, they were updated and sent in the same form to the supplier.

2.3.2. Software implementation. Although the AP function is not considered strictly a safety-critical function, the flight-resident software has been designed by the supplier in accordance with the guidelines and procedures developed for high-integrity fly-by-wire projects. This is because the AP function is integrated with the flight control basic function within the FCC.

Emphasis has been placed throughout the software development cycle on visibility, and this is achieved by the use of:

- simple software structures;
- a clear requirements definition by design audit and detailed documentation;
- rigorous production and configuration control.

The other aspect to be taken into account was the choice of the software instruction set. The Z8002 microprocessor has a comprehensive Assembler instruction set and addressing modes, but not all the instruction and addressing modes are compatible with the guidelines established by the supplier (GEC Avionics) for the development of the high-integrity software, which requires use of a simple instruction set and basic addressing modes. The indirectly addressed instruction, for example, has been prohibited.

2.3.3. Software qualification. The software qualification phase consists of two separate procedures, one performed by the supplier and the other performed by Alenia on the AMX FCS rig. Following completion of these procedures a software flight clearance is released.

(1) *Software supplier qualification.* The software integrity analysis concentrates on identifying potential software defects. This analysis is carried out using a bottom-up approach. The software is subjected to a series of independent design audits on the module design/code. The module design audits consisted of checks on:

  - compatibility of module design specification against software requirements;
  - adherence to codes of practice for design and coding;
  - overflow protection;
  - accuracy and completeness of documentation;
  - strict control of change incorporation;
  - completeness of test.

(2) *Software qualification by Alenia rig test.* In addition to the supplier software qualification, Alenia performs system-level tests on the AP software on its AMX FCS rig. The rig is a complete representation of the aircraft in terms of actual aircraft components. Avionic equipment is not installed in the rig, but is simulated by a dedicated host computer. In addition a bus analyser is provided in order to monitor the data transmission on the 1553 data bus between the FCC and the host computer. A logic analyser is also provided

for the microprocessor internal code monitoring in order to verify the correct computation of the received data, and the software module input and output data.

The aim of the rig tests is to verify the correct implementation of the following:

- AP control laws;
- modes selection logic;
- failure detection capability;
- system behaviour in case of failure;
- confidence test.

Each of these tests involves end-to-end checks to identify unexpected software behaviour or incorrect software implementation of the requirements.

3. Design of autopilot control laws

The design of the AMX AP was guided by two conditioning factors:

(a) The fulfilment of the general design criteria of the AMX FCS which established that no safety-critical functions shall depend on the FCC.

(b) The extension of the operational requirements at a mature stage of the design. The AP initially included only holding modes conceived to alleviate pilot workload during navigation. The acquire modes were requested at a later stage.

The above factors resulted in a challenging task for the designers as they had to cope with increased complexity of the control laws against a progressive reduction of the available memory size and throughput capability.

Present activities deal with tuning the acquire modes. The basic modes have already been successfully completed and flight tested.

The final design of the AP control laws was achieved through several iterations between the following steps:

- requirement analysis
- control law definition
- stability assessment
- performance assessment
- manned simulation
- flight trials

With each iteration of the process, a configuration that showed good agreement with stability and performances requirements was aimed for.

The AP control law design has been carried out according to the traditional design practice. Achievement of appropriate stability margins has been the basic driver in the AP control law synthesis. Extensive use was made of the classical tools, such as root locus and Nichols plots.

Once the required stability margins were achieved, the AP performances were evaluated through the non-real-time and manned simulation. Time-response criteria have been expressed in terms of static accuracy for the holding modes and of time-to-acquire and overshoot characteristics for the high-level modes. Non-compliance with requirements may lead to changes in gains or filters or to the introduction of non-linear elements, with the necessity of a new stability check.

Manned simulation played a significant role in the establishment of the mode logic and the AP functions associated with navigation. During early assessment work, there was a need for more precise performance assumptions, initially left up to the designer, but then discussed with pilots to get their point of view.

Later on, during the flight trials, additional tuning of the AP control laws dealt mainly with aspects such as aircraft sensitivity to steering commands.

3.1. *Basic modes*

These are the default modes entered on engagement of the AP. Schematic diagrams of the longitudinal and lateral AP control laws are shown in Fig. 4 and Fig. 5. PAH and RAH or HH are the holding modes. They operate to maintain, respectively, the longitudinal attitude, the lateral attitude and the heading existing at the time of AP engagement. Accuracy and performance requirements are derived from MIL-9490D.

PAH is the pitch attitude hold mode, and is engaged if θ is initially between $\pm 30°$. The accuracy is $\pm 5°$. Around the reference, a $5°$ attitude error has to be reduced to zero (taking in account the accuracy) in less than 3 s with a maximum θ overshoot not to exceed 20% of the disturbance. In the presence of turbulence, the error RMS has to be less than $5°$.

Root locus is used to verify the θ loop stability. An example is presented in Fig. 6. Proportional–integral compensation is applied to the θ error in order to reduce the residual error in the loop. A lag filter was introduced to attenuate high-frequency response. In Fig. 7, we can see a non-linear simulation of the aircraft response to a $5°$ θ input showing that the disturbance is in fact reduced to the accuracy threshold within the desired time.

RAH is the roll attitude hold mode and may be engaged within $\pm 60°$ of bank. Accuracy is of $\pm 1°$ around the reference and it is required that a $5°$ bank error has to be reduced to zero (taking in account the accuracy) in less than 3 s. In the presence of turbulence, the bank error RMS has to be less than $10°$. These requirements were satisfied without the necessity of using filters in the bank loop, while the gain value is scheduled with static pressure.

HH mode is selected if at engagement the bank angle is less than $\pm 7°$. In this case the present heading will be held; the requirements are similar to RAH mode. The heading error is converted to a bank error through groundspeed-dependent gain.

3.2. *High-level modes*

These are the modes that need preselection and the presence of a datum to be engaged.

- ALT is the barometric altitude hold mode. This mode maintains the altitude existing at the time of engagement.

- HDG is the heading acquire and hold mode.

- TRK is the track acquire and hold mode. A desired ground tack stored in the memory of the main computer is intercepted and followed selecting this

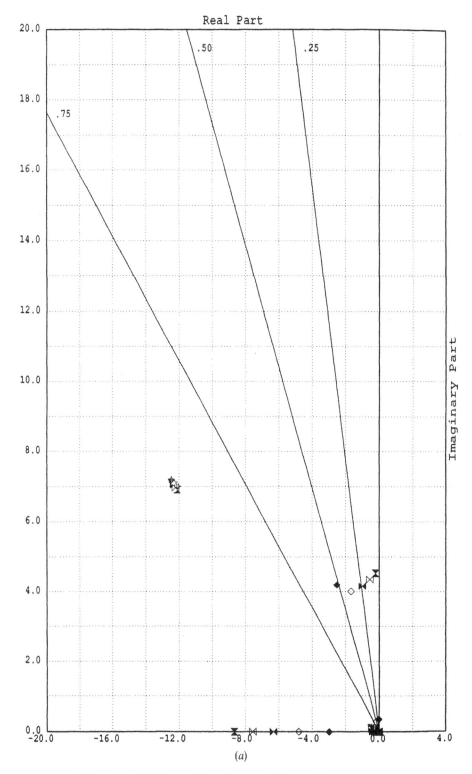

Figure 6. AMX autopilot – linear stability analysis (PAH mode).

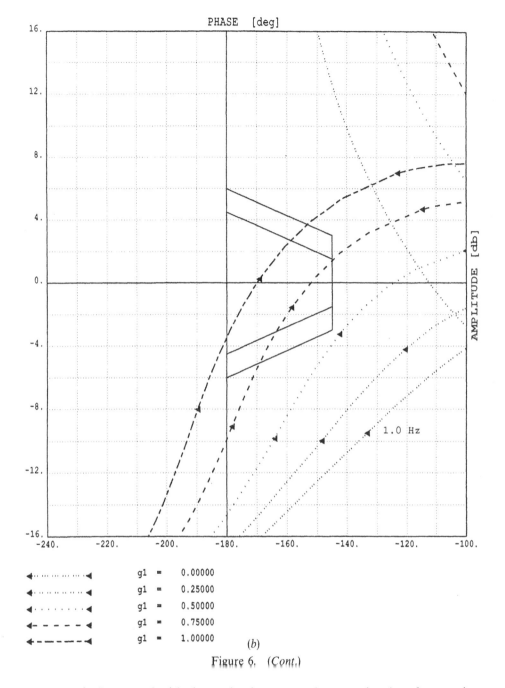

PHASE [deg]

AMPLITUDE [db]

1.0 Hz

	g1	=	0.00000
	g1	=	0.25000
	g1	=	0.50000
	g1	=	0.75000
	g1	=	1.00000

(b)

Figure 6. (Cont.)

mode. Integrated with the navigation system it steers the aircraft on a given course and operates to maintain it.

ALT mode accuracy and performance requirements are derived from the MIL-9490D requirement. The altitude error loop is assisted by an altitude rate loop to stabilize the phugoid mode. The altitude rate signal also makes the AP response faster against altitude disturbances. The altitude error is converted to a θ error by a

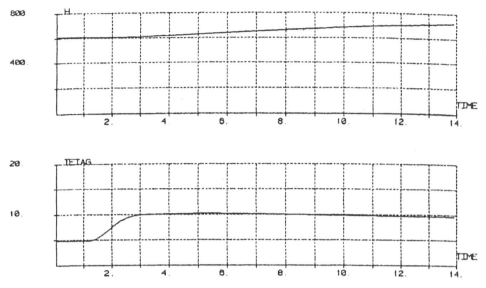

Figure 7.   AMX autopilot – non-linear simulation (PAH mode).

gain scheduled with the groundspeed and is processed by the θ loop of the AP control laws.

TRK mode performs a manoeuvre to null the distance and the angular error between the desired track and the actual track. The cross error and the heading error produce a bank demand that steers the aircraft onto the desired track. The mode may acquire only one track (manual/steering and radio/steering modes) or may follow a sequence of tracks that have been stored in the main computer before the flight or during the mission (auto/steering mode). In manual and radio modes the AP, after TRK selection, will engage the HDG mode until an adequate distance

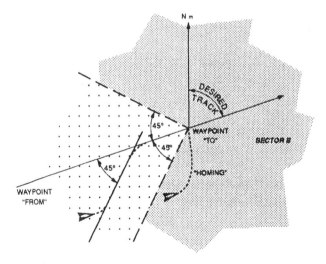

Figure 8.   AMX autopilot – TRK mode engagement.

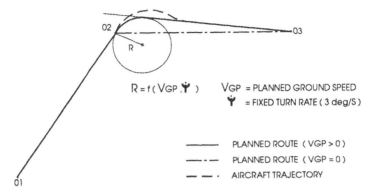

Figure 9. AMX autopilot – TRK mode leg changeover.

from the track is attained, and then track acquire will be performed. In auto/steering mode, after TRK engagement the nearer track will be selected and the AP will turn the aircraft to acquire a heading that intercepts this track at a 45° angle. Then, as with manual/steering and radio/steering modes, the AP will start the track acquisition when close to the track. If at TRK auto/steering mode engagement the angle between the heading and the track is more than 45°, the autopilot will drive the aircraft directly to the waypoint (Fig. 8). The next track will be acquired with the leg changeover defined in Fig. 9. The overshoot is due to the limited authority of the AP, through which the 3·0 deg s^{-1} target turn rate is not always attainable for every condition. This leads to waypoints which are defined by the minimum turn radius at that phase of the mission.

HDG requirements are similar to HH requirements. The heading error is converted to a bank error which is then sent to the bank demand loop. The maximum bank angle is 30° or 45°, depending on the dynamic pressure.

3.3. *Flight test results*

Flight tests resulted in several modifications to the AP control laws. This indicates, to some extent, a sometimes inadequate prediction of some aircraft behaviours.

The modifications fall into three basic categories:

(a) flight conditions in which the build-up of the acceleration plays a significant role;

(b) operational aspects;

(c) underestimating some external disturbances.

Examples referring to the above conditions are reported in the following sections.

3.3.1. *Acceleration onset at the pilot station.*
Keeping in mind the MIL requirements, 20 deg s^{-1} was chosen as the maximum roll rate for the HDG mode, in order to minimize the acquisition time. Pilots reported that 10 deg s^{-1} was a more desirable value, because the rise in comfort justified the performance reduction. Figure 10 shows a heading acquisition with the roll-rate limit set to the lower value.

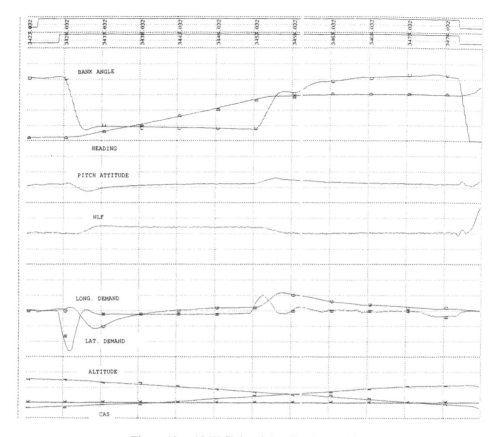

Figure 10.   AMX flight trials – HDG acquisition.

During heading acquisition, as the heading approaches the target value, the small heading error generates a small bank demand, giving a low turn rate and resulting in a long final phase. For this reason the heading error is forced to a constant value in the initial acquisition phase, faded to the computed error close to the datum. In the first issue of the control law, the reduced time to acquire the desired heading was achieved at the expense of a quite crisp lateral response. In accordance with the pilots' suggestions a reduced bank/heading ratio sensitivity was introduced. With this consideration a study is in progress and is intended to modify the logic by introducing a bank demand as a function of groundspeed and heading error, giving an open-loop signal that is faded to the heading error by a function of the bank angle and heading rate to reach the datum in the desired time. Figure 11 shows non-linear simulation plots of heading acquisitions, comparing the time response of the two control law versions.

Some more work is necessary to define when, in TRK mode, fast turn-in and turn-out manoeuvres are desirable, and when a gentle manoeuvre is better. A slow manoeuvre is necessary when acquisition starts at a low initial crosstrack error, and if the acquisition time increases. A compromise between time to acquire and comfort is under examination and the pilot's experience will be helpful in redefining the requirements.

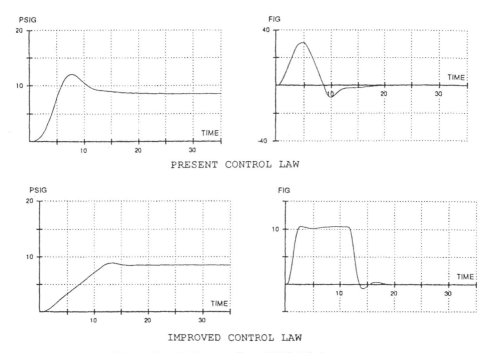

PRESENT CONTROL LAW

IMPROVED CONTROL LAW

Figure 11. AMX autopilot – HDG C/L improvement.

3.3.2. *Operational aspects.* In the first release of the AP control laws the system was designed to permit engagement of ALT mode only when the altitude rate was within ± 2000 ft min^{-1} (610 m min^{-1}). Beyond this range, the mode was not entered; thus the pilot action produced no effect. Our test pilots found this to be disagreeable, so they asked us to allow the mode selection even with high vertical speeds (up to $\pm 10\,000$ ft min^{-1} (3050 m min^{-1})) provided an appropriate control of the normal load factor build-up and the aircraft motion developed in the expected direction. For this reason the ALT logic was modified by introducing a non-linear element that reduces the altitude rate input signal at engagement, summing a constant opposite signal that is faded to obtain proper g and g-onset levels. Figure 12 illustrates the ALT flowchart with this modification.

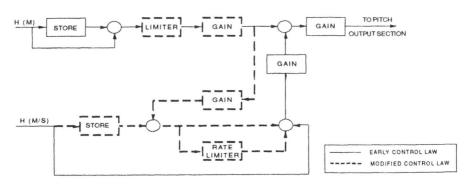

Figure 12. AMX autopilot – ALT C/L modification.

3.3.3. *Effect of external disturbances.* The radio/steering mode, in which a pilot-selected course to/from a TACAN station is captured and maintained, has led to some problems. The strong noise present in the radio signals used to generate the steering commands was underestimated. The situation is being readdressed with a massive filtering technique and lowering the gains close to the course.

4. EF 2000 FCS and AP description

EF 2000 is aerodynamically unstable in pitch and yaw axes. This comes from the necessity to obtain high performances and agile behaviour. A quadruplex full-time, full-authority digital FCS provides stabilization and control as basic functions. This FCS has no mechanical backup. In addition to the basic functions the FCS also provides carefree manoeuvring capability and has integrated autopilot/autothrottle functions.

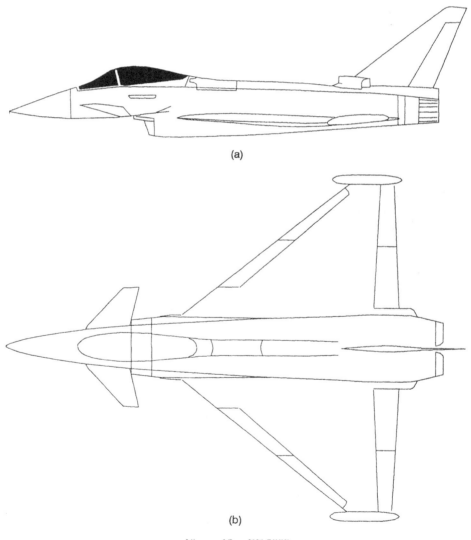

(a)

(b)

Figure 13. EF 2000.

The aircraft is twin engined and has a closely coupled delta–canard configuration (Fig. 13). Primary control surfaces are flaperons, canards and rudder; secondary controls are the leading-edge slats, air intakes and the airbrake.

Alenia has responsibility for the design of the basic AP control laws. The basic modes are:

- attitude hold
- altitude acquire
- heading acquire
- autoclimb

The AP requirements for stability and performance follow the MIL-F-9490D specifications.

In the current stage of development, the EF 2000 AP control law has been extensively verified by linear and non-linear simulation, while manned simulation is used to check the moding logic.

The AP control laws are divided into two loops: the longitudinal loop, where altitude or θ are controlled, sending a pitch-rate demand to the FCC; and the lateral loop that controls bank angle and heading and outputs a roll-rate signal (Fig. 14). AP inputs are introduced into the control law at virtually the same point

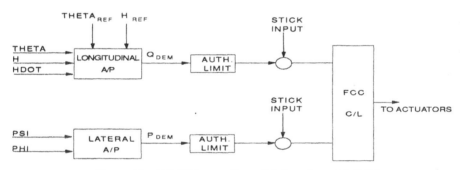

Figure 14.   EF 2000 autopilot – C/L interface with FCC.

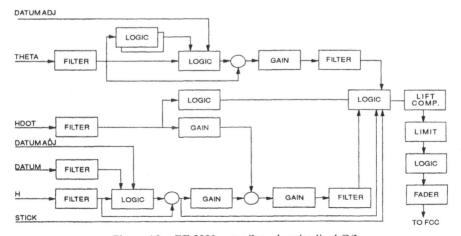

Figure 15.   EF 2000 autopilot – longitudinal C/L.

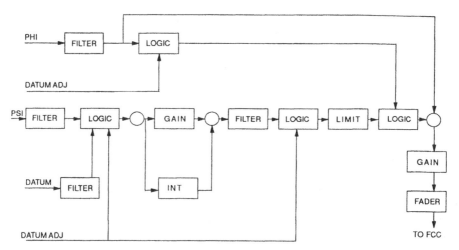

Figure 16. EF 2000 autopilot – lateral C/L.

as pilot inputs. In this way, the carefree functions and the response dynamics of the basic aircraft are maintained. AP signals have an authority limitation before the summing point. The block diagrams of the longitudinal and lateral AP control laws are illustrated in Fig. 15 and Fig. 16.

4.1. *Design of the EF 2000 autopilot*

Experience from the work on the AMX AP has been applied to EF 2000 AP design, so the layout of the control law structure has been made easier, starting with a precise view of the problems that had to be considered.

The addition of new control loops to the basic FCS control laws made it necessary to perform multiloop analysis (Fig. 17) to verify the stability margins of the outer loop (i.e. altitude loop) and the inner loop (basic FCS).

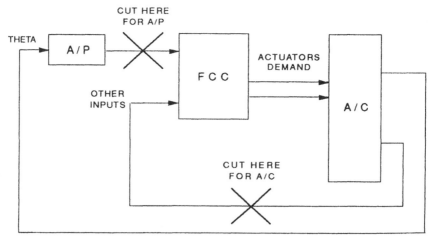

Figure 17. EF 2000 multiloop analysis.

The AP control laws were designed using classical design tools. The control law structure was defined starting with the analysis of the frequency and time response with a linearized model. Nichols and Bode plots were used to verify the stability requirements, while linear model response has been used to assess the system performance in the time domain. In the second step non-linear simulation was used to evaluate the system response to large manoeuvres and to modify the structure, adding non-linear elements where needed. The third step was to verify the moding logic through manned simulation. In this phase a large amount of non-linear elements were introduced to comply with moding requirements. Simulation model codes for batch and manned simulation are essentially the same, so their response is always in good agreement and model updating is easy.

4.2. Hold modes

The AP may be engaged in any of the following submodes, depending on the instantaneous aircraft attitude at the moment of engagement:

- θ hold (TH)
- altitude hold (AH)
- heading hold (HH)
- bank hold (BH)

If the absolute value of altitude rate is beyond a given threshold, TH mode will be present, otherwise AH mode will enter the aircraft in level flight. In the same manner, depending on whether or not bank angle is less than $\pm 7°$, BH mode or HH mode will be engaged.

When engaged, these submodes will tend to null error and maintain the datum value according to MIL requirements. Each reference datum may be modified by the pilot applying small corrections with the stick. A large stick deflection will deselect the AP and return full authority to the pilot.

AH may also be engaged by pressing the preselection button prior to AP engagement or during TH mode. The altitude at the time of engagement will be stored as the altitude datum. In a similar way HH mode may be preselected.

TH mode control laws have a path with a gain and a filter applied to the θ error. Gain is constant, while the filter time constants are scheduled with the altitude to give a smooth response to disturbances.

AH mode also needs the altitude rate signal for stabilization. The maximum normal load factor is limited to $\pm 0.5g$ if the altitude rate is within ± 2000 ft min^{-1} (610 m min^{-1}) at the moment of engagement. Outside this threshold, the maximum load factor authority is increased in order to minimize the overshoot of the datum altitude. However, moding and authority details are still under study.

BH mode stabilizes the aircraft in a steady turn at the desired bank angle. To compensate for altitude losses during the transient phase, the appropriate command is generated and fed to the pitch axis.

HH mode processes the heading error with a proportional–integral gain (introduced to null heading error in the presence of asymmetric configurations) and creates a bank demand, input to the bank loop.

4.3. Acquire modes

Acquire modes may be preselected prior to the AP engagement. These modes are:

- altitude acquire (AA)
- heading acquire (HA)
- autoclimb (AC)

AA mode is engaged when the altitude datum is present and valid, AP is engaged and the mode is selected. These inputs may be given in any sequence to activate the mode. The AP will initiate acquisition only if altitude rate conforms with the datum to be acquired, otherwise AH mode will be engaged, with AA mode pending an appropriate input to initiate the climb/dive in the proper sense. TH mode will be engaged during the climb (or dive) until the altitude error is small enough to start a blending manoeuvre to reach the datum at a fixed normal load factor. During the TH phase (the initial constant attitude climb/dive phase) the pilot may change the altitude rate by modifying the attitude datum with the stick.

The normal load factor during the blending TH–AH phase is scheduled with the initial altitude rate, giving a stronger manoeuvre when the attitude is high, to obtain a reasonably short transition time. Close to the datum AH mode will stabilize the aircraft. Figure 18 shows an acquisition starting from level flight.

Provision for an autolevel function is under study. This function will cause the aircraft to level off (provided the AP is engaged) at a certain altitude over the ground. If AH mode is engaged and the altitude rate is negative, the system will alert the pilot when altitude becomes low and will automatically engage AA mode to level off at an adequate predefined altitude over the ground.

HA mode may be engaged with the same logic as AA mode. Heading is acquired with a turn where its bank angle is a function of the airspeed. Just before the datum is reached, HH mode is engaged to minimize the transients.

AC mode is designed to perform an economy climb. When preselected, provided that throttle deflection is adequate, the mode will follow a climb profile that will

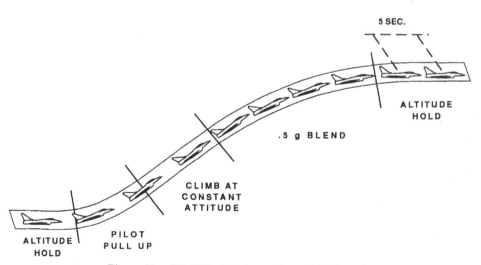

Figure 18. EF 2000 altitude acquire and hold mode.

maintain a speed schedule designed to minimize fuel consumption. Attitude control is used to keep the speed at the desired value. If AA mode is also preselected, the AP will disengage AC mode and perform the acquire when the altitude approaches the datum.

5. Autopilot control law design

The AP control law (C/L) has been designed to ensure two functions:

(*a*)   hold or acquire a datum;

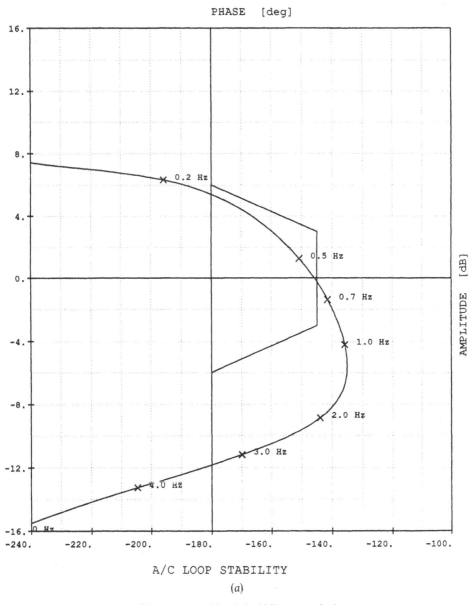

(a)

Figure 19.   Altitude hold linear analysis.

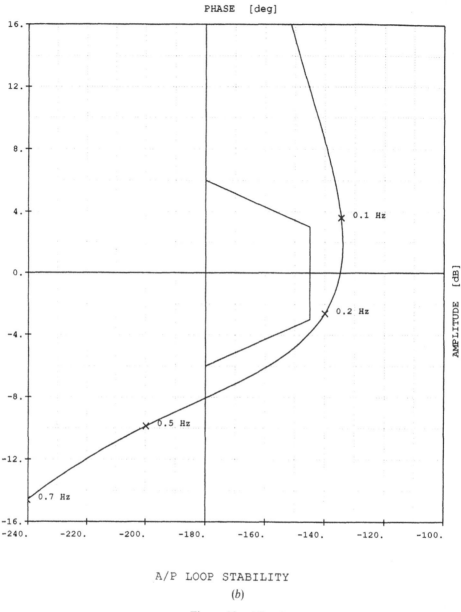

A/P LOOP STABILITY

(b)

Figure 19. (*Cont.*)

(b)   perform the reconfiguration of the C/L following a mode change request.

The code designed for point (a) for each mode performs the holding or the acquisition of a datum fulfilling stability and performances requirements. The code for the second point is designed to manage the point (b) code, selecting the proper subset of C/L, depending on the mode that has been selected. In general, (a) code and (b) code are rather independent, because the first is mainly generated by linear analysis, while

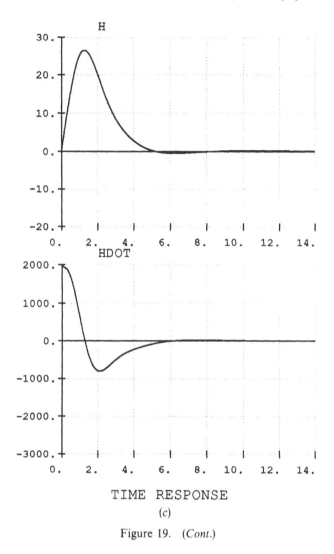

TIME RESPONSE

(c)

Figure 19. (*Cont.*)

the second is developed and checked by performing non-linear and manned simulation sessions.

During the definition of the C/L (code (a)), the designer has to cope with requirements and A/C configuration constraints. The solution of the flight C/L design problem comes from an iterative process targeted to find the best compromise to meet the requirements. For example, the stability of AP C/L requires the investigation of the loop relative to the mode engaged and also the examination of the A/C stability with AP engaged. We can see this process in Fig. 19, which shows the linear analysis of a benchmark A/C. In this figure the A/C has a good time response in an altitude acquisition phase. The frequency response of the AP, obtained by opening the stability loop in the altitude path, is also good, but the Nichols plot of the stability of the whole system, obtained by opening the control surface path, shows unsatisfactory stability margins. If the design was based only on the altitude loop analysis, it would enhance the A/C sensitivity to off-design conditions and therefore it could lead to potentially dangerous behaviour in the case of relatively small modelling errors (e.g. uncertainty of the centre of gravity).

Additional guidelines come from fulfilling the performance requirements. Again, concerning AH mode, MIL states that the period of the residual oscillations after the acquire phase must be greater than 20 s. In Fig. 20 we can see a root locus

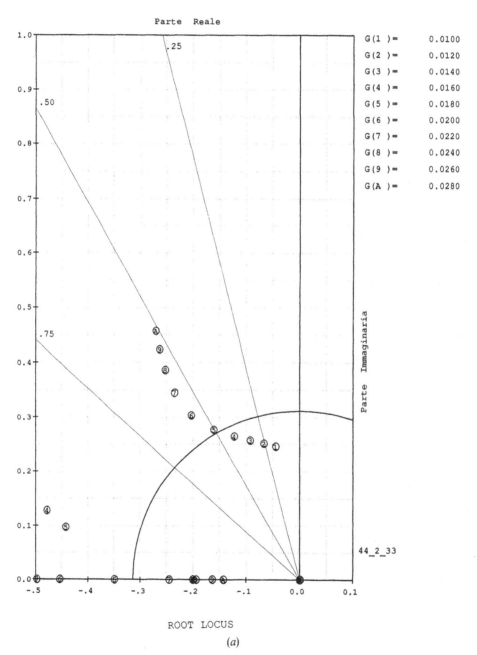

ROOT LOCUS

(*a*)

Figure 20. Altitude hold design.

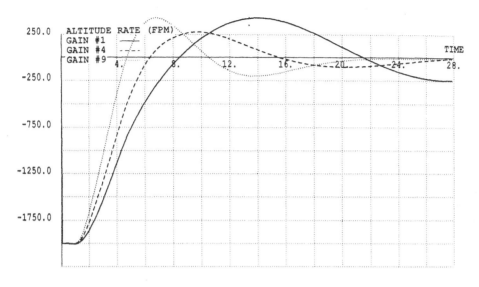

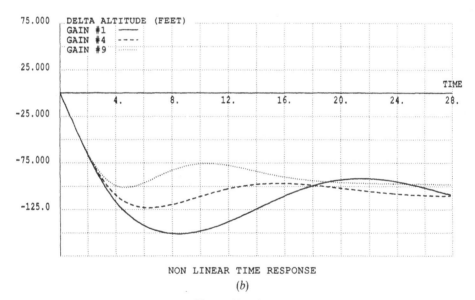

NON LINEAR TIME RESPONSE

(*b*)

Figure 20. (*Cont.*)

where the large circle sector discriminates the roots that have a period greater (inside the circle) than 20 s. The higher damped branch is relative to the short period. In this case gain designated as '4' is good, but stability margins have to be checked. The designer has to decide if the examined solution is good, or to explore other configurations for the C/L, like modifying the weight between altitude and altitude rate gains or adding a filter to obtain a better time response.

6. Conclusions

The use of digital techniques applied to the FCS design yields the following advantages:

319

- increase in computation capability;
- greater flexibility and more complex algorithms;
- no hardware requalification after software updating due to C/L modifications;
- complete integration between FCS and AP functions;
- reduction in weight and size of equipment.

In spite of the complexity of the software qualification procedure, the benefits mentioned above justify the choice of digital techniques, as confirmed by our experience on the development of the AMX AP. Margins for further improvement in the software implementation process in the FCC have been envisaged in the use of ADA code for C/L requirement. A programming technique accounting from the beginning for the specific characteristics of the FCC used is expected to speed up significantly the overall design/qualification process.

AMX experience was carried over to the development of the EF 2000 AP to a maximum extent, and it is proving very helpful in the prediction of software complexity and pilots' impressions of AP response when performances are not extensively defined by the requirements.

12

Development and flight experience of the control laws and the aeroservoelastic solution in the Experimental Aircraft Programme (EAP)

A. McCUISH and B. CALDWELL

1. Introduction

The Experimental Aircraft Programme (1983–1991) was jointly funded by government and industry. The aircraft first flew in 1986 and had its final flight in 1991.

The programme objectives were to demonstrate various technologies relevant to a future combat aircraft within the rigours imposed by having to achieve flight clearance and demonstration. Prime areas for demonstration were modern cockpit displays, avionics systems integration, advanced material construction, advanced aerodynamics and active flight control.

Nearly all future combat aircraft will have an unstable basic airframe owing to the advantages that are accrued: smaller, lighter, aerodynamically more efficient, etc. Necessary to such an aircraft is a full-time active control system. For BAe the flight control system developed under this programme was an evolution of the design used in the Jaguar fly-by-wire (FBW) programme (1977–1984). This FBW programme was aimed at identifying the design methodology and airworthiness criteria necessary for flight certification of a full-time digital active control system, with no mechanical backup. The control system was therefore treated as though intended for production. The programme successfully demonstrated, at different stages, a stall departure and spin prevention system and, progressively, the ability to control a highly unstable aircraft, ultimately achieving stability of a basic airframe exhibiting a time to double amplitude of $T_2 = 0·25$ s.

The requirements for the EAP were even more demanding than those of the Jaguar FBW:

- the airframe would be significantly more unstable;
- it must fly to high angles of attack and load factor;
- it should exhibit carefree handling.

Carefree handling was defined as no departure tendency or airframe overstressing no matter what the pilot's inputs are.

To achieve this required an upgrade to the performance of equipment compared with that of the Jaguar FBW and new equipment was manufactured in the areas of:

- faster computing;
- much improved actuator performance;
- significantly improved aircraft motion sensor units (AMSUs).

In a flight demonstration programme this equipment had to be manufactured, proven to be up to performance and proven to be flightworthy.

To encompass all the requirements of the flight objectives the control law architecture selected was a pitch-rate demand combined with an α demand system. In the final stage of the programme an N_z demand system was also combined with this. The pitch-rate demand system, sometimes known as a rate command attitude hold (RCAH), was a design feature retained from the Jaguar FBW programme where the pilots rated the handling qualities very favourably.

2. The aeroplane

The aeroplane, Fig. 1, being a technology demonstrator, rather than a pre-production or prototype aircraft, used existing components where this did not compromise the technologies being demonstrated. This included significant airframe sections and equipment resulting in an aircraft that would otherwise be considered

Figure 1.   The EAP aircraft.

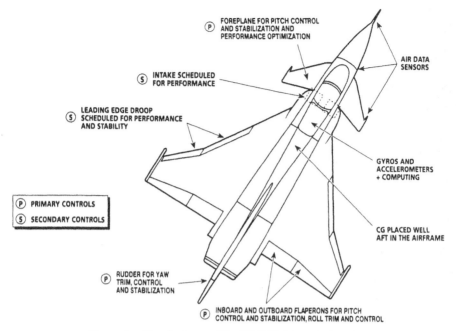

FOREPLANE FOR PITCH CONTROL
(P) AND STABILIZATION AND
PERFORMANCE OPTIMIZATION

AIR DATA
SENSORS

(S) INTAKE SCHEDULED
FOR PERFORMANCE

LEADING EDGE DROOP
(S) SCHEDULED FOR PERFORMANCE
AND STABILITY

GYROS AND
ACCELEROMETERS
+ COMPUTING

(P) PRIMARY CONTROLS
(S) SECONDARY CONTROLS

CG PLACED WELL
AFT IN THE AIRFRAME

RUDDER FOR YAW
(P) TRIM, CONTROL
AND STABILIZATION

INBOARD AND OUTBOARD FLAPERONS FOR PITCH
(P) CONTROL AND STABILIZATION, ROLL TRIM AND CONTROL

Figure 2. Key features of the EAP flight control system.

overweight; thus the absolute performance aspects were somewhat compromised. Extensive use was made of carbon fibre composites to minimize weight, e.g. wing and foreplanes, and indeed large-component carbon fibre composite construction was one of the critical technology demonstrations of the programme.

The aircraft layout, Fig. 2, was similar to that of the Eurofighter EF 2000 with the aim of achieving the best compromise of supersonic and subsonic performance: delta–canard, twin engine, side-by-side chin intakes, full-span trailing-edge flaps, leading-edge droops and a single fin and rudder.

The foreplane and leading edge were scheduled to optimize the lift/drag ratio in the trim state. This scheduling primarily favours trimming on the flaps. Pitch stabilization utilized all flaps and the foreplanes. Roll control was achieved via antisymmetric use of the flaps, while the rudder provided directional stability and control, the aircraft being directionally unstable at high Mach numbers and high incidences.

3. The cockpit and pilot inceptors

The general layout of the cockpit was that of the modern single-seat combat aircraft (Fig. 3). The primary forward panel was dominated by three multifunction displays and capped by a wide-angle head-up display (HUD). On the left were linear motion throttle inceptors which were borrowed from a Tornado. A departure from the current trend for a very small-movement sidestick controller was the retention of a more conventional centre-mounted stick. This stick, however, departed from the conventional by reducing the travels in both pitch and roll axes by around a half. The stick-force gradients chosen were known to provide comfortable 'feel' with constant maximum forces of 20 lb (9 kg) in pitch and 9 lb (4 kg) in roll.

Figure 3.   The EAP cockpit layout.

With no q-feel system (or similar) there were no control circuits attached to the stick which resulted in the stick having low inertia and a high natural frequency. Substantial viscous damping was applied to give a deadbeat response and provide the necessary solid feel. The damping was also known to alleviate any roll ratchet tendencies although the dominant factor in this phenomenon is the control law.

The rudder pedal travel was also reduced from the conventional with maximum forces of 75 lb (34 kg). In flight there proved little need for the rudder pedals as the coordination for rolling manoeuvres was provided by the control law. Available sideslip was constrained within limits by the control law so that its selection by pedal input was easily controlled.

The trim system was driven by a rocker button on the stick-top for pitch and roll, and by a thumbwheel, located on the left-hand panel, for directional trim. Series trim was effected by a software-commanded integrator signalled from the trim button such that the stick position was unaffected, remaining at the fixed spring centre position. Although pilots' preference for parallel trim was well understood it was not thought to justify the considerable mechanical complication within the limited space of the compact stick assembly and in view of the self-trimming nature of the pitch control law which provided neutral speed stability except at low airspeed. The software implementation of the pitch trimming is discussed later (§ 6.3.1) as this did cause some problems in flight. In general these inceptors proved highly satisfactory as an interface between the pilot and the FCS.

4. The control law philosophy

Much of the design method and philosophy built up through the Jaguar FBW programme was fed directly into the EAP. A prime component of this was the necessary visibility of the design. The design approach taken was classical using predominantly root-locus and frequency-response analysis. These methods are well understood and highly amenable to implementation and scheduling considerations. This approach also allowed for a simple and logical structure to be designed which isolated functions and non-linear characteristics such as the trim distribution function and non-linear control power compensation. The resultant design was specified in terms of gains, filters, integrators, switches, etc., each of which were defined as elements in an executable form. This became the formal specification of the control law and was the source code for any simulation and analysis. It was essential to the design process that the control law specification was executable in order that it was uniquely interpreted for the purposes of simulation, development and implementation into the flight control computer (FCC). As in the Jaguar FBW programme the control laws were to be treated as a production item meeting specific requirements on stability and probability of failures, etc.

The basic system architecture to meet the necessary integrity was quadruplex with a triplex source of air data and a triplex source of airstream direction detection data (ADD), i.e. incidence and sideslip data. This was coupled with a hierarchical control law architecture, in order to cope with multiple sensor failures, of the form:

- Full control law – all sensors available.

- Reversionary law – multiple failures of the ADD system.

- Fixed gain law – multiple failures of the air data system.

This last category was in fact not required to meet the integrity budget but, being a dropout of the 'Reversionary' laws, was made available as a matter of course.

5. Control law design method

The stability requirements for all the control laws developed were similar to MIL-F-9490D, an exception being the reduction of phase margin on rigid modes from 45° to 35°. However, additional robustness was introduced by requiring that the phase margin of 35° was maintained for loop gain variations of ± 3 dB.

The design process was composed of three elements: linear low frequency, non-linear and linear high frequency. The high-frequency design is aimed at avoiding structural coupling which is discussed in § 7. The separation of high- and low-frequency designs was achieved by including in the low-frequency (rigid-mode) design an allowance for the inclusion of the filters required to avoid structural coupling. This allowance was defined as a 'budget' of low-frequency phase lag (rigid-body modes) to the designers of the high-frequency filters within which the total lag of their filtering must remain.

A cornerstone of the design process was the inclusion, at the design stage, of the best available hardware models, e.g.

(*a*) rate gyro sensors;

(*b*) accelerometer sensors;

(*c*) α/β sensors;

(*d*) compute update rates/compute time delays.

Even with the very high quality of hardware used on the EAP aircraft, the combination of the hardware effects in any one loop will have a significant effect on the stability of the system and hence the inclusion of these effects is essential.

The linear behaviour of the control laws was derived using classical methods: root locus, Nichols plots, linear time and frequency response. This was evolved into a fully engineered controller via a significant amount of non-linear simulation which ensured that integration of the linear control law and the non-linear constraints (e.g. actuator position and rate limits, hydraulic supply limits, design manoeuvring limits, moding between different laws, etc.) was appropriate.

Figures 4 and 6 show simplified schematics of the pitch and lateral control law structures. In the pitch axis, Fig. 4, the basic concept of the control law was 'predetermined' to be a proportional plus integral demand with pitch-rate damper. From experience two first-order lead–lag filters in the feedback loops are all that is necessary to give the appropriate stability characteristics. Thus the basic controller is of

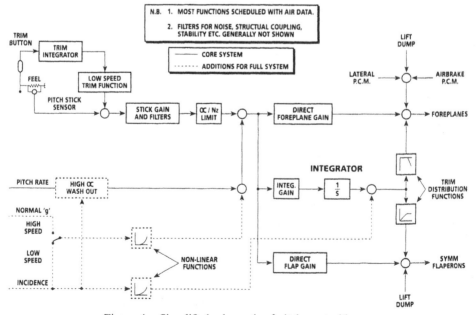

Figure 4. Simplified schematic of pitch control laws.

low order and readily understood. To maintain the visibility, however, an overall structure was imposed which isolated functions and non-linear characteristics, e.g. trim distribution, control power normalization, non-linear variation of stability. Without this structure the variation of the aerodynamics would appear directly in the schedules of the gains masking the function of the gain.

Having isolated these aerodynamic variations across the full envelope, the basic $P, I + q$ damper mimics the variation of speed and dynamic pressure, etc., explicitly, i.e. they reflect the basic flight mechanics parameters.

The stability feedbacks being thus defined, the next task was to design the handling of the aircraft. An important lesson learned during the Jaguar FBW programme was the benefits that may be gained from separating the regulator and the command path designs. For the EAP aircraft with its high levels of instability the regulator dynamics are dominated by the need to satisfy the stability requirements of gain and phase margin, etc. This usually leads to poor demand dynamics. This is overcome by prefiltering the command, typically with a forward feed component and shaping filters as shown generically in Fig. 5.

The flexibility introduced by this separation allows precise characteristics to be designed into the handling of the aircraft virtually independently of the stability loop dynamics. In particular this allowed a high level of PIO resistance to be achieved which resulted in there being no tendency for the aircraft to undergo PIO

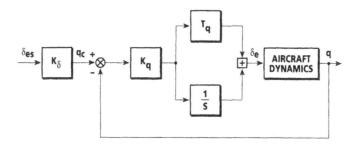

**REGULATION AND CONTROL THROUGH COMMON DYNAMICS CAN
LEAD TO POOR HANDLING DUE TO SYSTEM STABILITY CONSTRAINTS**

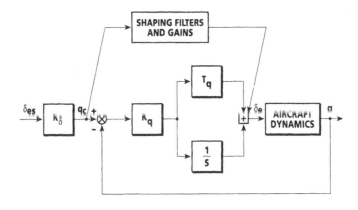

FEEDFORWARD SHAPING OPTIMIZES CONTROL WITHOUT
ALTERING OR DEPENDENCE ON REGULATION DYNAMICS

Figure 5.   Generic rate command/attitude hold (RCAH).

throughout the 259 flight programme. The prime, though not the sole, criterion to achieve this is to ensure a sufficiently large gain margin at sufficiently high frequency on the transfer function of pitch attitude to pilot input. No regard was paid to the so-called 'equivalent delay' of feel dynamics since it was firmly believed that pilots relate to the output displacement and not the force input. The latest criteria developed through both the Jaguar FBW and EAP programmes are given in Gibson (1990).

For the lateral/directional controller (Fig. 6) the challenge was to achieve a consistency of behaviour with aerodynamics which varied considerably in all axes and to allow for the variability in the augmented pitch axis. The basic architecture was conventional roll-rate demand with yaw-rate damping. Roll coordination to rudder was included with the inertial coupling control being via the foreplanes. Here again classical techniques have been used, relying predominantly on root locus and Nichols plots for the linear design. Experience has shown that the non-linear design of the roll coordination is dominated by the aerodynamic tolerances that must be considered. (This is not due to the tolerances being large; in fact they are significantly smaller than those considered in previous projects.)

The next major design area was the non-linear response where the objective was to provide agile carefree handling (i.e the ability to manoeuvre the aircraft rapidly without the pilot having to restrict inputs to respect a loading or incidence limit). The challenge was to retain as much agility as practicable since carefree handling can be 'readily' achieved if rates and accelerations are constrained but aircraft performance is then compromised.

At the lower airspeeds, where a combat aircraft must perform, the limiting factor to agility tends to be the position and rate limiting of the surfaces. The high surface rates required on a basically unstable vehicle require that hydraulic supply and

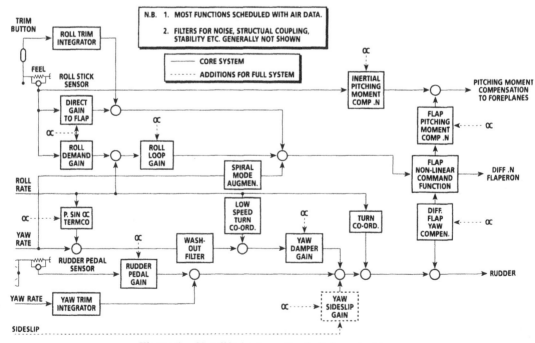

Figure 6.   Simplified schematic of pitch control laws.

demand aspects must be considered. On the EAP the number of actuators and their high rate would, if used continuously, soon exceed the hydraulic supply. This task required the extensive use of the real-time flight simulator to assess the near-infinite number of inputs that are possible – though much of the design may be achieved with the non-real-time simulation.

The integration of the pitch and lateral authorities gave priority to the pitch demand, i.e. at full back stick/maximum α if roll stick was applied the α demand was unchanged and the roll rate was limited to suit. The alternative would have been to reduce the incidence demand by some amount in favour of increased roll rates. Also in rolling manoeuvres where a change in pitch rate is demanded then the roll rate is reduced in favour of supplying the demanded pitch-rate change.

A necessary complexity in the design was to take into account the non-linearities in the aerodynamics, particularly pitch breaks in the stability data where these co-incide with the stressing limit of the aircraft.

It is also necessary to consider control law design items which, to some extent, do not exist in the design of a classical aircraft as the behaviour is inherent. For example, in rapid decelerations from high speed on a classical aircraft the pilot would gradually pull the stick aft to maintain the g. With this controller application of full aft stick demands the maximum allowable g. This is achieved at the start of the manoeuvre but a significant loss of g would result in the subsequent rapid slow-down had there not been a mechanism to assist the integrator in increasing the incidence. The mechanism simply measures the slowdown and demands the appro-priate pitch-up command. The alternative of increasing the integrator gain is not a solution as this is limited owing to linear stability requirements.

A feature added into the landing laws was the ability to employ a 'lift–dump' mode, the purpose of the mode being to increase the download in order to improve the braking and steering effectiveness. This was achieved by putting the foreplane and flaps to trailing-edge-up positions when the nose wheel was fully on the ground. The mode required pre-arming by the pilot and was automatically deselected if a go-around was initiated.

All of these functions form part of the control law specification and the integration/moding of these into the control law forms a large portion of the control law development.

6. Control law development phases and associated flight testing

Over the life of the EAP there were three main control law development phases. The first phase, utilizing rate feedbacks only, allowed flight over the majority of the subsonic envelope which verified the majority of the aerodynamics and the cali-bration of the ADD (incidence and sideslip) gauges. The second phase introduced the α and β feedback signals and the concept of carefree handling. The final phase expanded the carefree-handling envelope and was designed around a control law architecture proposed for the Eurofighter EF 2000.

6.1. *Phase 1: first flight standard*

6.1.1. *The reversionary control law.* This control law was designed for an aircraft ballasted to constrain the instability to a time to double amplitude of 0·25 seconds. This was the instability ultimately demonstrated on the Jaguar FBW programme. Starting at this instability level with unproven aerodynamics could be done with confidence as the hardware performance on this aircraft was improved over that of the Jaguar FBW and the control law design was similar. The control system was scheduled with air data (dynamic pressure, static pressure and Mach number) but utilized only the high-integrity angular rate information from the motion sensor units; no incidence or sideslip data was available. This restricted the incidence to within the more linear region below $C_{L_{max}}$, constrained the ability to coordinate rapid rolling and generally limited the performance of the pitch and lateral axes. No carefree handling was available at this control law standard.

A subset of this scheduled controller was also available within the control law which would cater for the remote possibility of failure of the air data system. This simply set the air data parameters in the scheduling if the air data system was determined to be invalid. In fact two sets of values were used, distinguished by undercarriage selection (up or down). This allowed the largest speed envelope to be cleared without the system becoming overgeared owing to the aerodynamic gain effects.

The pitch control law was pitch-rate demand using proportional rate damping and a filtered pitch-rate integral demand. This rate-command, attitude-hold (RCAH) system results in automatic trimming of the aeroplane since with the stick at neutral zero pitch rate is demanded. At low speed, where it is necessary to give the pilot a tactile sense of the reducing velocity, perceived static stability was introduced by biasing the stick neutral command with the reducing airspeed. This then required the pilot progressively to apply aft stick in the slowdown at these flight conditions.

As indicated previously the separation of the feedback design and the pilot command path design were achieved by utilizing a feedforward path and command shaping. This allowed a high level of PIO resistance to be designed into the control laws. The forward feed path allows sufficient pitch acceleration to be generated while the overall command shaping can, with only two or three first-order filters, achieve the correct pitch rate and α dynamics while ensuring that the transfer function of pitch attitude to stick displacement is suppressed at the PIO frequency.

The lateral/directional control law was roll-rate demand plus yaw-rate damping. The aim was to perform wind-axis rolling. There is no integrator to ensure the roll rate demanded is achieved and the resultant roll rate is a balance of the gains and aerodynamics. For handling purposes the lateral stick to roll-rate demand was designed to be near constant around the centrestick position for any flight condition. The demand was increased using a power law to give the appropriate full-authority roll rate at full stick deflection.

The directional control was a scheduled gain of pedal-to-rudder angle. The schedule, a function of airspeed, was such that the demanded sideslip was constrained to permitted levels. Wind-axis yaw rate was derived from body-axis yaw rate and roll rate and then fed back to the rudder via a wash-out filter to give the optimum dutch roll damping characteristics.

Within the non-linear design a network was constructed which applied a small transient input to the rudder when the lateral stick was centred. The purpose of this was to remove any residual sideslip on the roll exit. This, combined with the excellent coordination provided by the control law, proved very effective in practice, stopping the aircraft roll quickly and cleanly.

6.1.2. *Flight testing.* There were 52 flights with this standard of control law which verified the basic pitch-rate demand system, aerodynamics and the ADD calibrations. The flight envelope flown covered up to 500 knots CAS and Mach 1·3. Mach 1·1 was achieved on the first flight. The aircraft was incidence limited to a pilot-observed value of 18°, this giving a safe margin from $C_{L_{max}}$ where the use of pitch-rate demand is no longer feasible owing to stall characteristics. The extreme non-linearities of the aerodynamics at incidences greater than 18° require a knowledge of α to schedule the control system gains.

The early flying was aimed primarily at aerobatic manoeuvres as a work-up to the 1986 SBAC Airshow at Farnborough. The pilots were impressed by the handling qualities of the aircraft in this phase. The aspect of most concern was in performing loops where, owing to the speed variation through the manoeuvre, the incidence required careful monitoring so as not to exceed the limit. The other aspect subject to criticism was a slight oversensitivity in the roll axis for small-amplitude inputs.

Typical pilot comments for this phase of flying were:

- very natural and easy to fly;
- solid and very stable, but still responsive to control;
- takeoff, approach, landing and information flying were accurate and easy;
- steady attitude in turbulence and still easy to fly;
- no roll ratcheting or pilot-induced oscillation tendency.

6.2. *Phase 2: incidence limiting and carefree flying*

6.2.1. *The α limiting control law.* Having proved the basic performance of the FCS, the aerodynamics and the ADD system in the first phase, some of the ballast was removed resulting in a time to double amplitude of $T_2 = 0.18$ s for the basic airframe. The incidence signal was now used for limiting in the pitch axis and extensive scheduling in the lateral/directional axes. The pitch axis still used a core based on pitch-rate demand and this was blended across into the α demand system as the demand reached its limit. The limiting incidence was scheduled with flight condition to respect the stress limits of the airframe, and thus an N_z*w control system was in fact the result; at high mass the normal load was low and at low mass it was high. This maximizes the use of the strength available in the airframe.

The roll-demand and pilot rudder authorities were reduced as a function of incidence and flight condition in order to respect the stressing limits and prevent pilot-induced departures at high incidences.

The carefree handling of the aircraft was arrived at through the careful integration of the pitch and lateral/directional control laws. This design was only possible with a good deal of effort in pilot-in-the-loop simulation as well as off-line simulation in order that the maximum capability and suitable harmony between axes were made available.

As part of the simulation exercises the pilots were requested to try and 'beat the system' and cause the aircraft to depart. There were no restrictions to the pilot inputs even if these did not correspond to 'sensible' piloting or combat manoeuvres. The pilots proved very adept at finding the deficiencies in the development control laws and could do so much more quickly than engineers using the off-line facilities.

6.2.2. *Phase 2: flight testing.* The EAP aircraft was demonstrated with these control laws at the Paris Airshow of 1987. This was achieved after a very short flight test programme which proved the high-incidence aerodynamics and the effectiveness of the controller at preventing departure. The high-incidence, high-altitude testing, where the aircraft was fitted with an anti-spin gantry, was completed in 3 weeks with 20 sorties flown. This was followed by a period of testing where the altitude was progressively reduced to prove the carefree handling and the manoeuvring capability. There were no departures and the similarity to the modelling was verified with no changes being required to the control laws or aerodynamic model.

The flight envelope covered here was restricted to 400 knots Mach 0·9. Some 64 flights were flown in this phase with carefree manoeuvres tested throughout this envelope. Incidences of 34° were achieved and speeds down to 100 knots. There were no signs of departure although below 200 knots the pitch handling was rated as ragged and cyclic pitch stick inputs resulted in overshoot of the angle-of-attack limit. The handling qualities seen in phase 1 flying were repeated here primarily owing to the same type of pitch-rate demand controller being retained at $1g$ flight conditions.

The pilot comments from this phase were very positive:

- exceptional roll acceleration and damping;
- excellent handling and control in slowdowns to high angles of attack;
- flies extremely well;
- handling in loops and rolls extremely pleasant;
- impressive control of angle of attack, sideslip and roll rate in carefree manoeuvres.

The carefree handling was considered remarkable with, e.g., the pilot able to apply full roll stick and then full aft stick with the roll stick still applied. This feature was seen as particularly useful in situations where the pilot needs to be looking out of the cockpit, e.g. combat or low-level flying.

The aircraft also exhibited good tracking behaviour against ground targets and also in air-to-air tasks although the air combat experience was limited to a few flights. The aircraft also behaved well in turbulent conditions, the only criticism being that the flight path was unsteady which required attention when flying at low levels. This is a natural response of the pitch-rate demand system.

In an extension programme to these control laws the rate demand components were developed to allow flight across the full supersonic flight envelope. There was no extension to the carefree capabilities as the purpose was to gather aerodynamic data. During this flight test programme, where there was greater time spent in cruise or cruise climb getting to and from the supersonic test range, a slight lateral/ directional mis-trim was detected in the aircraft. There was also a distinct trim change in the transonic range which required retrimming by the pilot. This detracted from otherwise excellent handling qualities and although considered to be a slight nuisance it was raised at nearly every flight debriefing. In response to this behaviour it was decided to introduce a wings-level hold mode in the next flight phase.

6.3. *Phase 3: the 'full system' – AoA/N_z limiting*

A third and final phase was introduced to the EAP in order to supply some risk reduction to the EF 2000 project. To this end the control laws were to be designed around a structure destined for the EF 2000. This in fact was a development of the EAP structure involving the introduction of N_z and attitude feedback signals. This control law aimed to provide N_z and AoA limiting across the complete envelope.

The N_z signal was used at high-speed flight conditions where the accuracy of the AoA probes become critical. Low-wing-loading aircraft such as the EAP can easily generate load factors of greater than $1g$ per degree of incidence. Thus the accuracy becomes significant at such flight conditions when flying to the stressing limits. In the phase 2 flying which used incidence limiting the aircraft had been flying to higher g than intended. This resulted from the incidence correction term for fuselage bending due to normal g being underestimated. The aircraft achieved the measured incidence but this was less than the true value.

A significant development introduced at this time was the use of inertially derived incidence and sideslip. The inertial incidence was derived from the body rates and accelerations available within the FCC and was mixed with the probe-generated signal. The motivation for this was to reduce the sensitivity to probe disturbances when flying at incidence in the low-speed envelope. The sensitivity results from the high levels of incidence feedback required for stabilization. By utilizing inertially derived signals the high-frequency disturbances (e.g. turbulence) are eliminated. An inertial sideslip signal was used at high speed and here again the motivation was to reduce the effect of disturbances seen on the probes, in this case deriving from shock-wave effects seen in the transonic and supersonic envelope.

In the pitch command path an amplitude-dependent filter was introduced in order to incorporate particular response characteristics in the air-to-air tracking task. For small-amplitude inputs, as required for air-to-air gun tracking, pilots prefer a zero-attitude dropback response. For larger inputs, used during the target acquire phase, then a more aggressive flight path response is required which in turn results in attitude dropback.

As mentioned earlier a wings-level hold mode was introduced to relieve the pilot of the nuisance retrimming task as a result of small-aircraft asymmetries. This was a 'quick-fix' mode with very limited authority but it was included as an inherent component of the control laws, i.e. there was no pilot selection or deselection; this was fully automatic. This was not a true autopilot system, there was no track control or even directional coordination. The mode would simply hold the wings at zero roll angle when in cruise conditions, i.e. bank angle within $3°$ of level and minimal roll rate. The moding was designed such that operation was completely transparent to the pilot.

6.3.1. *Phase 3: flight testing.* The flight testing proved that the blending of N_z limiting to α limiting was transparent to the pilot. The only time this became noticeable was when flying at the manoeuvring limit in an extremely light mass state where the change from N_z limiting to $N_z * w$ (AoA) limiting was accompanied by an increase of around $1g$. This could be remedied by supplying the N_z control law with the mass state of the aircraft. This will be required for a service aircraft as otherwise performance will be seriously affected at masses below the stressing mass.

Following criticisms of the pitch gross manoeuvring when the aircraft was inverted, attitude scheduling was introduced to the command path in order to boost the manoeuvre. This extra complication results from the architecture of using pitch-rate demand blending to an N_z (or AoA) limiting system. In the inverted state the required change in incidence is larger by $2g$ compared with the upright case. For zero pitch rate, inverted or upright, the stick trim position is neutral; hence the pitch stick travel to manoeuvre to the N_z limit will be the same (zero to full aft) in both cases. In the previous flight testing this extra manoeuvring was taken up by the integrator which is a slow-acting device.

The assessment of the non-linear amplitude-dependent pitch-tracking filter was carried out during two flights dedicated to air-to-air combat manoeuvring with a target aircraft (a Hawk). This provided some interesting results and showed the basic tracking to be very good but spoiled by two faults. The more adverse of these was a result of the non-linear roll-damping network briefly described at the end of § 6.1.1. This imparted a kick to the rudder when the lateral stick was returned to centre. This in turn caused the tracking sight to swing off the target in the yaw axis just as the pilot had settled the pipper on the target. This highlights the risks of including highly non-linear effects in a control law; it may work in the case it was designed for but it must produce satisfactory results in all cases. In this particular instance the effect may be remedied by introducing an amplitude dependence, possibly similar to the non-linear tracking filter as in the pitch axis. The second effect was the occasional appearance of a 2–3 Hz bobble on the pitch tracking. This was surprising as the attitude to stick response is lowly geared at these frequencies and certainly free from PIO tendency. The problem was subsequently attributed to neuromuscular inputs being detected by the stick transducer and suggests that greater attitude attenuation is required for this type of task. The basic pitch gun tracking was rated as Cooper–Harper 2 with the lateral tracking as seen rated C–H 6 but was estimated to be C–H 2 if the non-linear effect could be removed since it was very easy to bring the pipper onto the target.

This flying also highlighted the need for a $C_{L_{max}}$ indicator on this aircraft as it was too easy for the pilot to pull maximum α to bring a solution on the target aircraft. Against a more equal target aircraft the loss of performance incurred for exceeding $C_{L_{max}}$ (i.e. a significant increase in drag) would be critical. There are no external clues for the pilot with there being very little buffet or canopy noise to indicate that the aircraft is close to or above $C_{L_{max}}$.

A criticism voiced at this phase of flying was that the normal g onset rate was too low. Throughout the programme this had been designed to a limit of $8g\ \mathrm{s}^{-1}$ and had not been criticized previously. The lack of criticism was probably due to the fact that the aircraft was strength limited to $6g$ and thus maximum g was achieved within 1·5 s. At low airspeeds the maximum capability was available, this being constrained by actuator rate and position-limiting characteristics. Increasing the g onset rate at the higher speeds could be achieved simply by increasing the direct feed gain from the pilot stick to surfaces.

For this standard of control law the roll acceleration performance had been increased to a level which the pilots desired based on the previous flying and flight simulation assessments. This led to a design with a consistently very crisp roll entry and exit across the medium- to high-speed flight envelope. Initial experience of these characteristics elicited very favourable comments but after a few sorties the acceleration was thought to be slightly too high. In some instances the pilot's head tended to strike the canopy owing to the lateral acceleration being generated at this point. From the flight results it is believed that a limit of $0·7g$ laterally, at the pilot's head position, would be a sensible limit for roll acceleration. This aspect may require further consideration for a two-seat aircraft where the second crew member may not be prepared or braced for the manoeuvre. Also the rear crew member, in a tandem

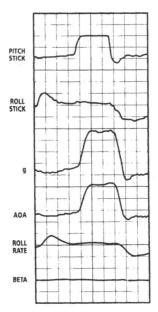

Figure 7. Snatch pull 1·7 Mach, 36 000 ft (10 975 m).

cockpit configuration, will be subjected to higher lateral accelerations being positioned higher above the roll axis than the front seat.

The improvement to the pilot workload brought about through the introduction of the wings-leveller was claimed to transform the aircraft. The mode proved completely transparent to the pilot. As a result of the removal of this nuisance factor from the roll axis, deficiencies in the yaw and pitch axes became more apparent.

The pitch axis, although self-trimming as a result of pitch-rate demand, did allow changes in the flight path to occur as a result of Mach accelerations or decelerations, particularly across the transonic range. Similarly there is a tendency with pitch-rate demand for the flight path to change as a result of turbulence. This was seen in medium to high turbulence and most notably at low level where flight path is more important. The aircraft did not pitch very much but the *g* spikes were noticeable. A tradeoff can be made here between aircraft attitude and flight path disturbance by adjusting the control law feedbacks, i.e. introducing incidence feedbacks. Alternatively a flight-path-hold or altitude-hold autopilot function would alleviate this problem. There was, however, no autopilot on the EAP as these functions were considered outside the control law demonstration objectives.

The deficiencies in the yaw axes were slight and resulted from the roll-damping 'rudder kick' and the wings-level-hold mode. On rolling out on to a heading the 'rudder kick' seen on centring the roll stick had the tendency to back off the selected heading by around 1°. The wings-level mode, which was a 'quick-fix' solution to roll mis-trim, was not coordinated with the yaw axis. As a result, if the yaw trim was not adjusted by the pilot and the pilot waited long enough, it was possible to detect that the aircraft was performing a flat turn. Neither of these characteristics interfered with the pilot's tasks and were very minor deficiencies.

The pitch trimming characteristics were significantly aggravated by implementation of the pitch trim button mechanization. As described previously this added a bias into the stick demand requiring the pilot to hold the stick position to maintain trimmed 1*g* flight. This was designed for low-speed approach conditions and it was not intended for use outside this envelope; however, pilots will 'by instinct' use the trim button if they perceive that the aircraft is out of trim. This perception occurs when achievement of steady-state 1*g* conditions was not acquired quickly enough. The pilots having used the button at these high speeds then found it difficult to reset the trim back to zero, there being no indicator for zero trim. (A function was available to reset the trim to zero but this also reset roll and yaw.) A simple solution to this characteristic is to mechanize the trim button input as a transient or as an input to the integrator, and thereby the pilot could assist the aircraft to achieve the steady condition in what would appear a natural use of trim control. Unfortunately there was no opportunity to implement this.

The carefree manoeuvring offered by this system was again considered impressive. Examples of the type of input and responses are given Figs 7–9. Figure 10 shows approximately 2 minutes of carefree handling where the pilot uses snatch pulls to the aft stop with and without rolling inputs, roll reversals, four-point rolls, etc. Also clear in this trace is the effect of loading and unloading the pitch stick while maintaining full roll command; the roll rate is seen to decrease and increase rapidly. The increase in roll rate was at times thought to be too rapid as it was not always possible to select the roll-out attitude; however, more conventional use of the controls may be more appropriate for such handling manoeuvres.

Pilot comments from this latter phase of flying were:

- carefree handling immensely impressive – copes effortlessly with brutal demands;
- impressive roll acceleration and damping;
- takeoff behaviour is excellent;

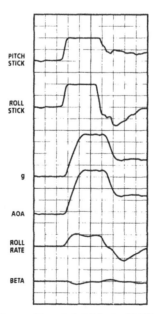

Figure 8.   Diagonal break 1·6 Mach, 40 000 ft (12 195 m).

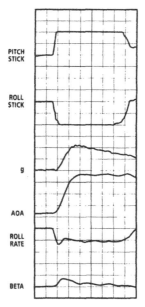

Figure 9.   Diagonal break 0·6 Mach, 30 000 ft (9150 m).

- air-to-air tracking generally very good;
- wings-level mode is an excellent feature and totally transparent.

7. Structural coupling aspects

The ongoing developments noted in the Introduction, in aircraft design requirements, aerodynamic configuration and associated control systems, necessitated

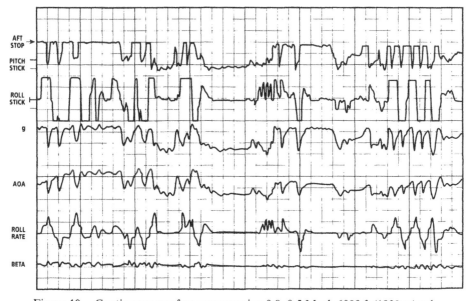

Figure 10.   Continuous carefree manoeuvring 0·8–0·5 Mach 6000 ft (1830 m) agl.

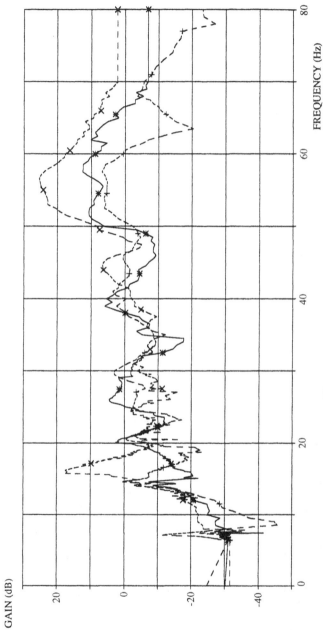

Figure 11. Ground test results.

338

parallel development of solutions to the problems associated with aero-servoelasticity, or, as it is more commonly referred to in the UK, structural coupling (SC).

The EAP aircraft combined for the first time in a British Aerospace combat aircraft all of the features leading to a 'difficult' SC problem:

- A flexible airframe, with resonant frequencies for 'significant' structural modes ranging in frequency between very close to the rigid aircraft modes and very close to the FCS sampling frequency (Fig. 11 shows typical pitch-axis SC ground test results).

- Removable stores mounted on sensitive parts of the structure (i.e. wingtips), leading to strongly configuration-dependent SC characteristics.

- Trailing-edge flying controls, with high, aerodynamics-dominated excitation forces at airspeeds where the FCS gains are high (Fig. 12).

- A powerful FCS (Fig. 4), featuring:
  - digital implementation;
  - high feedback gains at structural mode frequencies;
  - high-bandwidth sensors and actuators;
  - multiple parallel paths; and
  - a 'tight' phase budget for SC problem solution.

In view of this, and the overall degree of difficulty anticipated, SC was viewed as being a significant risk in the FCS programme. This was directly reflected in the SC analyses required and undertaken in comparison with previous projects.

The form of the SC analysis undertaken on the EAP derived principally from:

- the inadequacy of the augmented flutter model for the absolute calculation of SC characteristics;

- the need to consider in-flight conditions in detail;

- the requirement to cover potential variability in phase between the parallel paths of the FCS; and

- the need to account for digital effects on high-frequency structural modes.

7.1. *Aero/structural modelling*

The structural model was based on the standard EAP flutter model, i.e. a finite-element-based structural model coupled with unsteady aerodynamics to form the flexible aircraft matrix equations of motion. SC predictions made using the model failed to match the SC ground test results with sufficient accuracy for direct use in SC analysis. Rather than commit expenditure to match each predicted FCS sensor response/actuator excitation transfer function to test measurements, the EAP methodology was to combine the ground test results, representing zero-speed character-istics, with model-predicted incremental aerodynamic effects, to give a more reliable measure of the flexible aircraft transfer function at all flight conditions.

7.2. *FCS representation*

Preliminary frequency-response analysis indicated paths which, in combination with the structure and aerodynamics, had negligible gain at structural mode fre-

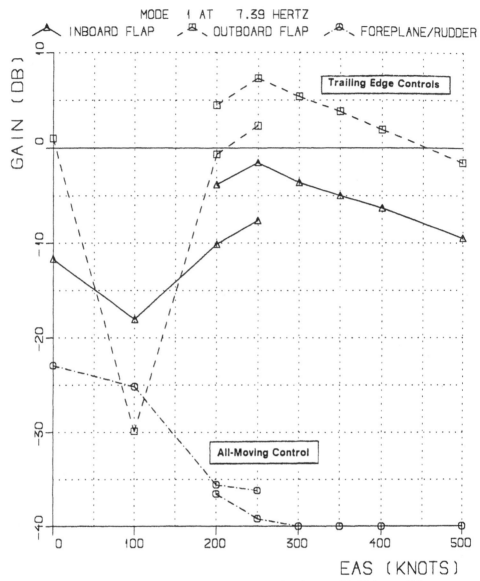

Figure 12.   Variation in control surface excitation forces with flight condition.

quencies. For the remaining paths a simplified representation of the FCS as a scheduled end-to-end high frequency gain facilitated combination with the aero/structural calculations.

7.3. *Combination of FCS and structural gains*

Addition of the FCS gain to the structural mode gain and frequency trends of the aero/structural modelling built up a mode-by-mode picture of the variation of the SC loop gain with flight condition. For each significant FCS path this indicated, in a very clear format, the critical flight conditions and the relative importance of modes and control surface contributions. To overcome the extreme sensitivity of the phase of the structural response, which was expected to be an unreliable quantity, individual FCS/structure paths were treated as scalars, and phase eliminated completely from the analysis.

7.4. *Treatment of digital effects*

Careful consideration of sampling theory led to a simple treatment which included all of the important effects, including:

- Rolling-average and downsampling processes in the aircraft motion sensor unit (AMSU) (i.e. the rate gyros and accelerometers).
- FCC sampling and zero-order hold (ZOH).

Fundamentally these processes led to some attenuation of high-frequency signal components, but also to significant aliasing of high-frequency response to augment low frequencies.

7.5. *Notch filter design*

The process discussed above resulted in a clear and concise picture of the requirements for the attenuation of SC effects. In line with the philosophy of non-reliance on phase, the SC design and clearance requirements, adapted from MIL-9490D, excluded the phase margin specification but required gain margins of 9 dB at structural mode frequencies. This was provided by simple notch filtering in the AMSU and FCC.

7.6. *Method developments*

The 'EAP method', i.e. the construction of the SC model and the resultant notch filter design satisfying a gain-margin-only requirement, was suitable for a demonstrator aircraft such as the EAP. It was recognized that this method would become increasingly costly if applied to a production aircraft with a range of stores to be catered for. In particular, the requirement to achieve a gain margin at low SC mode frequencies results in a notch filter set which is sensitive to the mode frequency. Thus a store configuration which alters the frequencies of these modes would require a different notch filter set.

The cost implications of this would be:

- the need to design multiple notch filter sets;
- significant design complexity in the FCS to allow switching between store configurations – which must be achievable in flight to allow for store release/jettison;
- increased complexity in, and reliability on, the stores management system which indicates if the store is on the aircraft or has been released;
- increased difficulty in introducing store configurations through the development of the weapon system.

It is therefore essential to maximize the robustness of the notch filter design to variations in low SC frequencies in order to avoid or minimize these complexities and costs.

With this in mind, various developments of method were implemented during the course of the EAP, the most significant development concerning the relaxation of the SC design and clearance requirements. This made allowable the consideration of the structural response phase for lower-frequency modes subject to the availability of additional flight test information which would confirm the structural response.

To provide this information, an in-flight structural mode excitation system

AMSU PITCH RATE / OUTBOARD FLAP

LOGLIN   FREQUENCY RESPONSE

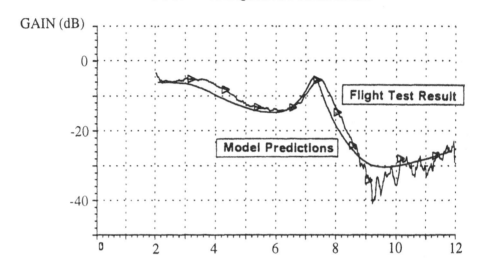

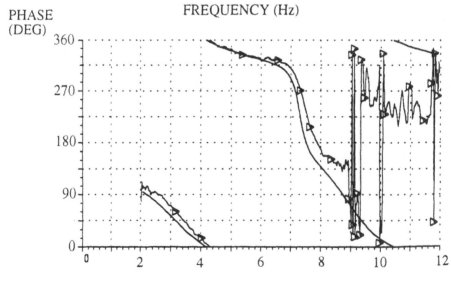

FREQUENCY (Hz)

Figure 13.   Typical results from EAP IFSME flight testing.

(IFSMES) was developed for flutter and SC flight test use; Ramsay (1992) describes the development of the flutter application of the system.

Development and use of the IFSMES proved the possibility of in-flight validation of model-predicted aerodynamic effects, and direct measurement of control power effectiveness as a function of aerodynamic incidence at structural mode frequencies, which was a major influence on the notch filter design.

Figure 13 shows typical results, demonstrating the quality of the data obtained.

8. Conclusions

Overall the programme has been highly successful and has proven the approach taken to design the control laws. The programme also demonstrated the basic control law structure intended for use in the EFA aircraft verifying the capability of achieving excellent handling qualities and gaining invaluable flight experience of a similar aircraft.

The SC clearance philosophy and analysis methodology employed on the EAP led to a clear understanding of the SC problem and of the strengths and weaknesses of the approach, from which a satisfactory flight clearance was derived. No flight envelope restrictions derived from SC considerations, and no ground or flight SC problems, were experienced throughout the 259 flights and the three FCS phase history of the aircraft.

The SC developments made during the programme, in particular the proving of the application of an IFSMES, have given confidence in the viability of the proposed SC clearance route for the EF 2000 project.

The highlights for the FCS control law design were the demonstration of an agile carefree-handling capability and the level of PIO exhibited by the aircraft. Indeed no PIO tendency was noted in any of the 259 flights by any of the 12 pilots who flew the aircraft. These results support the philosophy of retaining significant stick travels and the force and damping characteristics designed here. Valuable information on control sensitivity has been gathered particularly in the roll axis where limits to acceptable roll accelerations have been determined.

The success of the transparent, automatic wing-level-hold mode in transforming the basic long-term trimming/steady-state nature of the aircraft is a strong motivator to include such functions in the yaw and pitch axes. The reported reduction in pilot workload suggests that, in the absence of autopilot functions, the long-term trim behaviour is a significant contributor to workload.

A major highlight was that throughout the whole programme there were no great surprises due largely to the quality of the modelling of the aircraft and the control system. The quality of the aerodynamic model is an affirmation of the methods and wind tunnel analysis used, there being remarkably little adjustment to the model as a result of flight testing. There were also very few modifications to the basic control laws resulting in almost uninterrupted flight testing in each phase.

REFERENCES

GIBSON, J. C., 1990, Evaluation of alternate handling qualities criterion in highly augmented unstable aircraft. *Proceedings of the AIAA Atmospheric Flight Conference (90-2844), Portland, Oregon, USA*, pp. 435–444.

RAMSAY, R. B., 1992, In-flight structural mode excitation system for flutter testing. AGARD CP 519, Proceedings of the Flight Mechanics Panel Meeting, Crete, May 1992, Paper 15.

13

X-29 flight control system: lessons learned

ROBERT CLARKE, JOHN J. BURKEN, JOHN T. BOSWORTH and
JEFFREY E. BAUER

Nomenclature

ACC	automatic camber control (flight control system mode)
AOA	angle of attack
ARI	aileron-to-rudder interconnect
BMAX	rudder pedal command gain
$C_{L_{max}}$	maximum lift coefficient
CF	cost function
CP	constraint penalty
FCS	flight control system
FFT	fast Fourier transform
g	unit of acceleration ($32\cdot2\,\mathrm{ft\,s^{-2}}$)
G_{lim}	pitch stick limit gain
G_q	pitch rate feedback gain
$G_{\dot{q}}$	pitch acceleration feedback gain
G_{n_z}	normal acceleration feedback gain
GAIN1	pitch axis g-compensation gain
$GH(s)$	open-loop transfer function
GF1	symmetric flaperon gain factor
GS1	strake flap gain factor
GMAX	pitch-stick command gain
GYCWSH	rudder pedal-to-aileron washout filter time constant
Hg	the symbol for the element mercury
$\mathbf{HG}(s)$	loop gain matrix
$\mathbf{I}$	identity matrix
ISA	integrated servoactuator
j	$\sqrt{-1}$
K2	roll rate-to-aileron feedback gain
K3	$\dot{\beta}$-to-aileron feedback gain
K4	lateral acceleration-to-aileron feedback gain
K13	lateral stick-to-aileron forward-loop gain

K14   rudder pedal-to-aileron forward-loop gain
K17 $\dot{\beta}$-to-rudder feedback gain
K18   lateral acceleration-to-rudder feedback gain
K27   lateral stick-to-rudder forward-loop gain
LOF   left outboard flaperon
LVDT   linear variable differential transducer
M Mach number
MIMO   multi-input multi-output
n_y lateral acceleration (g)
n_z normal acceleration (g)
NACA   National Advisory Committee for Aeronautics
p roll rate ($\deg\,s^{-1}$)
P_s static pressure (inHg)
PCE   pilot compensation error
PMAX   lateral stick command gain
q pitch rate ($\deg\,s^{-1}$)
Q_c impact pressure (inHg)
r yaw rate ($\deg\,s^{-1}$)
RDM   return difference matrix
RM   redundancy management
RPE   resonance peak error
s Laplace transform variable
S_{aa} generic auto spectrum of a
S_{ab} generic cross spectrum of a to b
SF   scale factor
SISO   single-input single-output
T_t total temperature (°C)
TE   trailing edge
V_T true airspeed (knots)
XKI1   pitch axis forward-loop integrator gain
XKI3   lateral axis forward-loop integrator gain
XKP1   pitch axis forward-loop proportional gain
XKP3   lateral axis forward-loop proportional gain
XKP4   yaw axis forward-loop proportional gain
XPITCH   pitch axis input sequence used in fast Fourier transform
YPITCH   pitch axis output sequence used in fast Fourier transform
α angle of attack (deg)
$\dot{\alpha}$ angle of attack rate ($\deg\,s^{-1}$)
β angle of sideslip (deg)
$\dot{\beta}$ angle of sideslip rate ($\deg\,s^{-1}$)
δ_a differential flaperon deflection (deg)
δ_{a_p} lateral stick deflection (in)
δ_c canard deflection (deg)
δ_{e_p} pitch stick deflection (in)
δ_f symmetric flaperon deflection (deg)
δ_r rudder deflection (deg)
δ_{r_p} rudder pedal deflection (in)
δ_s strake flap deflection (deg)
θ pitch angle (deg)

σ_{FLT} flight-derived singular value
σ_{SSV} structured singular value
σ_{USV} unscaled singular value
τ time constant
ϕ bank angle (deg)
ω frequency (rad s^{-1})

1. Introduction

The Grumman Aerospace Corporation (Bethpage, New York) designed and built two X-29A airplanes under a contract sponsored by the Defense Advanced Research Projects Agency (DARPA) and funded through the United States Air Force. These airplanes were built as technology demonstrators with a forward-swept wing, lightweight fighter design. The use of tailored composites allowed the forward-swept wing design to be fabricated without significant weight penalties (Krone 1980). Both airplanes were flown at the NASA Dryden Flight Research Center to test the predicted aerodynamic advantages of the unique forward-swept wing configuration and unprecedented level of static instability (as much as 35% negative static margin; time to double amplitudes were predicted to be as short as 120 ms). Early on, the airplane designers recognized many potential advantages of this configuration. The forward-swept wing results in lower transonic drag as well as better control at high angle of attack (AOA) (Spacht 1980). The configuration was designed to be departure resistant and maintain significant roll control at extreme AOA. The typical stall pattern of an aft-swept wing, from wing-tip to root, is reversed for a forward-swept wing, which stalls from the root to the tip.

Through the eight years of flight test, over 420 research flights were flown by the two X-29A airplanes. These flights defined an envelope that extended to Mach 1·48, just over 50 000 ft altitude, and up to 50° AOA at 1g and 35° AOA at airspeeds up to 300 knots.

The flight experience at low AOA (below 20° AOA) with the initial flight control system (FCS) is covered in less detail since this design was done by Grumman Aerospace Corporation (Whitaker and Chin 1984, Chin *et al*. 1982). Several flight test techniques will be addressed. These techniques include the in-flight time history comparison with simple linear models and stability margin estimation (gain and phase margins) as well as new capabilities (structured singular value margins) which extend these single-loop stability measures to multiloop control systems. In addition, modifications to improve the FCS will be described; in particular, a technique used to improve the handling qualities of the longitudinal axis will be discussed.

The design of the high-AOA FCS modifications will be presented. Techniques used to expand the high-AOA envelope will be discussed, as well as the problems discovered during this effort. An FCS design feature was the incorporation of a dial-a-gain that allowed two control system gains to be independently varied during flight. This feature allowed many control system changes to be evaluated efficiently. These experiments allowed rapid incorporation of flight-derived improvements to the FCS performance.

2. Test airplane descriptions

The X-29A research airplane integrated several technologies, e.g. a forward-swept, aeroelastically tailored composite wing and a close-coupled, all-moving

Figure 1. The X-29A No. 2 airplane.

canard. Furthermore, the wing, with a 29·27° leading-edge sweep and thin, super-critical airfoil, is relatively simple, employing full-span, double-hinged, trailing-edge flaperons which also provide discrete variable camber. All roll control is provided by these flaperons, as the configuration does not use spoilers, rolling tail, or differential canard. The airplane has three surfaces used for longitudinal control: all-moving canards, symmetric wing flaperons, and aft-fuselage strake flaps. The lateral–directional axes are controlled by differential wing flaperons (ailerons) and a conventional rudder. The left and right canards are driven symmetrically and operate at a maximum rate of approximately 100 deg s^{-1} through a range of 60° trailing-edge (TE) up and 30° TE down. The wing flaperons move at a maximum rate of 68 deg s^{-1} through a range of 10° TE up and 25° TE down. The rudder control surface has a range of $\pm 30°$ and a maximum rate of 141 deg s^{-1}. The strake flaps also act within a range of $\pm 30°$, but have a maximum rate of only 27 deg s^{-1}.

The second X-29A (Fig. 1) was modified for high-AOA testing by adding a spin parachute which was attached at the base of the vertical tail. The spin parachute was installed to provide for positive recovery from spins, as spin-tunnel tests had indicated that the X-29A ailerons and rudder provided poor recovery from fully developed upright spins. The addition of an inertial navigation system and the spin parachute system increased the empty weight of the airplane by almost 600 lb.

3. Low angle-of-attack research

Research at low AOA was the focus of all flight testing on X-29A No. 1 throughout its four years of flight test (Gera *et al.* 1991, Bosworth and Cox 1989, Chacon and McBride 1988). Initially, the focus (from a flight control designer's viewpoint) was on proving adequate stability margins and fixing problems which impacted stability or redundancy management. In the last year of flying, the focus was shifted to make improvements to the FCS to overcome the deficiencies which had been identified by the pilots. The X-29A No. 2 airplane was used primarily to examine high-AOA characteristics, but was also used to study the stability margins of the lateral–directional axes using multi-input multi-output (MIMO) techniques at low-AOA conditions.

3.1. *Description of the flight control system*

The X-29A airplane is controlled through a triplex fly-by-wire FCS, which was designed for fail-operational, fail-safe capability. A schematic of the FCS is shown in Fig. 2. Each of the three channels of the FCS incorporates a primary mode digital system using dual central processing units along with an analogue reversion mode system. Both the digital and analogue systems have dedicated feedback sensors. The digital computers run with an overall cycle time of 25 ms. The commands are sent to servoactuators that position the aerodynamic control surfaces of the airplane.

The initial longitudinal axis control laws were designed using an optimal model following technique (Chin *et al.* 1982). A full state feedback design was first used with a simplified aircraft and actuator model. The longitudinal system stability was significantly affected as higher order elements, such as sensor dynamics, zero-order hold effects, actuators and time delays, were added to the analysis (Whitaker and Chin 1984). To recover the lost stability margins, a conventional design approach was taken to develop lead–lag filters to augment the basic control laws. Even after the redesign work, the stability margin design requirements were relaxed by the contractor to 4·5 dB and 30° (if all of the known high-order dynamics were included in the analysis). The government flight test team

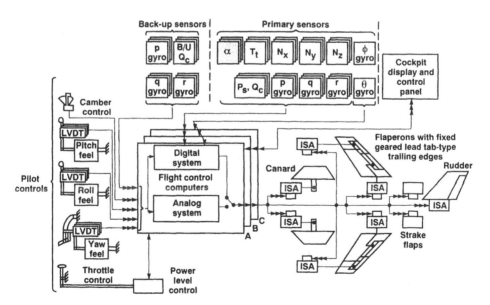

Figure 2. The X-29A flight control system. (Note: number of arrowheads designate level of redundancy and highlighted blocks represent changes made for high AOA.)

decided to require the use of flight-measured stability margins and set minimum margins at 3 dB and 22·5°.

Figure 3 is a block diagram of the longitudinal control system. Short-period stabilization is achieved mainly through pitch rate and synthesized pitch acceleration feedback. Normal acceleration feedback is used to shape the stick force per *g*. The proportional-plus-integral compensation in the forward loop improves the short-period response and steady-state response to pilot inputs. Positive speed stability, which is important during powered approach, is provided by either automatic engagement or pilot selection of airspeed feedback.

In addition to the short-period stabilization function, the primary mode includes an automatic camber control (ACC) function which, in steady flight, generates commands to the symmetric flaperons and strake flaps to optimize the overall lift-to-drag ratio of the airplane. The dynamic characteristics of the ACC feedback loops were designed to be significantly slower than those of the basic stability augmentation loops.

Figure 4 is the lateral–directional control-law block diagram. The bare airframe lateral–directional characteristics of the X-29A are stable, and the multivariable FCS is conventional (Whitaker and Chin 1984). Roll rate is proportional to the lateral stick deflection through a nonlinear gearing gain that enhances the precision of small commands while still enabling the pilot to command large roll rates with larger stick deflections. A command rate limiter is implemented in the roll and yaw axis control systems to minimize the potential of control surface rate limiting, caused by large commands. Another feature is the forward-loop integrator in the roll axis which provides for an automatic trim function and helps to null steady-state roll errors. Synthetic sideslip rate feedback is used to provide dutch roll damping and to assist in turn co-ordination. An aileron-to-rudder interconnect (ARI) is also used to help co-ordinate rolls commanded with lateral stick deflections alone.

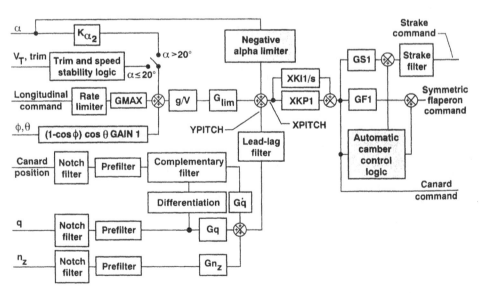

Figure 3. The X-29A longitudinal control system. (Note: the highlighted blocks represent changes made for high AOA.)

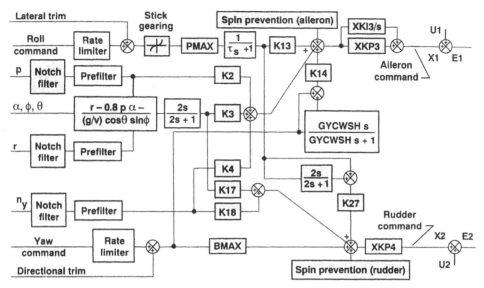

Figure 4. The X-29A lateral–directional control system. (Note: the highlighted blocks represent changes made for high AOA.)

3.2. *Flight test techniques*

The following section presents some of the tools used to analyse the X-29A aircraft. The three tools used for flight data analysis were single-input single-output (SISO) stability margins, time history comparisons, and MIMO or multivariable robustness margins. The first two tools were applied in near real time during envelope expansion of X-29A No. 1, while the last one was only used in postflight analysis on X-29A No. 2. Both stability margin analyses, SISO and MIMO, obtained the desired frequency responses without physically opening any of the feedback loops.

3.2.1. *Single-input single-output gain and phase margins*. Aircraft that have a high degree of static instability, like the X-29A, require close monitoring in the early envelope expansion stages of flight test. Fast Fourier transformation (FFT) techniques were used to measure the longitudinal open-loop frequency response characteristics over the entire flight envelope.

The X-29A longitudinal control system architecture lends itself to classical SISO stability margin analysis. As shown in Fig. 3, the feedback paths reduce to a single path, allowing for traditional gain and phase margin analysis, such as Bode analysis. To excite the vehicle dynamics, a series of pilot pitch-stick commands or computer-generated frequency sweeps were used. Briefly, the technique collects the time domain variables XPITCH and YPITCH driven by the sweep, and uses an FFT to estimate the open-loop transfer function $GH(s)$. The open-loop frequency response is displayed on a monitor, and gain and phase margins are determined. The details of the near-real-time SISO frequency response technique can be found in Bosworth (1989).

The frequency response of an SISO system can be estimated from the auto- and cross-spectra of the input and output. An estimate of the open-loop response is defined as

$$GH(j\omega) = \frac{S_{xy}}{S_{xx}} = \frac{S_{yy}}{S_{xy}} \qquad (1)$$

where S_{xy} is the cross-spectrum of the input XPITCH with the output YPITCH, S_{xx} is the auto-spectrum of the input, and S_{yy} is the auto-spectrum of the output. The overall procedure is shown in Fig. 5.

This test technique revealed much lower than expected margins (below the established flight test minimum margins) at a low-altitude transonic flight condition. As a result, the overall longitudinal loop gain was reduced to recover adequate stability margins. The actual amount of the reduction came directly from the comparison of predicted gain and phase margins with the analytical estimates. Once the control law change was made, the measured stability margins were greater than the requirement.

It was found that open-loop SISO frequency responses could be measured in flight without physically breaking the loop. The near-real-time capability enhanced the efficiency of the X-29A envelope expansion programme. Gain margins greater than 3 dB and phase margins higher than 22·5° were eventually demonstrated over the entire flight envelope.

3.2.2. Linear model time history comparisons. The real-time comparison of the airplane response with linearized models allows the flight test personnel to verify the aircraft is performing as predicted, to determine regions of nonlinear behaviour, and to increase the rate of envelope expansion (Bauer *et al.* 1987). Direct comparison of the measured aircraft responses to those generated by a simulation, driven with identical pilot inputs, provides timely information. It is an extremely useful test tool if the comparison between the actual and simulation responses can be made in real time.

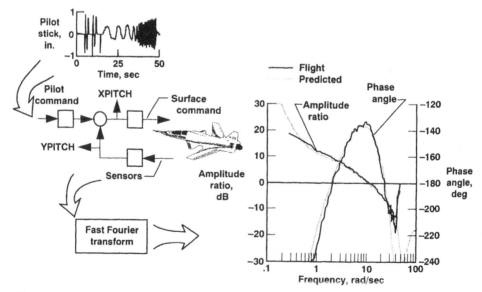

Figure 5. Near-real-time determination of longitudinal axis open-loop frequency response from flight data

Regions of nonlinear behaviour of the aircraft can easily be determined. For example, surface rate limits show up dramatically when the flight data are compared with the linear simulation response. Knowledge of this nonlinear behaviour can be useful in interpreting differences among results from other data analysis procedures, such as frequency response methods or parameter estimation techniques.

The success of the time history comparisons depends on a detailed and accurate mathematical model. For the X-29A airplane the models were obtained by linearizing the nonlinear equations of motion about a trimmed flight condition. The perturbation step sizes were $\pm 1\%$ of Mach number for total velocity, $\pm 2°$ for angle of attack, and $\pm 1°$ or $\deg s^{-1}$ for the remaining states and control surfaces. These step sizes provided reasonable estimates of the linear coefficients.

3.2.3. *Multi-input multi-output stability margins.*

The X-29A lateral–directional control system is a MIMO system, and the classical frequency analysis methods are inadequate for this type of control system. The classical methods, such as Bode or Nyquist analysis, do not allow for simultaneous variations of phase and gain in all of the feedback paths (Doyle and Stein 1981, Ly 1983, Paduano and Downing 1987). Recently, singular value norms of the return difference matrix $(\text{RDM} = I + GH(s)$ or $I + HG(s))$ have been considered as a measure of the system stability margins for multivariable systems (Doyle and Stein 1981, Ly 1983, Newsom and Mukhopadhyay 1985). However, singular value norms of a system can be overly conservative, and a control system designer could interpret the results as unsatisfactory when, in fact, the system is robust (Paduano and Downing 1987). A method for relieving the excessive conservatism is derived by structuring the uncertainties (Ly 1983, Doyle and Stein 1981).

To evaluate the stability robustness of a multivariable system, the Dryden Flight Research Center conducted a series of flight test manoeuvres on the X-29A No. 2 (Burken 1992). The flight-derived singular value (σ_{FLT}) was compared with predicted unscaled singular values (σ_{USV}), structured singular values (σ_{SSV}), and with the conventional single-loop stability margins. Although these flight-derived singular values were determined postflight, this analysis can be used for near-real-time monitoring and safety testing.

As the minimum singular value $(\underline{\sigma})$ of the input or output RDM approaches zero, the system becomes increasingly less stable. The flight singular values need to be determined by using frequency response techniques. The complex frequency response of a system can be estimated from the auto-spectrum and cross-spectrum of the input and output time history variables by transforming these time domain responses to the frequency domain using the FFT. The controller input-to-output transfer matrix, $\mathbf{X}_u(s)$, is defined as follows

$$\mathbf{X}_u(s)_{ij} = \sum_{k=1}^{N} (S_{x_j u_i}(s))_k (S_{u_i u_i}(s))_k^{-1} \qquad (2)$$

where $S_{xu}(s)$ is the cross-spectrum of the input u and output x, $S_{uu}(s)$ is the auto-spectrum of the input, and N is the number of time history arrays.

Using the relationship defined in (2), the open-loop gain matrix is

$$\mathbf{HG}(s) = \mathbf{X}_u(s)(\mathbf{I} - \mathbf{X}_u(s))^{-1} \qquad (3)$$

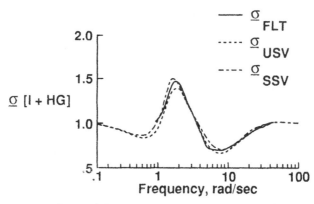

Figure 6. Flight and predicted minimum singular values of the X-29A at Mach 0·7, 30 000 ft altitude, with the baseline gain set.

Figure 6 shows the flight-determined minimum singular values, $\sigma[\mathbf{I} + \mathbf{HG}(s)]$, as well as analytical scaled structured and unscaled structured singular values. This plot shows that good agreement exists between the flight and analytical data. The analytical σ_{SSV} tend to agree slightly better with the flight data than with the analytical σ_{USV}.

For comparison purposes, the classical single-loop frequency response results (SISO) are shown in Table 1 along with the singular value (MIMO) analysis. These MIMO margins were obtained using the universal phase and gain relationship (Mukhopadhyay and Newson 1982). The MIMO analysis allows for simultaneous independent variations, while the SISO analysis allows for single-loop variation.

The minimum stability margin determined by the SISO method is 13 dB and 62°; whereas the flight singular value (MIMO) method resulted in a margin of 11·5 dB and 41°. As expected, the singular value method is conservative, but the results between the SISO and MIMO methods are similar. This is not surprising at this low-AOA condition since the X-29A lateral axis is largely uncoupled from the directional axis.

Extracting multiloop singular values from flight data and comparing the information with prediction validates the use of flight singular values as a relative measure of robustness. This comparison increases the confidence in

	Multiple-input multiple-output			Single-input single-output	
	σ_{USV}	σ_{SSV}	σ_{FLT}	Lateral	Directional
$\underline{\sigma}$	0·65	0·72	0·72		
Gain margin (dB)	8·5	11·5	11·5	18	13
frequency (rad s^{-1})	8	8	8	17	13
Phase margin (deg)	35	41	41	77	62
frequency (rad s^{-1})	8	8	8	2·5	4·5

Table 1. Comparison of SISO and MIMO lateral–directional stability margins at Mach 0·7 and 30 000-ft altitude.

using structured singular values for stability assessments of multiloop control systems. In addition, this technique extends the single-loop gain and phase margin concepts to multiloop systems.

3.3. *Control law modifications*

Several changes in the flight control system were required as a result of the high level of instability of the X-29A. A significant change was made to the airdata selection logic. The initial control laws used three equally weighted sources (a single noseboom and two side probes) for total pressure measurements. The most accurate source, the noseboom measurement, was almost never used by the flight control system since it was usually an extreme, not the middle value. To compensate, a change was made to use the noseboom as long as it was within the failure tolerance of the middle value. This change came back to haunt the test team as the failure tolerance was very large and it was discovered that a within-tolerance failure could result in such large changes in feedback gains that the longitudinal control system was no longer stable.

The flight data showed that reducing the tolerance to an acceptable level (going from 5·0 to 0·5 inHg) would not work as there was a narrow band in AOA from 7° to 12° where errors on the side probe measurements were as large as −1·5 inHg (airplane really faster than indicated). This large error was caused by strong forebody vortices which enveloped one or both of the airdata probes located on the sides of the fuselage. The solution was a 2·0 inHg tolerance and a bias of 1·5 inHg added to the side probe measurements. This worked since the sensitivity to the high gain condition (airplane faster than indicated) was much greater than that of the low gain condition (airplane slower than indicated). The airplane had been operated for almost three years before this problem was identified and fixed.

The high level of static instability of the X-29A caused the control law designers to stress robustness over handling qualities. During flight tests the aircraft models were refined, which allowed the control system to be fine-tuned to improve the handling qualities (Bosworth and Cox 1989). To keep it simple, there was a strong desire to fine-tune the control system without drastic changes in the system architecture. The process used to provide improved handling qualities involved four steps: selection of design goals, selection of design variables, translation of the design goals into a cost function, and iterative reduction of the cost function.

The Neal–Smith analysis provided a good quantitative method for assessing predictions of handling qualities. Unlike lower order equivalent system analysis, the Neal–Smith technique applies to systems that do not exhibit classical second-order behaviour. In addition, there is no ambiguity introduced by the goodness of fit of the higher order system to a low-order match. The Neal–Smith method takes the longitudinal stick position to pitch rate (or attitude) transfer function, and closes the loop around it with a simple compensator, representative of a simple pilot's transfer function. The compensator consists of a lead–lag filter with a gain and a time delay (Fig. 7). The application of the Neal–Smith criterion to the X-29A baseline control laws indicated a relatively large amount of lead required by the compensator to obtain the desired tracking performance. This correlated well with the pilot's comments, which indicated a desire for increased pitch responsiveness in tracking tasks.

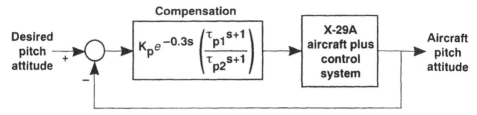

Compensation

Desired pitch attitude → + − ○ → $K_p e^{-0.3s} \left(\dfrac{\tau_{p1}s+1}{\tau_{p2}s+1} \right)$ → X-29A aircraft plus control system → Aircraft pitch attitude

Figure 7.   Closed-loop pitch attitude tracking task.

The design goals were to obtain a quicker pitch response without adversely affecting the stability, control surface activity, or introducing a pilot-induced oscillation problem. The point defined as the desired Neal–Smith criterion was nominally 0 dB and 10° (Fig. 8). A real scalar cost function (CF) was defined as follows.

$$CF = RPE + PCE/SF + CP \qquad (4)$$

where the resonant peak error (RPE) is the distance between the achieved resonant peak and desired peak (0 dB). The pilot compensation error (PCE) is the distance between the achieved pilot compensation and the desired compensation (10°). The constraint penalty (CP) is 10,000 if either the stability margin or surface activity constraints were violated, and 0 otherwise. The scale factor (SF) is 7·0, which is commonly used to compensate for the difference in magnitude of the units of decibels and degrees.

Bosworth and Cox (1989) covered the background and details of the improved handling qualities optimization of the X-29A airplane. The design goal of 10° of lead and 0 dB resonant peak was not achieved because of the design constraints; however, the amount of pilot lead was reduced by approximately 50%. The closed-loop resonant peak achieved by the modified gains was below

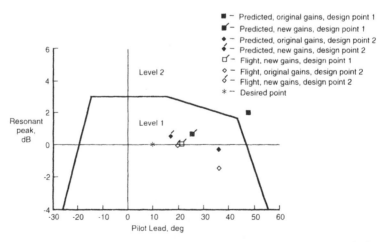

■ – Predicted, original gains, design point 1
◤ – Predicted, new gains, design point 1
◆ – Predicted, original gains, design point 2
◆ – Predicted, new gains, design point 2
◫ – Flight, new gains, design point 1
◇ – Flight, original gains, design point 2
◊ – Flight, new gains, design point 2
✳ – Desired point

Figure 8.   Neal–Smith analysis comparing the modified gains with the original gains for predicted and flight test results.

1·0 dB. This resulted in a Neal–Smith criterion that was well within the level-1 region of the Neal–Smith plane. The design process showed a definite trade-off between the constraints and the achievable Neal–Smith criterion. The modified design gains showed a slightly reduced level of stability margin and increased surface activity. In general, the pilot comments indicated a marked improvement in the performance of the new FCS software.

This design methodology provided a practical means for improving the handling qualities of the vehicle without excessive system redesign. The method provided a 100% increase in the pitch acceleration (from $16 \deg s^{-2}$ to $32 \deg s^{-2}$) with precise control. The final design for the X-29A exhibited a problem associated with rate limiting which resulted in a lower phase margin than predicted. Fortunately, the rate limiting problem occurred at frequencies higher than the range used by a pilot in handling qualities tasks.

4. High-angle-of-attack research

Research at high AOA was the focus of all flight testing on X-29A No. 2. The control laws were modified for high AOA and several airplane modifications were made to ensure that the programme could be conducted safely. The flight tests were conducted to discover the AOA limits of manoeuvring flight (which had been predicted to be up to 40° AOA) and symmetric pitch pointing flight (which was predicted to be as great as 70° AOA). The initial control laws were designed conservatively with provisions made for improvements if the flight data indicated that additional performance would not compromise stability.

4.1. *Description of the flight control system*

The control laws were designed to allow 'feet-on-the-floor' manoeuvring with the lateral stick commanding stability axis roll rate and the longitudinal stick commanding a blended combination of pitch rate, normal acceleration, and AOA. Rudder pedals commanded washed-out stability axis yaw rate, which allowed the airplane to be rolled using the strong dihedral of the airplane at up to 40° AOA.

This approach for controlling a blend of pitch rate and AOA at high AOA is different from the design philosophy of the F-18 High Alpha Research Vehicle and X-31 enhanced fighter manoeuvrability airplanes, which are primarily AOA command systems. The blended combination of feedbacks used in the X-29A FCS provided more of an $\dot{\alpha}$-type command system with weak AOA feedback. The pitch axis trim schedule provided small positive stick forces at $1g$ high-AOA conditions (approximately 1-in deflection or an 8-lb force required to hold 40° AOA).

The high-AOA FCS was designed using conventional techniques combined with the X-29A nonlinear batch and real-time simulations. Linear analysis was used to examine stability margins and generate time histories, which were compared with the nonlinear simulation results to validate the results. The linear analysis included conventional Bode stability margins, time history responses, and limited structured singular value analysis in the lateral–directional axes. In the pitch axis stability margins at high-AOA, conditions were predicted to be higher than the stability margins at the equivalent low-AOA conditions.

However, in the lateral–directional axes the unstable wing rock above 35° AOA dominated the response in linear and nonlinear analysis. Feedback gains that could stabilize the linear airplane models showed an unstable response in the nonlinear simulation driven by rate saturation of the ailerons. The control system design kept the feedback gains at a reasonable level and allowed the low-frequency unstable lateral–directional response.

4.1.1. *Design goals.* The FCS design was required to retain the low-AOA flight characteristics and control law structure which had been previously flown on X-29A No. 1. It was further required that the control system would ensure that spins would not be easily entered, and this required an active spin prevention system.

In the lateral axes the airplane was controlled with conventional ailerons and rudder. The ailerons had priority over symmetric flaperon deflections in the control laws. The control laws were designed in this manner since all roll control was provided by the ailerons and pitch control was provided by canards, strake flaps, and symmetric flaperons.

4.1.2. *Design of the longitudinal axis.* For the most part, the basic low-AOA X-29A normal digital longitudinal axis control laws remained unchanged at high AOA. No gains in the longitudinal axis were scheduled with AOA, but several feedback paths were switched in and out as a function of AOA. The following changes (which are highlighted in Fig. 3) were made in the design of the high-AOA control laws for the X-29A longitudinal axis.

(1) Modified ACC schedules that were designed to provide optimum lift-to-drag ratio symmetric flaperon and strake flap positioning at high AOA. These provide increased maximum attainable lift and reduced transonic canard loads.

(2) Fade-out of velocity feedback and fade-in of AOA feedback to control a slow divergent instability. Velocity feedback was not appropriate to control the instability which developed at high AOA; AOA provided the best feedback as the divergence was almost purely AOA.

(3) Active negative AOA- and g-limiters designed to prevent nose-down pitch tumble entries and potential inverted hung stall problems.

(4) Fade-out of single-string attitude-heading reference system feedbacks. The attitude information only provided gravity compensation for pilot inputs and n_z feedback. The single-string nature and relatively small benefit did not warrant the risk of system failure at high-AOA conditions.

(5) Symmetric flaperon limit reduction from 25° to 21°. Because high-gain roll-rate feedback would be required to prevent wing rock and because the wing flaperons are shared symmetrically and asymmetrically, 4° of flaperon deflection were reserved for aileron-type commands. The flaperons are commanded with differential commands having priority. At high AOA, the ACC schedule would otherwise command the wing flaperons on the symmetric limit and result in a coupling of the wing rock and longitudinal control loop through the symmetric flaperon.

During the accelerated entry high-AOA envelope expansion, the aft stick authority limit was reached earlier than expected (at 25° AOA full aft stick was reached for 160 and 200 knots). The original high-AOA FCS was limited, as were all previous X-29A FCS releases, to 5·4 incremental *g* command at high speed and 1·0 incremental *g* at low speed. The FCS modification increased this to 7·0 (+30%) incremental *g* command at high speed and 2·0 (+100%) *g* at low speed.

A functional check flight of the FCS change showed that although the stick sensitivity was changed it was still acceptable (since stick feel characteristics were unchanged, any change in command authority changes stick force per *g*). The X-29A pilots noted that during 1-*g* flight at 35° and 40° AOA '. . . the increased sensitivity of the longitudinal control was evident, but compensation by the pilot was easily accomplished.'

4.1.3. *Design of the roll–yaw axes.* In the lateral–directional axes the control laws were changed significantly at high AOA from the original low-AOA control system. The lateral–directional block diagram (Fig. 4) shows a full-state-type feedback structure. The high-AOA changes, for the most part, were simplifications in the control law structure flown on X-29A No. 1. The new control laws required many gains to be scheduled with AOA; several were just faded to constants while four command and feedback gains used three AOA breakpoints for table look-up. These three AOA breakpoints were the maximum allowed because of computer space limitations. Computer speed limitations required that a multirate gain look-up structure be incorporated since AOA (20 Hz) was expected to change more rapidly than Mach number or altitude (2·5 Hz). The control law changes and reasons for them include the following.

(1) The forward-loop integrator in the lateral axis was removed at high AOA. This eliminated a problem with the integrator saturating and causing a pro-spin flaperon command.

(2) Most lateral–directional feedbacks were eliminated. This left only high-gain roll rate feedback-to-aileron and washed-out stability axis yaw rate feedback-to-rudder. The high-gain roll damper was used to suppress the wing rock that developed near $C_{L_{max}}$. The washed-out stability axis yaw rate or β feedback helped control sideslip during airplane manoeuvres at high-AOA conditions.

(3) Pilot forward-loop gains were also simplified, leaving only lateral stick-to-aileron, lateral stick-to-rudder, and rudder pedal-to-rudder. The lateral stick gearing was changed from a second-order nonlinear gearing to a linear gearing at high AOA, and a wash-out filter was used in parallel with the ARI gain to provide an extra initial kick on rudder command.

(4) Spin prevention logic was added which commanded up to full rudder and aileron deflection if the yaw rate exceeded 30 deg s^{-1} with AOA $\geqslant 40°$ for upright spins and AOA $\leqslant -25°$ for inverted spins. The pilot command gain was increased to allow the pilot sufficient authority to override any of these automatic inputs.

At high AOA, vertical fin buffeting was encountered because of forebody vortex interaction. The control system was strongly affected through the

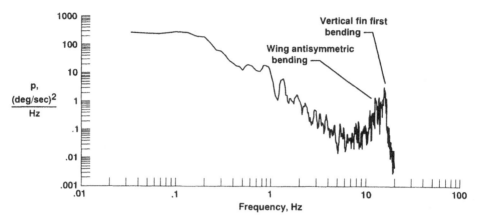

Figure 9. Roll rate gyro power spectral density.

excitation of several structural modes which were seen on the roll rate gyro signal. The buffet intensity was as high as $110g$ at the tip of the vertical fin.

The vertical fin excited the roll rate gyro and, through high-gain feedback, caused the flaperon actuators to attempt to track this high-frequency signal. Flight tests showed that an unexpected hydraulic system problem resulted from this flaperon command. During a 360° full stick aileron roll, the left outboard flaperon (LOF) hydraulic logic indicator showed a failure of the control logic for this actuator. The most probable explanation was that a flow restriction existed in the hydraulic lines driving the LOF and that this restriction showed up when large, high-frequency demands were placed on the actuator. Postflight analysis also showed that the measured LOF rates were approximately 7 to 8 $\deg s^{-1}$ lower than for the right outboard flaperon.

Since the roll rate gyro signal did not originally use any structural notch filters, the vertical fin first bending (15·8 Hz), wing bending antisymmetric (13·2 Hz), and fuselage lateral bending (11·1 Hz) structural modes showed up in the commands to the ailerons. Figure 9 shows the response of the roll rate gyro. The figure shows that most of the vertical tail buffet is transferred to the roll rate gyro through the vertical fin-first bending mode. Analysis of the flight data showed that the g level increased proportionately with dynamic pressure.

Notch filters and a software gain reduction on roll rate feedback were used as the long-term FCS solution to the problem. Before these changes, 50% of the manoeuvres in the region of failures indicated LOF hydraulic logic failures. After the changes were made, these failure conditions occurred only rarely, and even more severe entry conditions and higher buffet levels were encountered without incident.

4.1.4. *Other changes.* To aid in research and to allow for unknown problems in flight testing, several additional changes were made to the control system. These changes included a dial-a-gain capability to allow the roll rate-to-aileron and the ARI gains to be independently varied (K2/K27). These two gains can each have five different values.

Concerns about the severe wing rock had led to a slow build-up in AOA using the dial-a-gain variations. The airplane roll response was found to be heavily damped and the dial-a-gain system was used to examine reductions in feedback gain. The response of the airplane was significantly quicker (approximately 20%) with the reduced roll rate feedback-to-aileron gain. No objectionable wing rock developed because of the lower gain.

Wing rock was predicted to be a major problem with the X-29A configuration at high AOA. These predictions had been made based upon wind-tunnel (Croom *et al.* 1988, Murri *et al.* 1984) estimates and supported by drop model flight tests (Fratello *et al.* 1987). Early simulation predictions were that wing rock would effectively limit the useful high-AOA envelope to approximately 35° AOA. Wing rock was predicted to deteriorate quickly as the roll damping became unstable and the limited aileron control power was insufficient to stop it. Roll rate-to-aileron gains as high as $2 \cdot 0$ deg deg^{-1} s^{-1} were required to damp wing rock and prevent roll departures on the NASA Langley Research Center X-29A drop model. Flight tests with the dial-a-gain system showed that the K2 (roll rate-to-aileron) gain could be reduced from the initial maximum value of $0 \cdot 6$ to $0 \cdot 48$ deg deg^{-1} s^{-1} with no significant wing rock. Decreases of this gain were required to ensure that a rate limit-driven instability in the ailerons at high dynamic pressures would not occur.

The second use of dial-a-gain was to increase the roll performance of the X-29A at high-AOA conditions. The stability axis roll rates were almost doubled in the AOA region from 20° to 30° at 200 knots (from 40 to 70 deg s^{-1}) with only a small degradation in the roll coordination. Above this AOA, uncommanded reversals were seen because of control surface saturation and a corresponding lack of coordination (predominantly increased sideslip.) This performance improvement was accomplished with a 75% increase in the K13 (lateral stick-to-aileron) gain and an 80% increase in the K27 (lateral stick-to-rudder) gain. Increases in rates were possible throughout much of the high-AOA envelope, but were usually not as dramatic as in this example. The dial-a-gain concept proved to be a valuable research tool to test simple control law changes before the full FCS changes were made.

4.1.5. *Airdata system.* Airdata issues were addressed early in the development of the high-AOA control laws. Measurement of accurate AOA was very important as this would be a primary gain scheduling parameter as well as a feedback to longitudinal and lateral–directional axes. Accurate airdata were also important because of conditional stability of the lateral and longitudinal axes at high AOA. Stability margins would be compromised if airdata errors were large.

Two of the three AOA sensors, located on each side of the airplane, had a range limited to 35°. The location and range were considered inadequate and two options were considered to solve the problem. The first option, to mount two additional AOA vanes on the noseboom, was mechanized and flown on X-29A No. 1 for evaluation and was found to have excellent characteristics. The second option was to install NACA booms and AOA vanes on the wing tips. This option would have resulted in additional problems as the FCS did not rate-correct its AOA measurements. With large lateral offsets, roll rate corrections would have to have been included, which would have made AOA measurements sensitive to airspeed measurement errors. In the end, simplicity drove the decision to install two additional AOA vanes on the noseboom.

The second issue concerned the accurate measurement of airspeed at high-AOA conditions caused by local flow effects. Several unsuccessful alternative pitot probe locations were investigated. Belly probes were tested on a wind-tunnel model and were found to change the aerodynamics, while swivel probes proved unable to be flight qualified for installation forward of engine inlets on a single-engine airplane. Since an alternative location could not be found, and the side probes were expected to have poor high-AOA characteristics, the decision was made to use a single string noseboom pitot-static probe at high AOA.

The high-AOA control system design had to be highly reliable. In general, multiple (three or more) sensors were used to provide redundancy, but for impact pressure at high AOA the FCS relied on a single noseboom probe with two independent sensors.

The airdata system was carefully monitored during the high-AOA expansion. Differences between the noseboom and side probes were tracked as a function of AOA and compared with redundancy management trip levels (which were 1·5 inHg for static pressure, 2·0 inHg for total pressure and 5·0° for AOA). Predictions about the airdata system made during the FCS design were found to be pessimistic because the system worked better than expected. With the exception of the known problem with the total pressure measurements in the 7° to 12° AOA region, the maximum error at all other conditions was less than 0·5 inHg. In hindsight, the flight data showed that the side probes performed adequately for FCS gain scheduling purposes and the system did not need to be made single string on the noseboom probe.

One in-flight incident occurred because of the airdata system. Figure 10 shows the time history of a recovery manoeuvre. The figure shows that as sideslip exceeded 20°, the left rear AOA vane exceeded the sensor tolerance and

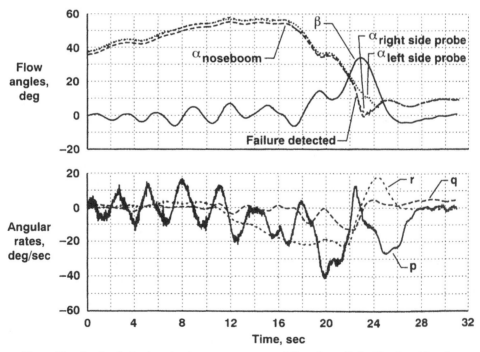

Figure 10.   Angle-of-attack redundancy management failure from flight 27—time histories.

was declared failed. This incident occurred during a recovery from 50° AOA. The airplane continued to operate on the two remaining sensors and was in no danger. Once the failure was known, the personnel in the control room were able to examine the individual AOA channels and discover that all vanes appeared to be functioning properly.

4.2. *Envelope expansion technique*

To increase the rate of envelope expansion, an incremental simulation validation technique (Gera *et al*. 1981) from the F-14 high-AOA ARI test programme was refined and used. This analysis technique was used for postflight comparisons and model updating. It allowed the simulation aerodynamic model to be updated between flight days with local improvements (e.g. changes to lateral control power) derived from previous test points. This allowed the pilots to train with a simulation which matched the most important aerodynamic characteristics and provided engineers with a method to track the aerodynamic differences that were discovered in the flight test. The magnitudes and types of changes to the aerodynamic model provided assurance that the airplane could be safely flown to the next higher AOA test point.

The updated aerodynamics were applied primarily in the lateral–directional axes as almost no changes were required in the longitudinal axis. The updates were constructed with mostly linear terms, but some local nonlinearities were also included.

The most important characteristics to match were the magnitude, frequency and phase relationship of the airplane response. At first, attempts were made to use all six degrees-of-freedom in the simulation, but longitudinal trim differences caused the simulation to diverge from the flight measurements before the manoeuvre was complete. Since the lateral–directional dynamics were of primary importance, the longitudinal dynamics were separated from them. The simulation-matching technique then used the measured longitudinal parameters and lateral–directional pilot stick and rudder pedal measurements as inputs to the batch simulation. This forced the airspeed, altitude, canard deflection and AOA to track the flight measurements, while allowing the lateral–directional axes complete freedom. Some work was also accomplished using an alternative technique which bypassed the control system and used the measured aileron and rudder as inputs, but the wing rock instability eventually made this too difficult. Without including the control system in the simulation the system is an unstable process, since the flight control system stabilizes the wing rock.

Several techniques were used to determine the aerodynamic model updates that would be made to the simulation model (Pellicano *et al*. 1990). These updates were maintained in a separate aerodynamic delta mathematical model, which allowed quick and easy modification. Once the aerodynamic models were updated, sensitivity studies on the real-time simulation were used to predict the airplane response at the next flight test expansion points.

4.3. *Pitch rate limitations*

An example comparison of a full-stick pitch axis maneuver with the complete six degree-of-freedom baseline simulation is shown in Fig. 11. The flight manoeuvre required the pilot to trim the airplane in level flight at 10° AOA at 20 000 ft altitude

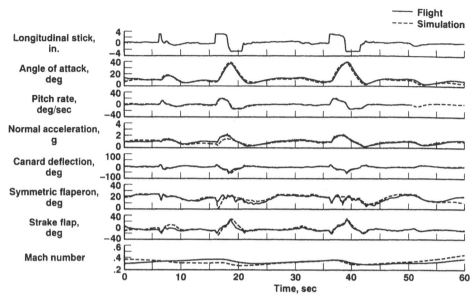

Figure 11.   X-29 No. 2 flight/simulation comparison for large-amplitude stick manoeuvre. Flight control software was the final-AOA software.

(approximately 0·3 Mach). The simulation was matched to the initial trim condition and then driven with the pilot stick and throttle inputs. The figure shows close agreement between the flight data and simulation which allows a high confidence to be placed on simulation analysis of the X-29 pitch rate capability.

The X-29 pilots consistently found the maximum pitch rate capability of the airplane inadequate. Figure 12 shows the predicted maximum nose-up and nose-down pitch rates of the X-29 as a function of Mach number (altitude varied from 10 000 to 20 000 ft). Several flight data points (both nose up and nose down) from the manoeuvre shown in Fig. 11 are also included as well as F-18 pitch rate data for comparison purposes. The simulation manoeuvres consisted of two types of manoeuvres: a full aft stick step input and a doublet-type input which consisted of a full aft stick input followed by a full forward stick input timed to try to force the control surfaces to maximum rates.

It is clear from the data that the X-29 requires approximately 50% higher rates to be comparable with an F-18 at low-speed conditions. Examination of the peak canard actuator rates shows that the X-29 was using nearly all of the capability (104 deg s^{-1}, no load rate limit) with the current control system gains. Increases in the canard actuator rates commensurate with the increases in pitch rate would be required for any improvement.

The simulation showed that most of the actuator rate was used in controlling the unstable airplane response. Figure 11 shows this in detail. Close examination of the canard response shows that during the full aft stick input the initial response of the canard is trailing edge down (and trailing edge up for the flaperon and strake flap). As is typical for an unstable pitch response the surfaces then move quickly in the opposite direction to unload and control the unstable response. The second motion is typically much larger than the initial motion and in most cases is more demanding of the actuator rates, especially at low dynamic pressure where large

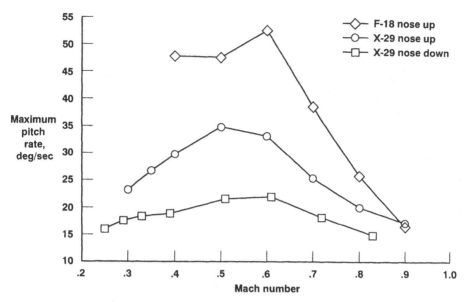

Figure 12. X-29 nose-up and nose-down pitch-rate capability using final high-AOA flight control software.

control surface motion is required. This agrees very well with data presented in Chin *et al.* (1982).

5. Conclusions

The X-29A airplanes were evaluated over the full design envelope. The flight control system successfully performed the tasks of stabilizing the short-period mode and providing automatic camber control to minimize trim drag. Compared with other highly augmented, digital fly-by-wire airplanes, the X-29A and its flight control system proved remarkably trouble free. Despite the unusually large, negative static margin, the X-29A proved safe to operate within the design envelope. Flight tests showed the following lessons.

(a) Adequate stability to test successfully a 35% statically unstable airframe was demonstrated over the entire envelope in a *flight test research environment*. Extrapolations to a production–operational environment should be made carefully.

(b) The level of static instability and control surface rate limits did impact the nose-up and nose-down maximum pitch rates. At low airspeeds, to achieve rates comparable with an F-18, new actuators with at least 50% higher rate are required.

(c) Testability of a flight control system on an airplane with this level of instability is important, and big payoffs can be made if provisions are provided for real-time capabilities.

(d) Airdata are critical for highly unstable airframes and extra analysis is required to ensure adequate stability. Typical fighter-type airplane airdata redundancy management tolerances do not apply. Tight tolerances must be used even at the risk of nuisance failure detection.

(*e*) The dial-a-gain concept proved a valuable aid to evaluate subtle predicted differences in flying qualities through back-to-back tests. It was also useful to flight-test proposed gain adjustments before major flight control system gain changes were made. This concept might not be easily applied to full state feedback designs, but forward-loop gains are good candidates for this use in any design.

(*f*) High angle of attack with high feedback gains will create problems with structural modes and require notch filters to eliminate flight control system response.

REFERENCES

BAUER, J. E., CRAWFORD, D. B. and ANDRISANI, D., 1987, Real-time comparison of X-29A flight data and simulations data. AIAA Paper No. 87-0344; see also: 1989, *AIAA Journal of Aircraft* (Feb.), 117–123.

BOSWORTH, J. T., 1989, Flight-determined longitudinal stability characteristics of the X-29A airplane using frequency response techniques. NASA TM-4122.

BOSWORTH, J. T. and COX, T. H., 1989, A design procedure for the handling qualities optimization of the X-29A aircraft. NASA TM-4142.

BURKEN, J. J., 1992, Flight-determined stability analysis of multiple-input–multiple-output control systems. NASA TM-4416.

CHACON, V., and McBRIDE, D. 1988, Operational viewpoint of the X-29A digital flight control system. NASA TM-100434.

CHIN, J., BERMAN, H., and ELLINWOOD, J., 1982, X-29A flight control system design experiences. AIAA Paper No. 82-1538.

CROOM, M. A., WHIPPLE, R. D., MURRI, D. G., GRAFTON, S. B. and FRATELLO, D. J. 1988, High-alpha flight dynamics research on the X-29 configuration using dynamic model test techniques. SAE 881420.

DOYLE, J. C., and STEIN, G., 1981, Multivariable feedback design: concepts for a classical modern synthesis. *IEEE Transactions on Automatic Control*, **26**, 4–16.

FRATELLO, D. J., CROOM, M. A., NGUYEN, L. T. and DOMACK, C. S., 1987, Use of the updated NASA Langley radio-controlled drop-model technique for high-alpha studies of the X-29A configuration. AIAA Paper No. 87-2559.

GERA, J., BOSWORTH, J. T., and COX, T. H., 1991, X-29A flight test techniques and results: flight controls. NASA TP-3121.

GERA, J., WILSON, R. J., ENEVOLDSON, E. K., and NGUYEN, L. T., 1981, Flight test experience with high-*α* control system techniques on the F-14 Airplane. AIAA Paper No. 81-2505.

KRONE, JR., N. J., 1980, Forward swept wing flight demonstrator. AIAA Paper No. 80-1882.

LY, U.-L., 1983, Robustness analysis of a multiloop flight control system. AIAA Paper No. 83-2189.

MUKHOPADHYAY, V., and NEWSOM, J. R., 1984, A multiloop system stability margin study using matrix singular values. *Journal of Guidance, Control and Dynamics*, **7**, 582–587.

MURRI, D. G., NGUYEN, L. T. and GRAFTON, S. B., 1984, Wind-tunnel free-flight investigation of a model of a forward-swept-wing fighter configuration. NASA TP-2230.

NEWSOM, J. R., and MUKHOPADHYAY, V. 1985, A multiloop robust controller design study using singular value gradients. *Journal of Guidance, Control, and Dynamics*, **8**, 514–519.

PADUANO, J. D., and DOWNING, D. R., 1987, Application of a sensitivity analysis technique to high-order digital flight-control systems. NASA CR-179429.

PELLICANO, P., KRUMENACKER, J., and VANHOY, D., 1990, X-29 High angle-of-attack flight test procedures, results, and lessons learned. SFTE 21st Annual Symposium, Garden Grove, California.

SPACHT, G., 1980, The forward swept wing: a unique design challenge. AIAA Paper No. 80-1885.

WHITAKER, A., and CHIN, J., 1984, X-29 digital flight control system design. AGARD CP-384: *Active Control Systems — Review, Evaluation and Projections*.

14

Practical aspects of the design of an integrated flight and propulsion control system

DAVID J. MOORHOUSE and KEVIN D. CITURS

1. Introduction

Control theory and system design methodologies are becoming more sophisticated each year; but at the same time the aircraft systems are becoming increasingly complex. This progress is not without steps backward, as recent crashes of the Grippen in Sweden (January 1989 and August 1992) and the YF-22 in the United States (April 1992) indicate that sometimes 'the dragon eats the knight'. The STOL and Maneuver Technology Demonstrator (S/MTD) program was a Wright Laboratory development of comparable flight control complexity. This program was structured to develop and validate through analysis, experiment and flight test, specific technologies intended to provide current and future high-performance fighters with both short takeoff and landing (STOL) capability and enhanced combat mission performance. The contract with McDonnell Douglas Aerospace (MDA) began on 1 October 1984. Figure 1 shows the major modifications that were made to an existing F-15B.

The primary technologies for consideration here were the integration of two-dimensional thrust vectoring and reversing exhaust nozzles into an all-new digital integrated flight/propulsion control (IFPC) system. Together with all-moving canard surfaces, this made a complex system. The overriding requirement of the IFPC system was to be 'capable of functionally integrating all aspects of flight, engine, and nozzle control including aerodynamic control surfaces, engine thrust, thrust vectoring, thrust reversing and differential efflux modulation, control and stability augmentation, high lift system, steering and braking'. The intent was that integration be an objective of the demonstration program, not just a means to achieve requirements as necessary. This is in contrast with a system acquisition in which the mission requirements dictate the integration that is needed.

With integration as a program objective, every possibility for a control effector was considered. The complete list is presented in Fig. 2, showing effects in all degrees of freedom. There was some uncertainty that a classical approach would optimize use of all these available effectors. At the same time, there was no desire to commit to a totally multivariable design approach.

All the bidders on the S/MTD contract were strongly encouraged to use multivariable control theory, although it was not an absolute requirement. The initial design effort compared results of classical techniques performed by MDA and

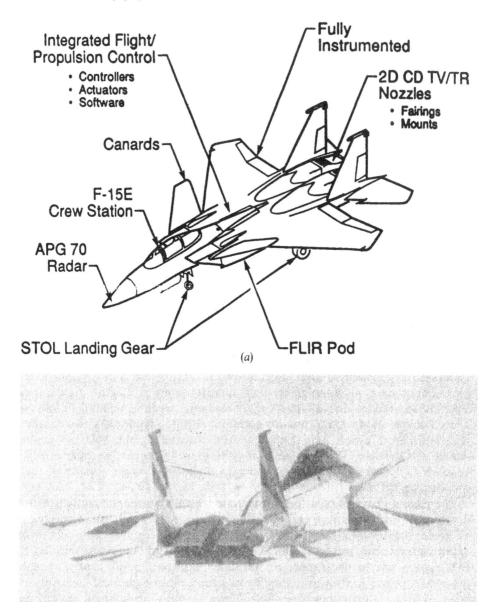

Figure 1. (*a*) S/MTD aircraft modifications and flight test vehicle with two-dimensional nozzles; (*b*) S/MTD flight test vehicle on its first flight, with the two-dimensional nozzles.

multivariable techniques performed by Honeywell Systems Research Center. This early comparison provided a rational basis to choose which design method to continue for implementation into the aircraft.

The objective of this chapter is to document the complete S/MTD control development experience. It is not intended as a theoretical exposition, however, but as a more practical example and guide to the design of a successful control system.

Effector	Pitch	Roll	Yaw	X-Force	Y-Force	Z-Force
Stabilators	√	√				
Canards	√		√		√	√
Ailerons		√				√
Flaperons		√				√
Rudders			√		√	
Gross Thrust				√		
Vectoring	√	√				
Reverser Vanes	√	√	√	√		
Main Gear			√	√		
Nose Gear			√			

Figure 2. Available control effectors.

Experiences with the two different design methodologies are presented. Flight test results are presented of the different modes implemented in the test aircraft. Lastly, overall results are presented in the form of lessons learned during the design development.

2. Design requirements

A new version of the US military flying-qualities specification had been issued in 1980 with substantiation documented by Moorhouse and Woodcock (1982). This specification, therefore, formed the basis for the S/MTD program requirements. In addition, however, since this program was the first application of the specification the opportunity was taken either to validate or improve individual requirements.

2.1. *System requirements*

The IFPC system was required to provide 'good inner-loop stability and positive manual control in all axes of the air vehicle throughout its intended operating envelope both in flight and on the ground (satisfying the intent of specification MIL-F-8785C)'. This requirement was intended to achieve good flying qualities over the whole envelope guided more by the intent than the letter of the specification. This recognized that meeting the letter of the specification was no guarantee of satisfactory flying qualities. The 'intent' was met by using additional, second-tier design guidelines. In addition, the requirement for 'positive manual control' was intended to preclude consideration of automatic landing systems. One flying-qualities requirement that was explicitly called out in the Statement of Work was a maximum equivalent-system time delay of 100 ms. From the development of MIL-F-8785C it was felt that this should be an explicit, hard requirement in any control system to be designed for any precise task, regardless of the method of design or implementation.

The system was also required to meet the intent of MIL-F-9490D with the stability margins as design goals, clarified by: 'Such single-input/single-output parameters may be too restrictive or too lenient for different aspects of the IFPC system in achieving the desired compromise between stability and performance. The contractor shall analyze and document deviations from the MIL-F-9490D requirements.' This was interpreted as a requirement to validate or modify the classical 6 dB gain margin and 45° phase margin for a complex multi-input/multi-output system.

Specific flight control modes were required with the rationale:

In order to provide the ability to assess task performance and minimize pilot workload in the flight vehicle, the integrated system shall also provide the flexibility to permit inflight selection of mission task oriented control modes as determined by analysis and simulation. Mode switching transients shall not produce unsafe aircraft responses. As a minimum, the following modes are required:

A CONVENTIONAL mode shall be designed for satisfactory performance over the flight test envelope, including conventional landing, without the use of the added technologies. This mode will serve as a baseline for performance evaluation and as a backup in the event of multiple failure of the new technology components.

A STOL mode shall be designed to provide precise manual control of flight path trajectory, airspeed and aircraft attitudes. The integrated control system and other technologies shall be combined to provide short field performance in weather and poor visibility. The purpose of this mode is to minimize pilot workload during precise manual landings, high reverse thrust ground operations and maximum performance takeoffs.

A CRUISE mode shall be designed to enhance normal up-and-away and cruise task performance. The purpose of this mode is to use the integrated control system and other technologies to optimize appropriate measures of merit representing an improvement over the cruise capability of the baseline aircraft.

A COMBAT mode shall be designed to enhance up-and-away maneuverability. The purpose of this mode is to use the integrated control system and other technologies to optimize appropriate measures of merit representing an improvement over the combat maneuvering or weapon delivery performance of the baseline aircraft.

These modes were specified in this form for technology demonstration purposes, with the above general guidance as to the intent of each mode. Once the benefits of each mode were identified, it was expected that they could be implemented in a production application partly or fully as necessary. During the development, the required STOL mode evolved into three separate modes – one for takeoff and approach, one for precision landing and one for ground handling.

2.2. Detailed requirements

Both the MDA and the government project offices were committed to a principle of designing to analytical flying-qualities requirements. There was complete agreement that the most critical part of a design is the definition of requirements. Any discussion of design methodology is meaningless without good requirements. The starting point was lessons learned, documented in Walker and LaManna (1982), such as making the command paths as direct as possible by placing filtering and shaping functions in feedback paths. The next step was a definition of preferred regions, i.e. design points, within the specification boundaries, which was documented in Moorhouse and Moran (1985). Many refinements and additions were made during a series of piloted simulations; however, no parameter was changed by 'simulation tuning'. When the need for a flying-qualities change was identified during the early simulations, the actual change was defined analytically and validated with further simulation.

Detail requirements for the precision-landing mode were defined independently of the military specification. Airspeed and pitch responses were required to be decoupled, removing the piloting task of coordinating stick and throttle inputs. It was also required that this feature be accomplished by feeding back airspeed to in-flight thrust reverser angle modulation to produce high-bandwidth speed control. In this mode, throttle commanded airspeed with no change in flight path and stick commanded pitch rate with no change in airspeed. When the aircraft was at the desired approach speed, glideslope angle was controlled with only stick inputs – a simpler task. In order to improve glideslope control even further a minimum phase flight path response to pitch stick inputs was also required.

3. Design methodologies

3.1. *Classical*

In general terms, a classical design develops in a step-by-step process to satisfy textbook inner- and outer-loop requirements in the military flying-qualities specification (illustrated conceptually in Fig. 3). Second-order effects such as adverse coupling are eliminated with crossfeed, but details of these requirements are frequently a fallout as the design progresses. A significant advantage is the insight into the system that is provided by the process itself of adding complexity in steps to

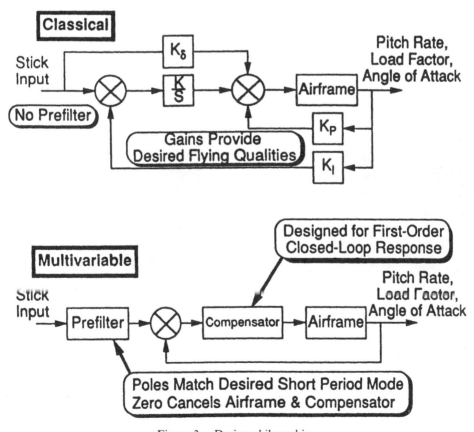

Figure 3. Design philosophies.

enhance system performance. On the other hand, the result is dependent on the fidelity of the aircraft model to an unknown extent. Robustness is verified by analysis or by trial-and-error sensitivity studies after the design. There is frequently not a well-defined end to the process (other than schedule).

A completely new design approach was used in the classical development of the flaps-up longitudinal control laws in the CONVENTIONAL mode. Figure 4 illustrates the process used in the design of this mode. The design guidelines were reformulated into low-order equivalent-system form complete with an allowable time delay. An initial architecture representing a classical proportional plus integral form was defined. A database of trimmed aerodynamic derivatives was computed. Using the low-order models, the initial architecture and the aerodynamic derivatives, an 'inverse equivalent-system' method was used to define the control system gains which the designer selected to be varied. The principles involved are very similar to those used in the generation of equivalent-system fits for the identification of flying-qualities parameters. The difference is that the 'high-order' system is adjusted to fit the frequency-domain representation of the desired flying-qualities characteristics. Successive applications at each flight condition result in a series of point designs from which gain schedules are developed. Equivalent-system techniques were used to verify the resulting flying-qualities characteristics. Robustness was checked by calculation of gain and phase margins. Nonlinear design and analysis was conducted using off-line six-degree-of-freedom flight simulation to produce the final control law design.

The inverse equivalent-system method is easy to use. The user specifies the low-order system dynamics and desired time delay, a frequency range over which the

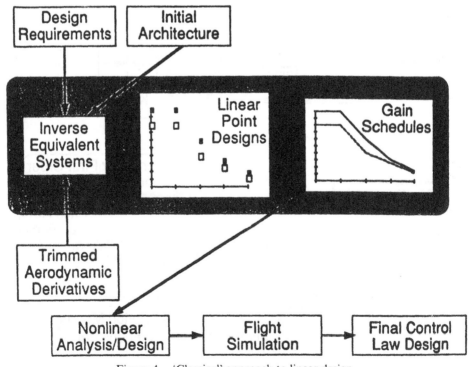

Figure 4. 'Classical' approach to linear design.

matching is calculated, and the control system gains to be optimized. Initial values and allowable limits are specified for the selected control system gains. The program then varies the selected gains to minimize the mismatch in gain and phase over the specified frequency range of matching. The user evaluates the resulting gains to determine if they are satisfactory. The next design point is then selected. During the S/MTD design effort, gains were calculated every tenth of a Mach across the design envelope at a given altitude. Each successive design used the gain values from the previous design point as initial values. A Mach sweep was conducted at 10 000 ft (3050 m) altitude increments. Use of this 'envelope expansion' technique usually resulted in smooth gain trends. Curve fitting of the point designs to produce the final gain schedules was simplified by the use of a system architecture in which the designer had physical insight of the impact of gain selection or aircraft parameter variations. The key to successful application of this method is a good understanding of the desired response characteristics.

3.2. *Multivariable*

In principle, multivariable approaches start with some form of full-state feedback and 'perfect' performance, also illustrated in Fig. 3. A compensator is designed to provide a specified first-order closed-loop response when cascaded with the airframe plant. A prefilter is then added to satisfy flying-qualities requirements. 'Perfect' performance, however, implies that all the requirements must be specified up front, which is not a simple task. Then this formulation must be simplified to a practical implementation while maintaining performance benefits.

The multivariable designs were performed by Honeywell Systems and Research Center using the well-known linear quadratic Gaussian (LQG) synthesis method. Free parameters were chosen in such a way that the entire formal design process could be reinterpreted not as a least-squares error minimization problem but as a 'loop-shaping' problem, i.e. a problem of designing feedback compensators to achieve desirable sensitivity and complementary sensitivity transfer functions at critical loop-breaking points in the system. This interpretation depends on a 'loop transfer recovery' property. The basic theory of this LQG/LTR technique is well documented, e.g. Doyle and Stein (1981) and Stein and Athans (1987), and will only be summarized here with emphasis on the more practical aspects.

The feedback design problem for integrated control reduces to one of specifying requirements in the form of loop shapes and then using the LQG/LTR method to realize the compensator. The design tradeoffs between the sensitivity matrix and the complementary sensitivity matrix are conducted by changing the specified loop shapes and computing new compensators. The method yields a compensator of order equal to the order of the open-loop plant model plus the order of the appended loop-shape dynamics. The complexity of a full-state compensator is not practical, so that the next step is to simplify to a reduced-order compensator. Modal truncation was used to remove asymptotic roots, followed by a procedure of successively balancing the compensator and truncating one state at a time. It is emphasized that compensator reduction is an art and good engineering judgement must be used to arrive at the best practical solution. Since the performance and robustness of the compensation degrade as states are removed, it is important to check stability margins and closed-loop transfer functions at each stage in the process. Robustness was analyzed by means of the upper bound on structured singular value (see Morton 1985, Morton and McAfoos 1985). Lastly, the required prefilter characteristics are defined explicitly by flying-qualities requirements.

3.3. *Combined approach*

Specific characteristics for each mode were established from the general requirements above. In each case, MDA performed a classical design and Honeywell performed an LQG/LTR design to the same requirements. The original hope was to implement both design approaches for evaluation in a flight test. As the control laws evolved, which as always means expands, it became obvious that there would be insufficient computer power to implement two complete sets. The more practical approach then was to make a choice of which design to implement in each case. The final choices, by mode and axis, are as shown in Fig. 5.

A simple example of the benefit of the combined approach comes from the CONVENTIONAL mode, which was designed to have good flying qualities using

Combination of Multivariable and Classical Analysis Methods Used to Design Control Laws

Axis/Mode	Honeywell LQG/LTR	MDA Classical
Longitudinal		
CONV		√
Cruise/Combat	√	
STOL-Land	√	
STOL-TOA	√	
STOL-GH	√	
Lateral/Directional		
CONV		√
Cruise/Combat		√
STOL-Land		√
STOL-TOA		√
STOL-GH		√
Thrust		
CONV		√
Cruise/Combat		√
STOL-Land	√	
STOL-TOA		√
STOL-GH		√

Figure 5.   Choices of design approach.

only the aerodynamic control surfaces. This mode used conventional controls to satisfy conventional requirements; it was felt that the multivariable approach provided no practical benefits and therefore classical techniques were used. The first consideration in either design approach, however, is controls-fixed static instability. With the addition of the canards to the F-15, controls-fixed static stability is approximately 10–12% unstable subsonically. The first step in the classical approach was to schedule canard deflection as a function of angle of attack to provide positive stability. This change also eliminated the problem of an unstable pole from the multivariable formulation. A simple plant was defined by including the stability contribution from the scheduled canard deflection.

4. Design experience

4.1. *Classical*

Classical methods were used in the design of the lateral and directional axes in all control system modes and in the longitudinal axis of the CONVENTIONAL mode, both flaps up and flaps down.

The lateral/directional control system was designed using root-locus and Bode plot design methods. The control effectors included differential canard and rudders for directional control, ailerons and differential stabilator for roll control (plus differential flaps in the flaps-down modes). Stabilator rate and position commands from lateral control system inputs were limited to ensure pitch priority.

4.1.1. *Flaps down.* The development of the flaps-down lateral and directional control laws was driven by the heavy emphasis on the crosswind landing requirements. Careful blending of differential canard and rudder produced a direct sideforce capability which offset nearly 50% of the crosswind effect, whether in a slipped or a crabbed approach. The direct sideforce was blended with a sideslip command to provide a natural-appearing response to pedal input. Full-authority roll-rate feedback augmented the roll-mode dynamics. Directional augmentation was provided by a blend of lateral acceleration and estimated sideslip rate. In addition to the proportional feedbacks, a directional integrator was implemented to provide uniform sideslip response to pedal inputs. Lateral/directional interconnects from the differential ailerons and stabilators provided roll coordination. The interconnects are washed out at a frequency determined by the bare-airframe roll damping. Hence, they provided a yawing moment command based on roll acceleration. An additional roll-rate feedback path commanded the yawing moments necessary to coordinate sustained roll rates.

The CONVENTIONAL-mode flaps-down longitudinal control system used pitch stick and proportional plus integral feedbacks of pitch rate and angle of attack. The blend of pitch-rate and angle-of-attack feedback to the integrator was chosen to provide a constant pitch rate per inch of stick deflection. The angle-of-attack feedback was chosen to provide a stable steady-state variation of stick force per degree angle of attack at zero pitch rate. The ratio of proportional to integral gains was chosen to produce a classical short-period second-order response. The canard deflection was scheduled with angle of attack to produce the desired pitch stability characteristics. Collective stabilator deflection was the primary pitch effector.

Development of the ground-handling control laws was complicated by the adverse effects from thrust reversing. Special logic was needed to ensure the aircraft

Parameter	Original	Cruise	CONV/Combat
ω_n (rad/sec)	3.31	3.31	4.68
ζ	0.7	0.8	0.8
L_α (rad/sec)	1.2	1.2	1.2
$1/T_{\theta_2}$ (rad/sec)	1.2	1.2	2.0
τ_e (msec)	70	65	65
CAP	0.4	0.4	0.8†
			0.47‡
	† CAP Calculation Based on L_α ‡ CAP Calculation Based on $1/T_{\theta_2}$		

Figure 6. Pitch transfer function characteristics.

would achieve a three-point attitude in the presence of predicted nose-up pitching moments generated by thrust reversing in ground effect (see Moorhouse *et al.* 1989). When reverse thrust was selected by the pilot, a weight-on-wheels indication initiated a nose-down input, retracted the drooped flaps and ailerons, cancelled the lateral/directional interconnects and commanded maximum reverse thrust. In addition, the pilot could select an AUTOBRAKE function to command maximum anti-skid braking. Ground track stability was enhanced by the addition of yaw-rate feedback to nose-wheel steering. The lateral directional control system architecture changed with weight on wheels to provide direct sideforce commanded by the lateral stick and yaw rate commanded by the pedals. Linear design and analysis was applied wherever possible. Owing to the nonlinearities and constantly changing conditions which occur during ground operations, the majority of the design was accomplished using knowledge of the forces and moments to be expected and extensive six-degree-of-freedom modeling. Problems with unexpected jet/ground-effect interference are documented by Moorhouse *et al.* (1993).

4.1.2. *Flaps up lateral and directional.* The lateral control laws use a conventional roll-rate feedback path and a limited roll-rate feedback path for gust rejection. The directional control laws incorporate lateral acceleration and estimated sideslip-rate feedbacks. Interconnects from the lateral control commands to the directional controls are used for roll coordination. Differential stabilator is used to provide additional directional stability augmentation at high angle of attack. Equivalent-system analysis verified that the resulting control system design provided the desired response characteristics. Sideslip excursions due to lateral stick were small, with no oscillatory roll component. All of the analytical parameters were within the Level 1 MIL-F-8785C boundaries; however, manned simulation testing resulted in Level 2 pilot ratings for fine tracking. After extensive analysis and simulation testing, a solution was identified which required modification of the roll/yaw phase angle relationship as a result of turn coordination. The investigation of this problem was thorough enough to result in a tentative criterion for use in future efforts (see Moorhouse 1990, Moorhouse *et al.* 1990).

4.1.3. *Flaps up longitudinal*. Manned flight simulation testing revealed a problem with the original MIL-F-8785C Level 1 design guidelines. The design guidelines were modified as a result of this analysis. Inverse equivalent systems were used to incorporate a subtle design change into the existing architecture. This can best be illustrated by reviewing the initial and modified designs at one flight condition.

The initial control law design effort began with the establishment of a set of flying-qualities goals based on MIL-F-8785C guidelines and past experience. These goals were defined in the form of classical second-order pitch-rate and load factor responses to stick input with the values designated as original in Fig. 6. These design goals were applied at airspeeds above 300 knots calibrated airspeed, approximately the 'corner speed' of the aircraft. These goals were of course modified for speeds below corner speed.

A simplified linear block diagram of the CONVENTIONAL-mode longitudinal control laws at speeds above 300 knots is shown in Fig. 7. This mode was designed to produce Level 1 flying qualities for precision air-to-air tracking. A forward-loop integrator and a constant gain on the n_z feedback to the integrator provided the desired steady-state stick force per *g*. The proportional gains on the stick, pitch-rate feedback, and normal acceleration are used to produce the desired frequency and damping characteristics. The canard is driven by an angle-of-attack feedback to increase the lift-to-drag ratio while also stabilizing the unstable bare airframe. The values of the gains were computed using the inverse equivalent-system method.

The resulting control law design met all the design goals. Pitch-rate and load factor responses were second order in the 0·1 to 10 rad s^{-1} bandwidth. To verify the design, the pitch-rate response was matched with a second-order equivalent system with L_α fixed at the bare-airframe value. The match had a low cost function and equivalent time delay was less than 70 ms. Unfortunately manned flight simulation results indicated the design produced Level 2 flying qualities during precision tracking. It was decided to modify the CONVENTIONAL and COMBAT modes to improve tracking characteristics, and to leave the CRUISE mode as originally designed. Analysis suggested that the pitch-axis response had insufficient bandwidth. The selected solution involved modification of the desired pitch-rate numerator lead term. More complete details of this redesign activity can be found in Bland *et al.* (1987).

For a classic aircraft L_α is a function of the bare airframe and true airspeed, and cannot be modified by the flight control system. However, considering L_α simply as the numerator time constant in the pitch-rate response, it can be tailored by proper selection of gains in the flight control system. The value of L_α in the pitch-rate numerator was increased by a factor of 1·67 to achieve a higher 'apparent L_α'.

The apparent L_α can be altered in the pitch-rate response, but not the 'true L_α' which determines the relationship between pitch and flight path. Therefore the flight path rate, or normal acceleration, will no longer be a classical second-order response. The result is a third-order *g* response with some '*g* creep.' While significant *g* creep is normally considered undesirable, our experience has shown a small

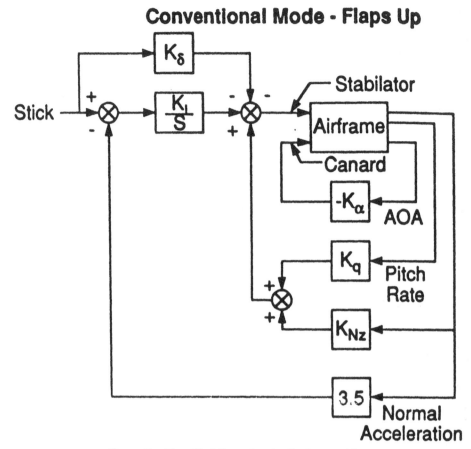

Figure 7. Simplified linear longitudinal control law.

amount to be acceptable. It was certainly an acceptable compromise to provide precise tracking. The pilots unanimously preferred the improved tracking over the CRUISE mode in all flight phases. The third-order g response can be seen in Fig. 8 which shows a step input with the original and apparent L_α design.

$$\frac{n_z}{\delta} = \frac{K(s + 1 \cdot 67 L_\alpha)e^{\tau s}}{(s + L_\alpha)(s^2 + 2\zeta\omega_{sp}s + \omega_{sp}^2)}$$

It is important to point out that, for the multivariable designed COMBAT mode, it was more effective to modify the prefilter that is a part of the methodology. It might also seem that a prefilter would be a better method of attaining the desired response for the CONVENTIONAL mode. The additional schedule complexity, time delay and potential aerodynamic uncertainties made this an unacceptable choice.

It was decided, therefore, to use the new approach being presented here. A new set of design goals was defined to incorporate the apparent L_α design as also indicated in Fig. 6. The magnitudes of the gain changes required are shown in the following table:

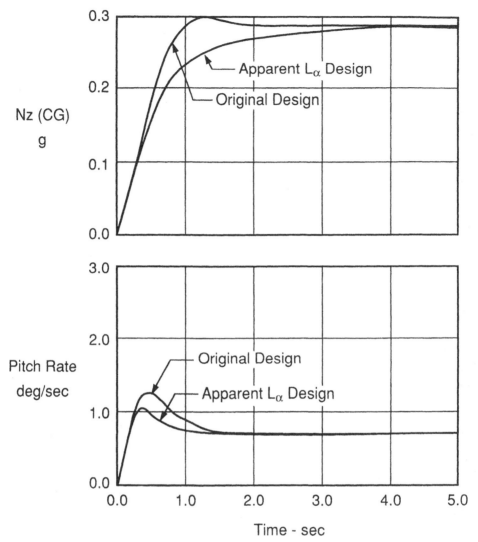

Figure 8.   Apparent L_α system time response.

Gain	Original	Apparent L_α
N'_δ	1 660	2 227
K_I	1·0	2·255
K_q	0·552	0·630
K_{nz}	1·622	2·427

Flight   condition: $M = 0·7$,
Alt = 20 000 ft (6100 m).

Using an inverse equivalent-system design approach allowed the team to implement an unusual design change while retaining the original architecture. This eliminated the potential problems involved in incorporating a prefilter and simplified the incorporation of nonlinear control system elements following the basic gain selection. The basic insight and understanding of the original design were maintained.

4.2. *Multivariable*

4.2.1. *Flaps down longitudinal.* Requirements for the precise landing mode included decoupling airspeed from the pitch axis in order to provide the precision for minimizing touchdown dispersion. High-bandwidth augmentation of airspeed stability is achieved because of the high response rates of the reverser vanes compared with changing thrust by means of engine speed. Airspeed is fed back to vane deflections with different schedules top and bottom to give zero pitching moment. Throttle deflection then commands airspeed. With this effective speed hold, pitch-rate command by the stick becomes flight-path-angle-rate command. With additional requirements to produce a minimum phase flight path response using direct lift control, the multivariable approach was the clear choice. For this mode, various regulated variables were considered. First, flight path rate was approximated by subtracting high-passed angle of attack from pitch rate. This was discarded because a high-enough high-pass break frequency caused sensor noise problems, and also the gust response was deficient. Both pitch rate alone and a blend of low-passed load factor and pitch rate met stability and robustness requirements. Piloted simulation showed that pitch rate alone, together with airspeed as the regulated variable for the thrust axis, provided superior flying qualities.

One term that is a frequent topic of discussion is what is meant by the 'performance' of a design methodology. In this case, control system performance was measured by the amount of cross-coupling. This showed a measurable increase as the order of the compensator was reduced from fifth order to second order, as shown in Fig. 9. It can be seen that the biggest effect is a significant increase in speed response due to a pitch-rate command. The primary responses are completely unaffected by this reduction in order of the compensator. Because there are no criteria with which to judge how much degradation can be allowed, piloted simulation was required to verify that a second-order compensator was satisfactory.

Longitudinal control laws in the short takeoff and approach mode differ primarily in the use of pitch vectoring to augment the control power of the stabilator. The regulated variable was a linear combination of angle of attack at low frequency and pitch rate at high frequency, consistent with the control effectiveness of the effectors. A command-shaping prefilter was used to control angle of attack precisely. Constant stick force per angle of attack was provided at lower angles of attack, with increased stick force required at high angle of attack. A wash-out filter was inserted into the pitching-moment commands to the nozzles and canards so that all steady-state trimming would be accomplished using the stabilator. During ground operations, the wash-out filter to the nozzle was set to behave like a fixed gain. This allowed the use of thrust vectoring to improve short/rough-field operations.

4.2.2. *Flaps up longitudinal.* Pitch rate was chosen as the regulated variable for the CRUISE and COMBAT modes, with a blend of low-passed load factor at subsonic speeds. The stabilator, canard and thrust vectoring were used together as a single moment effector. A 2 second wash-out was used on the canard and nozzles so that the stabilator provided trim. This is misleading, however, because all three are on schedules to provide stability and minimum drag with trim stabilator near neutral.

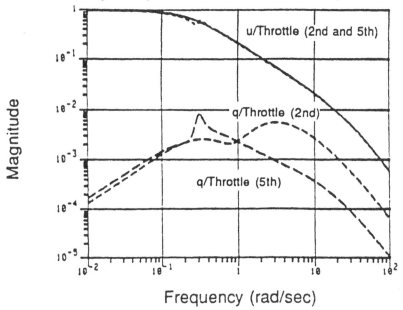

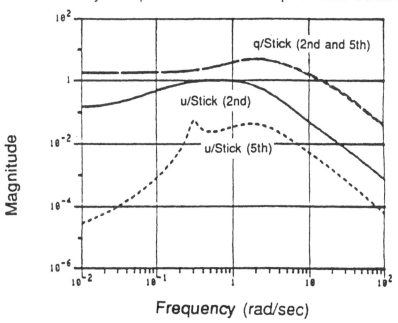

Figure 9. Decoupling with five- and two-state compensators.

For the CRUISE and COMBAT modes, additional requirements were imposed. Canard deflection and nozzle vector angle were coordinated to minimize steady-state drag at 1*g* in the CRUISE mode and at elevated load factors in the COMBAT mode. The canards were also used for short-term pitch control. In these cases, the benefits of the multivariable approach, guaranteeing stability and robustness, made it the best choice. The compensator order could be reduced to a practical proportional plus integral structure without significant degradation in stability margin. A more significant effect of compensator order reduction may be deviation from the ideal form of the compensated plant, which is discussed later.

5. Flight test results

The most important result, of course, is how the flying qualities were rated in flight. All modes/axes were rated Level 1 where predicted by simulation. For instance, the CRUISE mode was rated Level 2 for tracking (as expected) but Level 1 for a flight path capture task. The landing mode on a very gusty day received Level 2 ratings 'because of high turbulence', as allowed by the specification.

The simplest and most direct form of presentation of handling-qualities flight test results would be tables or graphs of Cooper–Harper ratings. There is a lot more information behind the simple numbers. Here, we present the flying-qualities results mostly in the form of pilot comments. These comments (with ratings) show excellent agreement with the analytical and simulation development results. They also provide an interesting diversion in comparing different likes of the different pilots, which would be masked by a simple presentation of Cooper–Harper ratings.

Handling qualities during tracking (HQDT)

Mach 0·6/10 000:	Gross	Fine
CONV HQR	2	2
COMBAT HQR	3	3

- 'CONV stopped when I put it there.' (Pilot A)
- 'COMBAT HQDT improves with G increases.' (Pilot A)

Mach 0·7/20 000:	Gross	Fine
CONV HQR	3	3
CRUISE HQR	4	4
COMBAT HQR	3	3

- 'Very nice pitch control in COMBAT, less pitch sensitive than CONV. This was the best mode of all for tracking.' (Pilot C)

Mach 0·9/20 000	Gross	Fine
CONV HQR	3	2
CRUISE HQR	4	3
COMBAT HQR	3	3

- 'HQDT in all three modes was very nice.' (Pilot B)
- 'The biggest delineator for all tasks in the three modes was the relatively poor gross acquisition in CRUISE. The initial overshoot was large, 25 + mil, but I usually only had one overshoot in CRUISE. In COMBAT and CONV I generally had a much smaller overshoot, better predictability, but often more than one overshoot. One small difference was noted in that the COMBAT mode seemed to have less pitch acceleration in the gross acquisition task than CONV mode – to get the same performance I had to pull harder and this was evident as apparent higher stick forces.' (Pilot B)

Note that all of the ratings are Level 1 for the CONVENTIONAL and COMBAT modes, and Level 1/borderline Level 2 for the CRUISE mode. Very little discrimination would be implied by the ratings alone. The comments reflect perceived differences and preferences even when the ratings are identical. Thus Pilot B says that COMBAT gross acquisition is better than CONV, but assigns a rating of 3 to both. Pilot C says that COMBAT was the best mode of all for tracking, but assigns the same rating as the CONV. Also, it can be seen that Pilot A consistently preferred the CONV mode while Pilot C chose the COMBAT mode. No explanation will be attempted to explain these results. The 'engineering evaluation' of the different pilots is that both Pilots A and B are high gain relative to Pilot C. We can certainly rationalize their preferences on this basis, and it supports the technique of requiring multiple pilot opinions. In a development program, however, do we have to satisfy all pilots? If not, whose opinion is given precedence? In the S/MTD program we are lucky – the differences are within the Level 1 range and the distinction is academic. This aircraft is unique in having the capability to switch modes at will, and fly the modes back to back. Fine distinction within Level 1 characteristics is not normally a problem in a development program. What is important is that these comments do indeed repeat the comments that were noted during the piloted simulation program.

Landing configuration

- Pitch captures – 'Nice and stable.' (Pilot B)
- Flight path captures – 'Sluggish just like normal.' (Pilot B)
- 'Everything was fine. I didn't see anything I didn't like.' (Pilot B)

5.1. *Final validation*

The real purpose of the last flight of the program was to evaluate the SLAND mode in conjunction with autonomous landing guidance (ALG) in a severe environment, i.e. night. In the pilot's own words:

In this, the S/MTD had its finest hour. Using SLAND made a considerable reduction in pilot workload, and was the perfect complement to ALG. The integration of the system was wonderful, and certainly the overall system is much better than the sum of its individual parts would indicate. I truly believe the testing of SLAND under day, VFR conditions is not indicative of its true worth. I felt it made my approach to a totally black airfield no more difficult than a simple video game; all I had to do was keep the velocity vector aligned with the carets.

During most flight testing, as pilots we typically only comment on those things that don't work the way we would like them to work. As a result, you seldom hear about those things that work well. Tonight's flight highlighted just how incredibly well the myriad of technologies incorporated in this airplane worked.

The Integrated Flight and Propulsion Control stands out in my mind as the most amazing. Integrating all the different control surfaces including the canards and the nozzles, making it all fly like a regular F-15 or better, and still only requiring normal stick and throttle inputs by the pilot is fantastic. During tonight's flight, the only obvious clue that we weren't flying a regular airplane was that no regular airplane could do what we were doing.

6. Overall results

The first result is not really a surprise. For modes or axes that have only conventional controls and requirements, the multivariable design technique does not offer any benefit. The modern control theoretician will cite the benefits of guaranteed robustness; however, this is not an overriding concern for a 'simple' system. One of the comparisons to be made when assessing two methodologies is the ease of implementing the design or, if necessary, the ease of correcting deficiencies. In this respect, the insight into the design process provided by the classical method of control law development/analysis has a distinct advantage over the multivariable techniques. For example, during the early flight testing of the CONVENTIONAL mode, it was discovered that the system damping was lower than desirable at low-altitude high-speed flight conditions. While the causes and fixes for the condition were being investigated, a simple patch to the software was installed, changing the feedback gains and providing sufficient damping to continue the flight test program. If those particular control laws had been developed using the modern method, a complete analysis of the system would have been required to define the changes required to improve the flying qualities. This is further support for the conclusion that multivariable theory is not warranted for a conventional design problem. The converse is probably also true – such a problem in one of the complex modes might not be so amenable to a simple fix.

For all flaps-up modes, the lateral and directional requirements are conventional so that the classical design approach was again chosen. Initially, only conventional lateral and directional requirements were considered for the STOL modes. After the choice of a classical design had been made, piloted simulations indicated that the crosswind landing imposed an additional requirement. This led to the implementation of direct sideforce (differential canard plus rudder deflection) as a function of rudder pedal input. This can now easily be formulated as a requirement to be included in the multivariable design approach, but in practice it was a fallout as the classical design progressed.

The second result follows from the previous discussion. With multiple control effectors, or complex requirements, a multivariable technique is preferred with one important qualification – the requirements must be specified *a priori*. The LQG/LTR technique was very effective in decoupling pitch and airspeed together with a minimum phase flight path response for the precision-landing mode. It is these authors' contention, however, that modified or additional requirements are more likely to be extracted from the classical process than from a multivariable technique, because of the insight gained by the step-by-step development. The lateral and directional axes of the STOL mode were designed classically. Well into the development it was realized that direct sideforce control was the way to meet the crosswind landing requirements. Once the requirement had been formulated, the design could have been completed using either technique.

An aspect of the development that gave totally unexpected benefits was the synergism of the parallel design process. One of the critical areas of multivariable control theory is to establish all the design requirements as the starting point. A full-performance design is then synthesized to satisfy them. The classical approach addresses the requirements individually, in principle, although an experienced designer uses past experience to approach the final solution efficiently. Both MDA and Honeywell used experienced control system designers. Even so, the classical MDA design benefitted from knowledge of the performance attainable by the multivariable design. Simultaneously, the Honeywell design simplified the high-order compensator of the full-performance design aided by the knowledge of the performance attainable with the simpler formulations of the classical design. The result was a very effective convergence on an optimum balance of performance and complexity (depicted in Fig. 10).

The program intent was to utilize multivariable control theory to the extent necessary for control integration. It was decided to do the dual development of the control laws using both the classical and modern methods, and compare the results prior to choosing the set to be programmed. As described above, the two methods essentially start from the opposite positions with respect to both performance and complexity, and evolve toward the same point, as shown in Fig. 10. The use of the flight simulator to compare the flying qualities obtained with each set aided in both the evolution and in the comparison. Considering the similarity of the resulting two sets of control laws, the consensus was that both the speed and accuracy of this evolutionary process, regardless of the method being used, depended more on the capability of the individual doing the work than on the process itself. In addition, the combined approach was more efficient than either by itself for the modes or axes with more than conventional requirements.

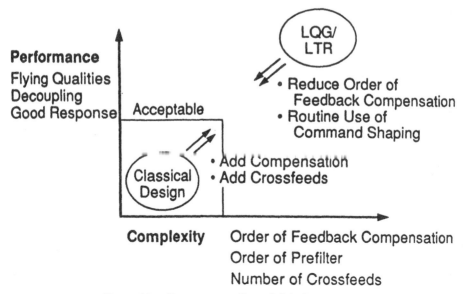

Figure 10.   Convergence of parallel design process.

While the contract Statement of Work called out compliance with the intent of MIL-F-8785C, it was recognized that strict compliance with those specifications was not sufficient to assure optimum flying qualities. This comment was not, and should never be, interpreted to imply that the specification boundaries can be ignored. It is critical to recognize, however, that not one flying-qualities parameter violated a specification boundary during the S/MTD development. The problem was either one of defining a satisfactory point within the boundaries, or one of augmenting inadequate requirements. In order to prevent misinterpretation, it is also critical to recognize that we are talking about small differences between Levels 1 and 2. Meeting the specification boundaries and guidelines will guarantee the avoidance of truly bad characteristics, such as pilot-induced oscillations. The only really inadequate requirements were judged to be for precision landing and roll/yaw phasing. Throughout the S/MTD flight test program, not one pilot rating or comment worse than Level 2 was received. At the risk of being too obvious, it is noted that problems defining or interpreting flying-qualities requirements apply to both classical and multivariable design techniques equally, and are very critical to the overall design process. The results from the S/MTD program applicable to requirements definition have been documented by Moorhouse (1990).

As stated earlier, one of the design goals was to meet the stability requirements of MIL-F-9490D, or to define recommended changes to the criteria. The design used the 6 dB gain and 45° phase limits as total loop design requirements. The flight test problem mentioned above was manifested first by the pilots complaining about the pitch axis 'ringing'. In other words, aircraft response to the normal flight test stick inputs did not damp out as expected. Analysis of the flight test data revealed that the gain margin had decreased to approximately 3 dB. The temporary fix that restored damping also restored gain margin to 6 dB and gave flying qualities satisfactory for completion of the flight test. This is not to imply that loss of margin always results in loss of damping. The physical manifestation will depend on the source of the problem. The S/MTD project has successfully used the MIL-F-9490D requirements for overall loop gain and phase margin. Although not addressed in this chapter, this also applies to the 8 dB requirement at structural frequencies. Although this does not constitute formal validation of the requirements, it does indicate that caution should be exercised before accepting less stability. In particular, the common perception that the specification allows 3 dB and 22·5° would have been unacceptable for this program. It is these authors' contention that, regardless of the methodology that is used, a design should still satisfy classical measure of stability.

This discussion of the practical aspects of control system design methodology must end with a reminder that we have discussed linear techniques. Starting with linear design methods is a universal approach but the steps needed to achieve a final solution can be easily neglected. All aspects of the design need to be evaluated, but especially any nonlinear behavior must be exposed before it is encountered in flight. This is most easily accomplished in a piloted simulation using large-amplitude maneuvers or disturbance inputs. A great deal of emphasis was placed on this aspect of the S/MTD development, especially landing in winds, windshear and turbulence. This was very successful in general: all pilot ratings were Level 1 in the flight test program. One landing did reach stabilator rate limiting, however, although there was no discernible change in the flying qualities. This was a surprise even though the gusty conditions were more severe than the design requirements.

The problem was investigated in the simulator, where it was determined that continuous turbulence could not replicate the flight results. It is recommended that a series of discrete gusts be used for flying-qualities evaluation (see Leggett *et al.* 1990).

7. Conclusions

This chapter has documented the development of a full-envelope integrated flight/propulsion control system. A combination of classical and multivariable design methodologies was used, including a unique inverse procedure to produce second-order equivalent systems defined by specified flying-qualities requirements. The implementation was based on a rational choice between the two methodologies. It is suggested that a parallel design approach in the beginning will produce efficient convergence on a practical optimum design.

Finally, all control law revisions should be done analytically and only evaluated by simulation. Regardless of the design technique used, the process begins by specifying detailed design guidelines selected to meet the flying-qualities specification. Once the designs were complete and analyzed, based on the design guidelines, manned flight simulation was used only to validate and demonstrate the flying qualities achieved prior to flight test. Problems encountered during simulation or flight testing were addressed first by reviewing the original design guidelines and evaluating the success achieved in implementing them.

REFERENCES

BLAND, M. P., CITURS, K. D., SHIRK, F. and MOORHOUSE, D. J., 1987, Alternative design guidelines for pitch tracking. *Proceedings of the Atmospheric Flight Mechanics Conference, Monterey, California* (AIAA Paper No. 87-2289), CP 876, pp. 40–48.

DOYLE, J. C. and STEIN, G., 1981, Multivariable feedback design: concepts for classical/modern synthesis. *IEEE Transactions on Automatic Control*, **AC-26**, 4–16.

LEGGETT, D. B., MOORHOUSE, D. J. and ZEH, J. M., 1990, Simulating turbulence and gusts for flying qualities evaluation. *Proceedings of the Atmospheric Flight Mechanics Conference, Portland, Oregon*, AIAA Paper No. 90-2845, CP 898, pp. 445–455.

MOORHOUSE, D. J., 1990, Lessons learned from the S/MTD program for the flying qualities specification. *Proceedings of the Atmospheric Flight Mechanics Conference, Portland, Oregon*, AIAA Paper No. 90-2849, CP 898, pp. 473–479.

MOORHOUSE, D. J. and MORAN, W. A., 1985, Flying qualities design criteria for highly augmented systems. *Proceedings of the IEEE National Aerospace & Electronics Conference, Dayton, Ohio*.

MOORHOUSE, D. J. and WOODCOCK, R. J., 1982, Background information and user guide for MIL-F-8785C, Military Specification – Flying Qualities of Piloted Airplanes. Air Force Wright Aeronautical Laboratory Report AFWAL-TR-81-3109.

MOORHOUSE, D. J., LAUGHREY, J. A. and THOMAS, R. W., 1989, Aerodynamic and propulsive control development of the STOL and Manoeuvre Technology Demonstrator. *AGARD Conference Proceedings, Madrid, Spain*, CP 465, Paper 1.

MOORHOUSE, D. J., CITURS, K. D., THOMAS, R. W. and CRAWFORD, M. R., 1990, Handling qualities of the STOL and maneuver technology demonstrator from specification to flight test. *AGARD Conference Proceedings, Quebec City, Canada*, CP 508, Paper 13.

MOORHOUSE, D. J., REINSBERG, J. A. and SHIRK, F. W., 1993, Study of jet effect and ground effect interference problems on a STOL fighter. *AGARD Conference on Computational and Experimental Assessment of Jets in Cross Flow, Winchester, UK, April*.

MORTON, B. G., 1985, New applications of mu to real parameter variation problems. *Proceedings of the 24th IEEE Conference on Decision and Control, Fort Lauderdale, Florida,* pp. 223–238.

MORTON, B. G. and McAFOOS, R. M., 1985, A mu-test for real parameter variations. *Proceedings of the American Control Conference, Boston, Massachusetts,* pp. 135–138.

STEIN, G. and ATHANS, M., 1987, The LQG/LTR procedure for multivariable feedback control design. *IEEE Transactions on Automatic Control,* **32**, 105–114.

WALKER, L. A. and LaMANNA, W. J., 1982, Development of the F/A-18A handling qualities using digital flight control technology. *Proceedings of the 26th Annual Symposium of Experimental Test Pilots, Beverly Hills, California.*

15

Control law design and flight test results of the experimental aircraft X-31A

H. BEH, G. HOFINGER and P. HUBER

Nomenclature

AoA	Angle of Attack
A/C	Aircraft
CPU	Central Processing Unit
DASA	Daimler-Benz Aerospace
EFM	Enhanced Fighter Maneuverability
FCC	Flight Control Computer
FCLs	Flight Control Laws
FCS	Flight Control System
IMU	Inertial Measurement Unit
IO	Input/Output
LVDT	Linear Variable Data Transducer
PST	Post-stall
RI	Rockwell Internatinal
TV	Thrust Vectoring
Φ	bank angle
Θ	pitch attitude
α	angle of attack
α_c	angle of attack command
β	sideslip
β_c	sideslip command
δ_{SF}	symmetrical trailing-edge command
δ_{DF}	differential trailing-edge command
δ_C	canard command
δ_R	rudder command
κ	thrust vectoring command yaw axis
σ	thrust vectoring command pitch axis
$\cos \mu \cos \gamma$	direction cosine
$\sin \mu \cos \gamma$	direction cosine
J	performance index
V	aircraft speed

g	gravity constant
n_y	lateral load factor
n_z	load factor
p	roll-rate body axis
q	pitch-rate body axis
r	yaw-rate body axis
x_p	pilot command roll axis
x_q	pilot command pitch axis
x_r	pilot command yaw axis

Indices

c	command
e	experimental axis
k	nodal line axis (flight path axis)
w	wind axis

Vectors and matrices

$\mathbf{p}$	pilot command vector
$\mathbf{u}$	surface command vector
$\mathbf{u}_c$	steady-state actuator command vector
$\mathbf{x}$	state variable vector
$\mathbf{y}$	output vector
$\mathbf{x}^T$	transpose of vector $\mathbf{x}$
$\mathbf{x}_k$	vector $\mathbf{x}$ at time k
$\mathbf{A}$	system matrix
$\mathbf{B}$	input matrix
$\mathbf{C}$	output matrix
$\mathbf{K}$	feedback matrix
$\mathbf{P}$	Riccati gain matrix
$\mathbf{Q}$	state variable weighting matrix of performance index
$\mathbf{R}$	surface command weighting matrix of performance index
$\mathbf{X}^T$	transpose of matrix $\mathbf{X}$

1. Introduction

The X-31A post-stall experimental aircraft (Fig. 1) was developed to demonstrate enhanced fighter maneuverability by using thrust vectoring to fly beyond the stall limit. The goal of the EFM program is to demonstrate the tactical advantage of a fighter aircraft being capable of maneuvering and maintaining controlled flight including the post-stall regime up to 70° AoA.

Two fighter-type X-31A aircraft were built by Rockwell International and Daimler-Benz Aerospace under contract to the Advanced Research Projects Agency (ARPA) and the German Ministry of Defense (GMoD).

Since the first flights of aircraft #1 on October 11th, 1990, and aircraft #2 on January 19th, 1991, the two aircraft have accumulated a total of 510 flights (280 on aircraft #1 and 230 on aircraft #2) or 380 flight hours till December 1994. First the conventional envelope was tested up to 0·9 Mach, 40 kft pressure altitude, 485 kcas and 30° AoA. Symmetrical loads were cleared between 7·2g and −2·4g. Shortly

Figure 1. Experimental aircraft X-31A.

after PST flight test was started, both aircraft were transferred from the RI flight test facility at Palmdale, California, to the NASA Dryden Flight Research Center (DFRC) at Edwards Air Force Base to continue the PST envelope expansion flight test (January 1992). Since then more than 200 PST flights have been accomplished and the PST envelope is now cleared up to 70° AoA, between 10 kft and 30 kft pressure altitude, a maximum of 4*g* during PST entry and a maximum of 225 kcas entry speed. An extensive tactical flight test to demonstrate the EFM capability of the X-31A aircraft was performed at DFRC in 1994 after the operational PST clearance was given. This flight test includes close-in-combat versus modern fighter aircraft.

Additional information on X-31 flight control law design philosophy is given by Beh and Hofinger (1994). Huber *et al.* (1991) give additional information on X-31 flight testing, and Huber and Galleithner (1992) present further flight test results on high angle of attack and post-stall flight.

2. X-31A flight control system

The X-31A aircraft is a longitudinal unstable (time to double amplitude as low as 200 ms) delta-wing A/C with canard configuration. The primary aerodynamic control surfaces are symmetrical trailing-edge flaps and canard for the longitudinal axis and differential trailing-edge flap and rudder for the lateral/directional axes. In addition a thrust vectoring system is added to the engine exhaust nozzle utilizing three paddles. Each paddle covers an angular section of 120° around the exhaust nozzle and can be deflected up to 35° into the plume, leading to a thrust deflection in the pitch and yaw axis of more then 10°. This TV system is used to augment the aerodynamic control power during low speed and PST flight.

The X-31A FCS is a full-authority digital fly-by-wire system. It consists of three identical FCCs (two CPUs each) supported by a so-called tie-breaker FCC. This tie-breaker is like the other FCCs but with just one CPU. It selects the healthy FCC lane in case of second FCC failure, which gives a quadruplex system reliability. The safety-critical flight control components are electrically quadruplex and connected to all four FCCs. These are the pilot inceptors (stick and pedal), the rate gyros, the accelerometers and the actuators of trailing-edge flap, canard and rudder. The safety-critical actuators (primary control surfaces) are hydraulically duplex. The other components are not considered safety critical, but are necessary to fulfill the EFM requirements and to be able to fly within the PST regime. These are AoA and sideslip sensors located at the nose boom, the air data computer, the inertial measurement unit and the actuators of the thrust vectoring paddles, leading-edge flap, speedbrake and engine air intake. These components are electrically duplex, except the simplex IMU which has a self-test monitoring feature. The nonsafety-critical actuators (secondary control surfaces) are hydraulically simplex. A failure of a nonsafety-critical component must be monitored by the redundancy management and reported to the flight control laws. Figure 2 gives an overview of the FCS architecture.

In the basic FCS mode all feedback signals are used to calculate the actuator commands. There are two basic modes, because the TV can be enabled and disabled by the pilot. But PST flight is prevented by the FCL as long as TV is disabled. For takeoff and landing TV is automatically disabled for safety reasons. Depending on the actual sensor failure situation, reversionary modes provide a step-by-step system degradation, i.e.:

- R1 – inertial measurement unit disengage mode
- R2 – flow angle disengage mode
- R3 – fixed-gain mode in case of an air data failure

The most degraded mode, R3, still has save 'fly-home' capability. In Fig. 3 the step-by-step degradation is shown; note that the more degraded mode includes the dis-

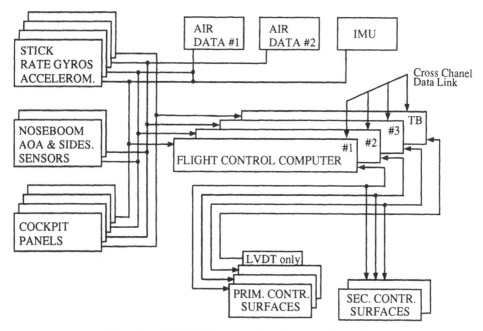

Figure 2.   X-31A flight control hardware configuration.

engagements of the less degraded modes, i.e. in R3 mode the IMU and flow angles are also disengaged. The arrows illustrate the possible degradations in case of a hardware failure. The degraded modes are also pilot selectable in a nonfailure situation for flight test purposes.

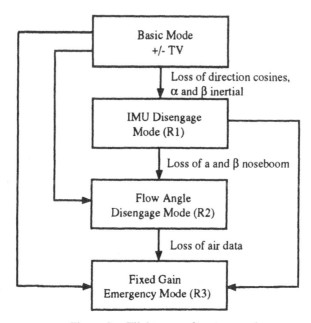

Figure 3.   Flight control system modes.

For safety reasons a spin recovery mode was introduced into the flight control laws. This mode must be selected by the pilot. There are classical direct links from stick and pedal to canard, trailing-edge flap, rudder and thrust vectoring, giving the pilot full surface deflection authority. A proportional and integral pitch-rate feedback is the only closed stabilization loop in the spin recovery mode.

3. Basic structure of the control laws

The X-31A FCLs have three main external interfaces, i.e.:

(a)   the pilot command vector $\mathbf{p}(x_p, x_q, x_r)$;

(b)   the sensed feedback vector $\mathbf{y}(p, q, r, \alpha, \beta)$; and

(c)   the actuation command vector $\mathbf{u}(\delta_{SF}, \delta_{DF}, \delta_C, \delta_R, \sigma, \kappa)$.

Within the PST flight envelope the pilot command vector consists of wind-axis roll-rate command (x_p), AoA command (x_q), and sideslip command (x_r). At high dynamic pressure the flight conditions load factor command replaces the AoA command.

Figure 4 shows the X-31A FCLs in a closed loop together with the aircraft dynamics. There are three subunits within the FCLs, the linear feedback unit K and the nonlinear feedforward units f_u and f_y. The feedforward unit f_u calculates the necessary steady-state command vector $\mathbf{u}_c$, i.e. the trimmed surface deflections, for the pilot command vector $\mathbf{p}$ depending on the actual flight condition and aircraft configuration data (e.g. c.g., weight). In parallel f_y calculates the corresponding

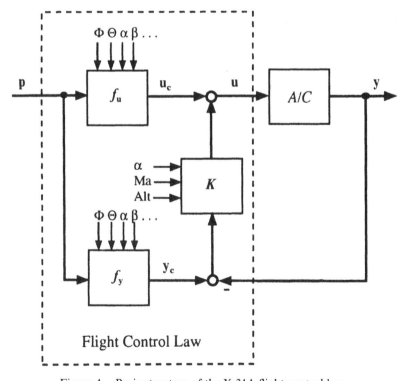

Figure 4.   Basic structure of the X-31A flight control law.

steady-state command vector y_c. This means that for all feedback signals an associated command signal must be calculated from the pilot input.

The actuator commands are calculated with the following vector equation:

$$u = K * (y - y_c) + u_c \qquad (1)$$

Thus the actuator command vector **u** is the sum of the steady-state command vector (trimmed surface deflections) and the feedback difference vector multiplied by the feedback gain matrix.

3.1. *Determination of the feedback gain matrix* **K**

The feedback gain matrix **K** is determined using the linearized aircraft model split into longitudinal, respectively lateral/directional, motion. All additional dynamics (e.g. actuation and sensor models) are not considered. This leads to fourth-order models. Using the Z-transform the vector difference equations are (Fig. 5)

$$x_{k+1} = Ax_k + Bu_k$$

$$y_k = Cx_k \qquad (2)$$

The feedback matrix is mathematically calculated using the optimal linear digital regulator design. Thereby the main task for the designer is the definition of the weighting matrices **Q** and **R** of the quadratic performance index J (eqn (3)):

$$J = \frac{1}{2} \sum_{k=0}^{\infty} (x_k^T Q x_k + u_k^T R u_k) \qquad (3)$$

$$K = (B^T PB + R)^{-1} B^T PA \qquad (4)$$

The minimization of the performance index for infinite time results in a time constant optimum feedback matrix **K**. This matrix is calculated (eqn (4)) using the system and weighting matrices and the matrix **P** which is the solution of the 'matrix Riccati equation' (eqn (5)). This equation is often referred to as the algebraic Riccati

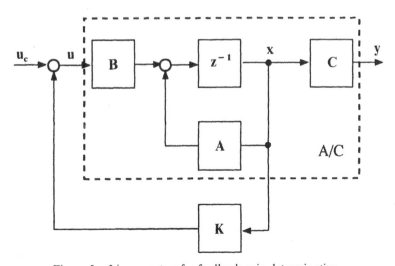

Figure 5. Linear system for feedback gain determination.

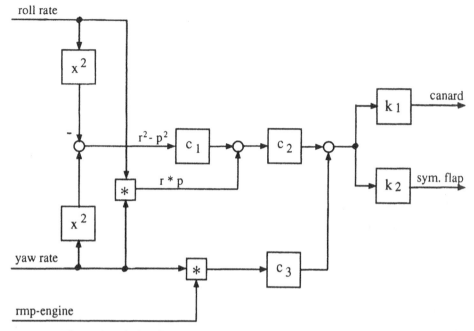

Figure 6.   Pitch-axis inertial and gyroscopic coupling compensation.

equation:

$$P = A^T[P - PB(B^TPB + R)^{-1}B^TP]A + Q \tag{5}$$

The stability and handling analysis is carried out with the full high-order system. If this check shows unsatisfactory results the weighting matrices have to be adjusted and the optimization procedure is repeated.

3.2. *Calculation of the feedforward paths*

The feedforward paths are calculated independently of the feedback path using the steady-state equations of motion of the aircraft. Steady state is interpreted in this context as the resulting stable flight condition with constant pilot inputs (e.g. steady-state wind-axis roll). When taking into account all influences in calculating the steady-state vectors y_c and u_c, the functions f_y and f_u describe the inverse steady-state model of the aircraft. They include all direct link and compensation paths (e.g. gravity effects, inertial coupling, speedbrake moment compensation, etc.) and are dependent on configuration and flight condition. Constraints such as complexity, model fidelity or computer power do not allow the full implementation of all paths; therefore some of them must be simplified or omitted. The steady-state feedback difference error is a measurement of this simplification.

3.3. *Inertial and gyroscopic coupling compensation*

The gyroscopic moments are square dependent on the angular rates and therefore not considered in the linearized model. At high angular rates these moments cannot be neglected. Uncompensated, these moments would lead to unacceptably large deviations and the aircraft reaction would be lagged by its dynamics. Intro-

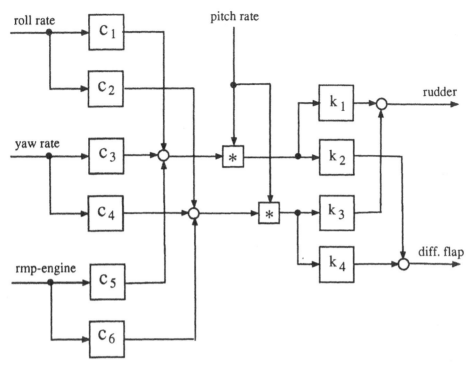

Figure 7. Lateral/directional axes inertial and gyroscopic coupling compensation.

duction of an integral feedback would help, but would introduce overshoots. A better solution is feedforward compensation acting instantly (just lagged by the sensor and actuation dynamics) against the disturbances. The small remaining deviations due to model uncertainties are controlled by the feedback loops. Figures 6 and 7 show the block diagram of the longitudinal, respectively lateral/directional, inertial and gyroscopic coupling compensation. The constants c are dependent on the inertias of the aircraft and the engine. They are used to calculate a normalized compensation moment from the rates in front of the gains k. These gains are functions of flight conditions. The outputs are the necessary surface deflections to compensate the inertial and gyroscopic moment.

3.4. *Gravity-effect compensation*

The simplest description of the force equation is in the flight path axes system. Here the forces in the y and z direction (n_y and n_z) consist of just the centripedal force and gravity:

$$n_{k_z} = \frac{V_k}{g} q_k + \cos \mu \cos \gamma \tag{6}$$

$$n_{k_y} = -\frac{V_k}{g} r_k + \sin \mu \cos \gamma \tag{7}$$

They will be used to calculate the flight path rate-command signals from the commanded g and gravity components:

$$q_{k_c} = \frac{g}{V_k} \left(n_{k_{z(c)}} - \cos \mu \cos \gamma \right) \tag{8}$$

$$r_{k_c} = \frac{g}{V_k} \left(-n_{k_{y(c)}} + \sin \mu \cos \gamma \right) \tag{9}$$

For this the body-axes acceleration commands are transformed into the flight path axis.

The dependency of the rates on gravity in eqns (8) and (9) introduces additional terms in the exact formulation of the aerodynamic damping. These additional aerodynamic terms are neglected.

The time differential of the gravity component leads to angular accelerations. These moments are compensated by a feedforward command.

3.5. *Additional control structure elements*

The simplified model used for the determination of the feedback matrix may lead to a high-order system with reduced stability and/or degraded handling qualities. With filters in the feedback and feedforward loops this can again be improved. Failed or missing feedback signals have to be substituted by observers.

If the calculation of the feedforward signals is missing some steady-state term (in case of a hardware failure), a steady-state error (the difference between the commanded and sensed signal) will remain. Wash-out filters in the feedback loop are used to drive this error to zero. Here just the high-frequency part of the feedback signal will be used.

Limitations due to loads, control power, sensors and pilot accelerations are retained by scaling and rate limiting the pilot command loop.

4. Longitudinal flight control laws

The pitch stick position is scaled in the FCLs from $-1\cdot0$ (max push) to $+1\cdot5$ (max pull). This position corresponds directly to an AoA or load factor command. At low dynamic pressure flight conditions the FCL is the AoA command mode. Here a command of $-1\cdot0$ corresponds to $-10°$ AoA, $+1\cdot0$ corresponds to $+30°$ AoA, and $+1\cdot5$ corresponds to $+70°$ AoA. If PST is disabled the AoA command is limited to $+30°$. A force detent in the stick feel system at $+1\cdot0$ gives the pilot information on whether he or she has pulled into PST or not. At high dynamic pressure $-1\cdot0$ commands approximately $-2\cdot4g$ and $+1\cdot0$ commands $7\cdot2g$. Pulling over the detent does not change the maximum command of $7\cdot2g$ (this is the aircraft's load limit). The switchover between these two command systems is at the flight condition where $30°$ AoA results in the maximum load factor of $7\cdot2g$. This is around 380 psf (2620 kPa). Depending on the stick command the associated command is calculated, i.e. in the AoA command flight regime the associated load factor command is calculated and vice versa. PST flight is only possible if the aircraft is in the AoA command mode.

Figure 8 presents a simplified block diagram of the longitudinal FCLs for the AoA command mode. On the left-hand side the pilot input (in this case the AoA command) and the main feedback sensor signals, AoA, and pitch rate are shown.

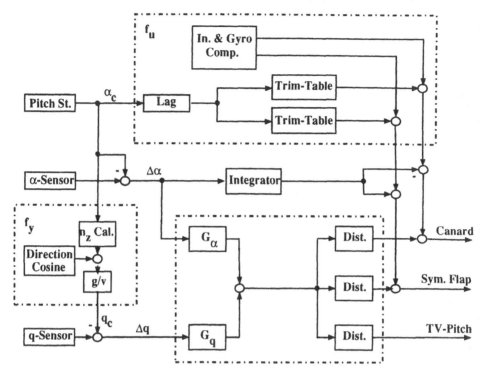

Figure 8. Simplified block diagram of the X-31A longitudinal flight control laws (basic mode with α command).

The calculated output canard command, symmetric trailing-edge flap command and the TV command for the pitch axis are on the right-hand side. Within the block diagram three areas are surrounded by a chain line. They reflect the main units as described above.

4.1. *Direct link longitudinal axis* (f_u)

The direct link path in the upper right corner of Fig. 8 calculates the steady-state canard and trailing-edge flap positions from the commanded AoA and the actual flight condition. Derived from trimmed conditions with wings level and maximum dry power, the appropriate canard and trailing-edge flap deflections are stored in a set of two trim tables. The additional degree of freedom with two control effectors is used to optimize minimum drag at low AoA and lateral stability (aerodynamic stability derivatives are dependent on canard position) at high AoA. This set is called the 'cruise' trim schedule. But this schedule results in high landing speeds, and therefore a second set of trim tables was calculated with the goal of maximum lift at landing AoA. This 'high-lift' schedule is also stored in the FCC. The pilot can switch from one schedule to the other, but the 'high-lift' trim schedule is restricted for takeoff and landing. For simplification only one set of trim tables is shown in Fig. 8.

A lag filter with a time constant corresponding to the AoA response of the aircraft is used in front of the trim tables to compensate the pilot command lead prior to the aircraft response. The canard and trailing-edge deflections to compensate the inertial and gyroscopic moments as defined in Fig. 6 are added to the output of the trim tables and this sum forms the steady-state trim commands. Since the aerodynamic control power is sufficient for trimming, TV is not used.

4.2. *Calculation of pitch-rate command* (f_y)

The pitch-rate command calculation is shown on the left side of Fig. 8. First the associated n_z command is derived from the AoA command, using a stored trimmed lift table, dynamic pressure, and an estimated aircraft weight. From this the direction cosine of gravity in the aircraft's z axis is subtracted and multiplication by the gravity constant divided by aircraft speed results in the pitch-rate command (eqn (8)).

4.3. *Feedback loop longitudinal axis*

The feedback loops are shown in the lower right corner of Fig. 8. Inputs are the differences between commanded and actual AoA and pitch rate. These deltas are multiplied by the gains stored in three-dimensional tables as functions of altitude, Mach number, and AoA, and the results are summed. This sum is the normalized pitching moment the feedback loops require. Further down, this pitching moment is distributed with factors to the canard, symmetrical flap, and pitch-axis TV command. These factors depend on the flight condition and the actual engine thrust level. They are calculated in the FCLs using stored tables. There are two different distributions stored in the FCLs: with and without TV. Without TV this factor is set to zero and the others are raised accordingly.

It is the designer's choice to select an appropriate distribution of the different pitching-moment producers during the optimization of the feedback gains. The $\mathbf{x}_k$ and $\mathbf{u}_k$ vectors of eqn (2) for the longitudinal axis are

$$\mathbf{x} = [V_k, \alpha, q, \Theta]$$

$$\mathbf{u} = [\delta] \tag{10}$$

In this equation δ is a linear combination of canard, symmetrical trailing-edge deflection, and pitch TV. For the elements of the performance index weighting matrices only their relative value to each other is important, and therefore the control weighting matrix can be set to one. For the state weighting matrix a diagonal matrix was chosen with zeros in the speed and pitch attitude rows. This results in small feedback gains for speed and pitch attitude which are neglected because of their small influence on AoA motion. The phugoid cannot be controlled with these feedbacks, but the resulting phugoid is within the handling-quality requirements. The frequency and damping of the AoA motion is chosen with the values of q_α and q_q:

$$\mathbf{Q} = \text{diag} \, [0, q_\alpha, q_q, 0]$$

$$\mathbf{R} = [1] \tag{11}$$

In addition an AoA, respectively load factor, error integral is implemented which does not much influence the AoA motion. The phugoid remains unchanged with the AoA integral and is more damped with the load factor integral.

5. Lateral/directional control laws

Roll stick position is scaled in the FCLs from -1 (max left) to $+1$ (max right). Depending on the flight condition a maximum wind-axis roll command $p_{c\,max}$ is calculated (up to 240 deg s^{-1} at low AoA and high dynamic pressure). The maximum roll rate is scaled with flight condition such that the available control power will be used as much as possible for the steady-state roll, with sufficient control power left for stabilization and departure prevention. This maximum command multiplied by the scaled roll stick input gives the wind-axis roll-rate command p_c.

The rudder pedal command is calculated in a similar way. Here $\beta_{c\,max}$ is calculated as a function of AoA and aircraft speed (up to 12° at low AoA and low speed). This maximum command multiplied by the scaled rudder pedal gives the sideslip command β_c. During rapid rolling the sideslip command is blended to zero, to use all the control power for rolling (roll priority).

Figure 9 shows a simplified block diagram of the lateral/directional FCLs. The large box in the center marked 'Gains' represents the matrix multiplication of a combined feedback and feedforward matrix (five columns, three rows) by the input vector (p_c, Δp, Δr, $\Delta \beta$, β_c) resulting in surface commands for rudder, differential trailing-edge flap, and yaw TV.

5.1. *Direct link lateral/directional axes* (f_u)

In the lateral/directional axis direct links exist from the wind-axis roll-rate command as well as from the sideslip command to the trailing-edge flaps

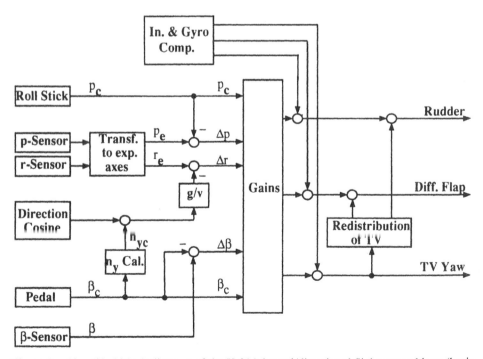

Figure 9. Simplified block diagram of the X-31A lateral/directional flight control laws (basic mode).

(differential), rudder, and TV. They are calculated by multiplying the commands with gains. The gains are stored in tables and interpolated depending on flight condition. The direct link commands correspond to the deflections calculated in a steady-state flight condition.

The direct link yawing moment is fed to the aerodynamic rudder at AoAs up to 30°. At PST conditions the rudder becomes ineffective. Thus direct link is blended to TV which takes over the full authority in yaw at 45° AoA.

As with the longitudinal FCL, the surface deflections to compensate the inertial and gyroscopic moments as defined in Fig. 7 are added to the surface commands.

5.2. Calculation of yaw-rate command (f_y)

The yaw-rate command is not directly a pilot input, and thus it has to be calculated in the FCLs. First the associated n_y command is derived from the sideslip command, using a stored sideforce table, dynamic pressure and estimated aircraft weight. From this the direction cosine of gravity in the aircraft's y axis is subtracted and multiplication by the gravity constant divided by aircraft speed results in the wind-axis yaw-rate command (eqn (9)).

5.3. Feedback loop lateral/directional axes

Similar to the longitudinal control laws the differences between the commanded and actual sideslip, wind-axis roll and yaw rate are multiplied by gains depending on AoA, altitude and Mach number. The TV gains also depend on estimated thrust. If TV is switched off this command is redistributed to the rudder and differential flap. TV nearly gives a pure yawing moment; in the stored redistribution tables a combination of rudder and differential flap is precomputed depending on flight condition, also giving a pure yaw moment.

The feedback gain matrix is determined by optimal control theory. The $\mathbf{x}_k$ and $\mathbf{u}_k$ vectors of eqn (2) for the lateral/directional axes are

$$\mathbf{x}^T = [\beta, p_k, r_k, \Phi]$$

$$\mathbf{u}^T = [\delta_{DF}, \delta_R, \delta_{TV}] \tag{12}$$

The weighting matrices are defined as shown below. The Φ column and row elements are set to zero and the Φ feedbacks which have only marginal influence on dutch roll and the roll mode are neglected. The diagonal elements of the matrix $\mathbf{Q}$ mainly define the eigenvalues of the lateral/directional aircraft motion. The other elements are used to decouple the yawing and rolling motion. The control power is weighted with the diagonal matrix $\mathbf{R}$. A spiral-mode stabilization needs a Φ feedback or the feedback of an equivalent signal, but the resulting spiral mode fulfills the handling-quality requirements.

$$\mathbf{Q} = \begin{bmatrix} q_\beta & q_{\beta p} & q_{\beta r} & 0 \\ q_{\beta p} & q_p & q_{pr} & 0 \\ q_{\beta r} & q_{pr} & q_r & 0 \\ 0 & 0 & 0 & 0 \end{bmatrix}$$

$$\mathbf{R} = \begin{bmatrix} r_{\delta_{DF}} & 0 & 0 \\ 0 & r_{\delta_R} & 0 \\ 0 & 0 & r_{\delta_{TV}} \end{bmatrix} \tag{13}$$

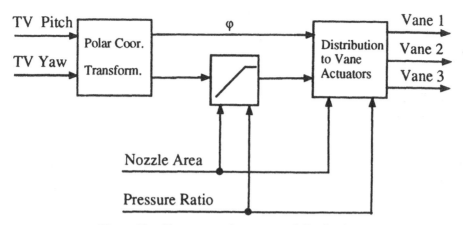

Figure 10.   Thrust vectoring command distribution.

6. Thrust vectoring command distribution

The longitudinal and lateral/directional FCSs command effective thrust deflections in pitch and yaw directions (Fig. 10). These have to be transformed into the associated vane actuator commands. For this the pitch and yaw commands are transformed into polar coordinates where the maximum effective deflection can easily be limited with engine parameters without changing the direction of the command. Stored TV tables depending on desired thrust deflection as well as engine parameters are used to calculate the vane deflections in two steps. First the plume boundary vane position is calculated, and then the thrust deflection vane commands are superimposed. At vane deflections larger than 26° geometric clearance is not guaranteed; therefore a software limitation is introduced which allows only one paddle to deflect more than 26°. When TV is switched off the two lower vanes (2 and 3) can be used as airbrakes.

7. Engagement/disengagement of thrust vectoring

The TV system can be switched on and off by the pilot. In case of a failure TV is automatically blended out. This blending is implemented in the flight control software in such a way that the aerodynamic surfaces get additional commands which produce the same overall moment as with TV. As long as sufficient aerodynamic control power is available there is no difference in the moments with and without TV. With small pilot commands this is valid in the whole conventional flight envelope and for the pitch axis even in the PST regime. In all these cases the linear handling qualities are nearly unchanged with TV in and out. In case of a thrust vector failure in PST no sufficient yawing moment is available. To keep the sideslip as low as possible in this case, the rudder and the differential flap command is blended out during recovery to low AoA. Owing to the reduced control power, the roll performance is also reduced with TV off.

8. Weight and thrust estimation

For calculation of the load factor command from the AoA command and vice versa, as well as for thrust estimation, information about the aircraft's weight is

405

necessary. For the calculation of the effective thrust deflection information about the aircraft's thrust is necessary, because the resulting moment is proportional to the thrust. No sensed signals for weight or thrust are available, so these values are estimated with the lift and drag equations which are transformed into body axes. To minimize the dynamic error of these steady-state models the following limitations are included:

- lag filters reduce high-frequency effects;
- the estimations are rate limited;
- the estimated values are bounded;
- during extreme maneuvers the estimation retains the last value.

The overall accuracy of the estimations depends on the accuracy of the aerodynamic model. With new software an engine model was introduced to improve thrust estimation.

9. Post-stall (PST) mode

The control law structure does not change with the introduction of the PST mode. Only the breakpoints in the gain tables and AoA-dependent scalings are extended to the larger AoA range. Flying into the PST is only possible if all of the given prerequisites are fulfilled. If this is the case the pilot can pull to AoAs larger than 30°. To prevent the pilot from unintentional PST entries a detent is introduced in the stick-force spring at 30° AoA command. If one or more of the prerequisites are not longer fulfilled, or in case of a failure, the AoA command is automatically reduced to 30°.

10. Flight test results

The achieved flight envelope encompasses an airspeed range from 485 kcas down to 40 kcas, an AoA range from −8° to 70°, and a high-altitude supersonic corridor up to Mach 1·3. The flight envelope is shown in Fig. 11. The aircraft can be flown entirely carefree within the subsonic flight regime. Tailside maneuvers, however, have not been demonstrated.

Along with the evaluation of steady-state flying-quality maneuvers, like accelerations/decelerations, wind-up turns, steady-heading sideslips and 360° rolls, a low-order equivalent-system identification method was used for flying-quality assessment throughout the envelope expansion process. Actual flight test results were compared to the handling-quality requirements of the Military Specification MIL-F-8785C. Figures 12 and 13 show short-period pitch-axis results obtained by equivalent derivative identification. Figure 14 shows dutch roll frequencies versus damping derived at various flight conditions. Flight data match predictions from linear analysis, depicted by full square symbols. Level 1 of 'conventional' flying-quality requirements could be achieved throughout the flight envelope. For the PST flight regime, however, no settled requirements exist and generally conventional requirements cannot be applied. Pilots rated the handling and performance of the aircraft at high AoAs as excellent, impressive and quite satisfactory.

Figure 15 depicts a time history of a velocity vector roll at 70° AoA. The roll about the velocity vector is one of the basic maneuvers of PST flight. It can be seen

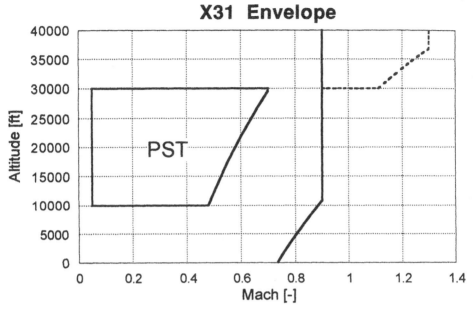

Figure 11.   X-31A flight envelope.

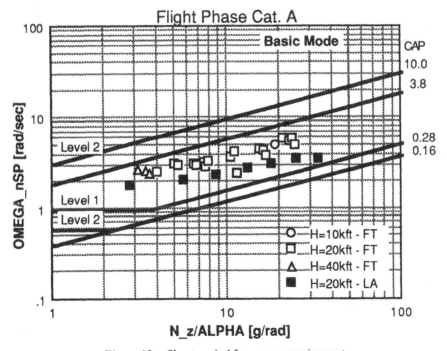

Figure 12.   Short-period frequency requirement.

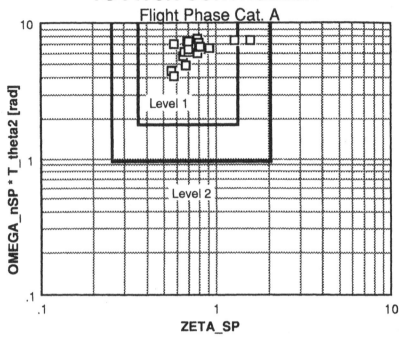

Figure 13.   Requirement for short-term pitch response.

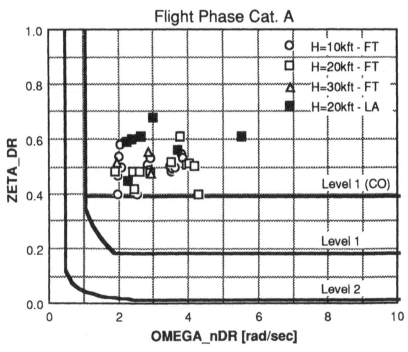

Figure 14.   Dutch roll frequency and damping requirement.

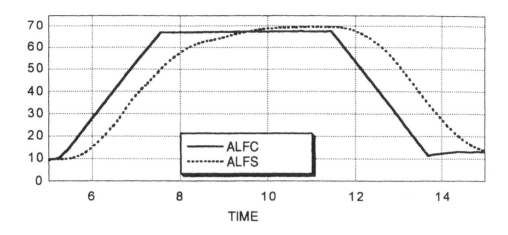

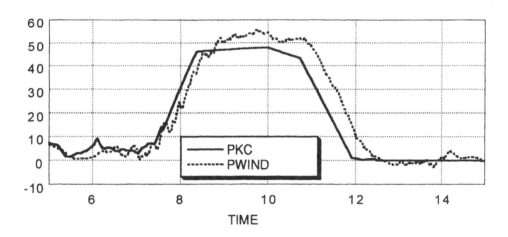

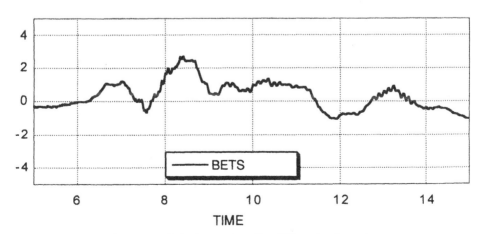

Figure 15.   Velocity vector roll at 70° angle of attack.

that the resulting roll rate (wind-axis roll rate) follows the command (PKC) precisely, while sideslip stays close to zero throughout the maneuver.

Figure 16 shows AoA and load factor versus airspeed during a 5*g* dynamic PST entry from 260 knots initiated from 1*g* flight by an abrupt full aft stick input.

The Neal–Smith criterion was applied to X-31A flight data to assess the aircraft's high AoA pitch-tracking qualities. The criterion indicates Level 2, which corresponds to qualitative pilot comments (see Fig. 17). It should be noted that this criterion was developed to assess pitch tracking in the low AoA regime.

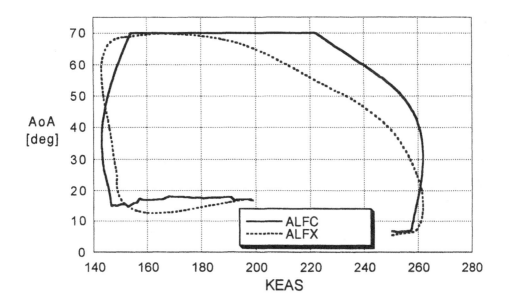

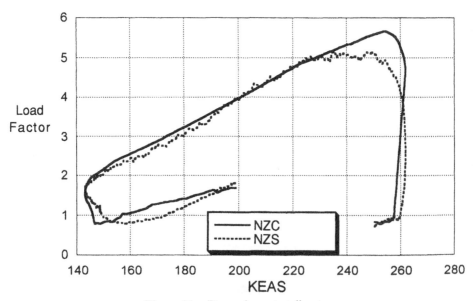

Figure 16.   Dynamic post-stall entry.

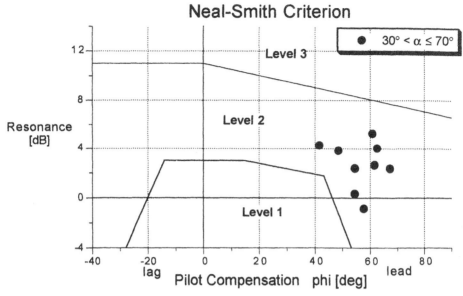

Figure 17. Neal–Smith criterion applied at high angle of attack.

Gross-acquisition and target tracking in the lateral axis (wind-axis roll) at high AoA were rated to be quite satisfactory. This could be proved during the tactical utility evaluation phase. More than 300 close-in-combat engagements were flown, during which the X-31A could demonstrate its superiority by taking advantage of exceptional PST maneuverability.

11. Concluding remarks

The flight test results confirmed that the design approach and the control law structure concept, including the TV system, were successful and all objectives could be met. The aircraft dynamics were rated by the pilots close to existing Level 1 handling-quality requirements throughout the whole conventional envelope. Within the PST envelope no settled requirements exist; however, the pilots rated the handling and performance of the airplane as excellent, impressive and quite satisfactory.

During the flight test phase no significant control law structure changes were necessary. The only major modification was the introduction of integral feedback of sideslip and roll-rate error above 30° AoA. This modification was introduced owing to unpredicted large asymmetries in the high AoA regime, particularly around 50°, resulting in unacceptably large steady-state error in roll rate and sideslip during PST maneuvers.

Two updates of the aerodynamic model shortly before first flight and during flight test showed the flexibility of the optimal control approach. This was demonstrated by a redesign of the feedback gains performed in less than 1 month.

ACKNOWLEDGEMENTS

The authors would like to express their thanks to the members of the International Test Organization at NASA Dryden for their support in preparation of this chapter.

REFERENCES

BEH, H. and HOFINGER, G., 1994, Control law design of the experimental aircraft X-31A. ICAS-94-7.2.1.

HUBER, P. and GALLEITHNER, H., 1992, X-31A high angle of attack and initial post stall flight testing. AGARD.

HUBER, P., WEISS, S. and GALLEITHNER, H., 1991, X-31A initial flying qualities results using equivalent modeling flight evaluation technique. AIAA-91-2891.

Index

Page numbers ending in n, e.g. 7n, indicate that the reference is in a footnote. Other abbreviations are explained in the index.

Printed and bound by CPI Group (UK) Ltd, Croydon, CR0 4YY

01/11/2024

01782599-0007